COMMITTTEE ON TRANSITIONS TO ALTERNATIVE VEHICLES AND FUELS

DOUGLAS M. CHAPIN, *Chair*, NAE,[1] MPR Associates, Inc., Alexandria, Virginia
RALPH BRODD, Broddarp of Nevada, Henderson
GARY COWGER, GLC Ventures, LLC, Bloomfield Hills, Michigan
JOHN M. DECICCO, University of Michigan, Ann Arbor
GEORGE C. EADS, Charles River Associates (retired), Washington, District of Columbia
RAMON ESPINO, University of Virginia, Charlottesville
JOHN M. GERMAN, International Council for Clean Transportation, Ann Arbor, Michigan
DAVID L. GREENE, Oak Ridge National Laboratory, Knoxville, Tennessee
JUDITH GREENWALD, Center for Climate and Energy Solutions, Arlington, Virginia
L. LOUIS HEGEDUS, NAE, Arkema, Inc. (retired), Bryn Mawr, Pennsylvania
JOHN HEYWOOD, NAE, Massachusetts Institute of Technology, Cambridge
VIRGINIA McCONNELL, Resources for the Future, Washington, District of Columbia
STEPHEN J. McGOVERN, PetroTech Consultants LLC, Voorhees, New Jersey
GENE NEMANICH, ChevronTexaco Corporation (retired), Scottsdale, Arizona
JOHN O'DELL, Edmunds, Inc., Orange, California
ROBERT F. SAWYER, NAE, University of California, Berkeley
CHRISTINE S. SLOANE, Sloane Solutions, LLC, Kewadin, Michigan
WILLIAM H. WALSH, JR., Consultant, McLean, Virginia
MICHAEL E. WEBBER, University of Texas at Austin

Project Staff

ALAN T. CRANE, Senior Scientist and Study Director
JAMES J. ZUCCHETTO, Director, Board on Energy and Environmental Systems
JONNA HAMILTON, Program Officer *(until December 2011)*
EVONNE TANG, Senior Program Officer *(beginning December 2011)*
DAVID W. COOKE, Associate Program Officer
ALICE V. WILLIAMS, Senior Program Assistant
LaNITA JONES, Administrative Coordinator
DANA CAINES, Financial Manager

Consultants

DAN MESZLER, Meszler Engineering Services
STEVE PLOTKIN, Argonne National Laboratory
MARC MELAINA, Consultant
MICHAEL P. RAMAGE, NAE, ExxonMobil Research and Engineering Company (retired)
JAMES R. KATZER, NAE, ExxonMobil Research and Engineering Company (retired)
GARY W. ROGERS, FEV, Inc.
DEAN TOMAZIC, FEV, Inc.
AARON BIRCKETT, FEV, Inc.

---

[1] NAE = Member, National Academy of Engineering.

Preface

The U.S. light-duty vehicle fleet is responsible for about half the petroleum consumed in this nation and about 17 percent of its greenhouse gas emissions. Concerns over national security and climate change have increased interest in alternative ways to power the fleet.

Many technologies, with widely varying levels of current capability, cost, and commercialization, can reduce light-duty vehicle petroleum consumption, and most of these also reduce greenhouse gas emissions. However, any transition to achieve high levels of reduction is likely to take decades. The timeframe of this study goes out to 2050. Projecting the cost and performance of technologies out that far entails many uncertainties. The technical issues alone are extraordinarily complex and interrelated. Further, its statement of task also asked the Committee on Transitions to Alternative Vehicles and Fuels to consider the related policy options.

The committee's analyses, while exploratory and not definitive, having significant uncertainty, indicate that the costs and benefits of large reductions in petroleum consumption and greenhouse gas emissions will both be substantial. Its work also suggests that policy will be an essential element in achieving these reductions. Alternative vehicles and some fuels will be more expensive than their current equivalents, at least for several decades, and advanced technology could be used for increased power or other purposes rather than be focused solely on reducing petroleum use and greenhouse gas emissions. Thus, it is critical to have a clear vision of the options and how they might be implemented if progress is to be made efficiently with a minimum of disruption and a maximum of net benefits. This report explores those options and the related issues, and it sheds light on the decisions the nation may be making.

The members of the study committee worked extraordinarily hard on this task. I am very grateful for their efforts. They represent a remarkably broad and accomplished group of experts. Given the complex nature of the task at hand, producing a report that was satisfactory in every detail to every member was challenging. Given the difficulty we have had in achieving consensus, I will not attempt to summarize the result here. The report speaks for itself.

The committee and I greatly appreciate the efforts made by our highly qualified consultants and the many others who contributed directly to our deliberations via presentations and discussions and the many authors on whose work we relied.

The committee operated under the auspices of the NRC's Board on Energy and Environmental Systems. We owe a special debt of gratitude to James Zucchetto, Alan Crane, Evonne Tang, David Cooke, and Alice Williams of the NRC staff. In spite of what must have seemed like an endless succession of in-person and conference call consultations among the full committee and working groups, meetings to gather information, and revision of the text, their energy and professionalism never wavered. The committee and I personally offer our heartfelt thanks.

Douglas M. Chapin, *Chair*
Committee on Transitions to
Alternative Vehicles and Fuels

Acknowledgments

The Committee on Transitions to Alternative Vehicles and Fuels is grateful to the many individuals who contributed their time and efforts to this National Research Council (NRC) study.

The presentations at committee meetings provided valuable information and insights that enhanced the committee's understanding of the technologies and barriers involved. The committee thanks the following individuals who provided briefings:

Patrick Davis, U.S. Department of Energy,
Phillip Patterson, U.S. Department of Energy,
Jacob Ward, U.S. Department of Energy,
David Howell, U.S. Department of Energy,
Jay Braitsch, U.S. Department of Energy,
Diana Bauer, U.S. Department of Energy,
Sunita Satyapal, U.S. Department of Energy,
Fred Joseck, U.S. Department of Energy,
David Danielson, ARPA-E,
Austin Brown, National Renewable Energy Laboratory,
Andy Aden, National Renewable Energy Laboratory,
David Green, Oak Ridge National Laboratory,
Steve Plotkin, Argonne National Laboratory,
Bill Charmley, U.S. Environmental Protection Agency,
Robert Fri, Consultant,
Mike Ramage, Consultant,
Robbie Diamond, Electrification Coalition,
Mark Finley, BP,
Alan Krupnick, Resources for the Future,
Virginia McConnell, Resources for the Future,
Linda Capuano, Marathon Oil Company,
Sascha Simon, Mercedes Benz,
Ben Knight, Honda,
Dan Sperling, University of California, Davis, and
Reiko Takemasa, Pacific Gas and Electric Company.

The committee owes special thanks to Michael Ramage (NAE) and James Katzer (NAE), who generously volunteered their time and expertise to assist in many complex and difficult issues. This report has benefited greatly from their contributions. The members of the committee and the staff deeply regret the death of Jim Katzer in November 2012.

The committee also appreciates the contributions of the following personel from FEV, Inc., who helped in reviewing the methodology and results of the vehicle analysis: Gary Rogers, Dean Tomazic, and Aaron Birckett.

This report was reviewed in draft form by individuals chosen for their diverse perspectives and technical expertise, in accordance with procedures approved by the NRC's Report Review Committee. The purpose of the independent review is to provide candid and critical comments that will assist the institution in making its published report as sound as possible and to ensure that the report meets institutional standards for objectivity, evidence, and responsiveness to the study charge. The review comments and draft manuscript remain confidential to protect the integrity of the deliberative process. We wish to thank the following individuals for their review of this report:

Menahem Anderman, Advanced Automotive Batteries,
Paul N. Blumberg, NAE,[1] Independent Consultant,
Andrew Brown, NAE, Delphi Corporation,
Lawrence D. Burns, NAE, University of Michigan,
Robert Epperly, Independent Consultant,
Albert R. George, Cornell University,
Chris T. Hendrickson, NAE, Carnegie Mellon University,
Jason D. Hill, University of Minnesota, St. Paul,
Maryann N. Keller, Maryann Keller & Associates, LLC,
Joan M. Ogden, University of California, Davis,
John M. Reilly, Massachusetts Institute of Technology,
Bernard I. Robertson, NAE, DaimlerChrysler Corporation (retired),
Gary W. Rogers, FEV, Inc., and
R.R. Stephenson, Independent Consultant.

Although the reviewers listed above have provided many constructive comments and suggestions, they were not asked to endorse the conclusions or recommendations, nor did they see the final draft of the report before its release. The review of this report was overseen by Elisabeth M. Drake, NAE, Massachusetts Institute of Technology (retired), and Trevor O. Jones, NAE, ElectroSonics Medical. Appointed by the National Research Council, they were responsible for making certain that an independent examination of this report was carried out in accordance with institutional procedures and that all review comments were carefully considered. Responsibility for the final content of this report rests entirely with the authoring committee and the institution.

---

[1]National Academy of Engineering.

Contents

APPENDIXES[1]

---

[1]Note that Appendixes D through H appear only in the electronic version of this report, available at http://www.nap.edu/catalog.php?record_id=18264.

Select Acronyms and Abbreviations

AEO	*Annual Energy Outlook*
AFV	alternative fuel vehicle
bbl	barrel
bbl/d	barrels per day
BEV	battery electric vehicle
Btu	British thermal unit
CAA	Clean Air Act
CAFE	Corporate Average Fuel Economy
CCS	carbon capture and storage
CNG	compressed natural gas
CNGV	compressed natural gas vehicle
CO_2	carbon dioxide
CO_2e	carbon dioxide equivalent
CTL	coal to liquid (fuel)
EERE	Office of Energy Efficiency and Renewable Energy
EIA	Energy Information Administration
EISA	Energy Independence and Security Act of 2007
EOR	enhanced oil recovery
EPAct	Energy Policy Act
ETA	Energy Tax Act
FCEV	hydrogen fuel cell electric vehicle
FFV	flex fuel vehicle
gge	gallon of gasoline equivalent
GHG	greenhouse gas
GREET	Greenhouse Gases, Regulated Emissions, and Energy Use in Transportation model
GTL	gas to liquid (fuel)
H_2	hydrogen
HEV	hybrid electric vehicle

ICE	internal combustion engine
ICEV	internal combustion engine vehicle
IHUF	Indexed Highway User Fee
ILUC	indirect land-use change
IPCC	Intergovernmental Panel on Climate Change
LCA	life-cycle assessment
LCFS	Low Carbon Fuel Standard
LDV	light-duty vehicle
Li-ion	lithium ion
LT	light truck
$MMTCO_2e$	million metric ton(s) of CO_2 equivalent
mpg	miles per gallon
mpgge	miles per gallon of gasoline equivalent
NAAQS	National Ambient Air Quality Standards
NEMS	National Energy Modeling System
NHTSA	National Highway Traffic Safety Administration
NO_x	mono-nitrogen oxides, including nitric oxide (NO) and nitrogen dioxide (NO_2)
PEV	plug-in electric vehicle
PHEV	plug-in hybrid electric vehicle
quad	quadrillion British thermal units (of energy)
RFS	Renewable Fuel Standard
RFS2	Renewable Fuel Standard, as amended by EISA
RIN	Renewable Identification Number
tcf	trillion(s) of standard cubic feet
VMT	vehicle miles traveled

NOTE: A more complete list of acronyms and abbreviations is given in Appendix E of the electronic version of this report, available at http://www.nap.edu/catalog.php?record_id=18264.

Overview

This National Research Council report assesses the potential for reducing petroleum consumption and greenhouse gas (GHG) emissions by the U.S. light-duty vehicle fleet by 80 percent by 2050. It examines the technologies that could contribute significantly to achieving these two goals and the barriers that might hinder their adoption. Four general pathways could contribute to attaining both goals—highly efficient internal combustion engine vehicles and vehicles operating on biofuels, electricity, or hydrogen. Natural gas vehicles could contribute to the additional goal of reducing petroleum consumption by 50 percent by 2030.

Scenarios identifying promising combinations of fuels and vehicles illustrate what policies could be required to meet the goals. Several scenarios are promising, but strong and effective policies emphasizing research and development, subsidies, energy taxes, or regulations will be necessary to overcome cost and consumer choice factors.

All the vehicles considered will be several thousand dollars more expensive than today's conventional vehicles, even by 2050, and near-term costs for battery and fuel cell vehicles will be considerably higher. Driving costs per mile will be lower, especially for vehicles powered by natural gas or electricity, but vehicle cost is likely to be a significant issue for consumers for at least a decade. It is impossible to know which technologies will ultimately succeed, because all involve great uncertainty. It is thus essential that policies be broad, robust, and adaptive.

All the successful scenarios combine highly efficient vehicles with at least one of the other three pathways. Large gains beyond the standards proposed for 2025 are feasible from engine and drivetrain efficiency improvements and load reduction (e.g., weight and rolling resistance). Load reduction will improve the efficiency of all types of vehicles regardless of the fuel used.

If their costs can be reduced and refueling infrastructure created, natural gas vehicles have great potential for reducing petroleum consumption, but their GHG emissions are too high for the 2050 GHG goal.

Drop-in biofuels (direct replacements for gasoline) produced from lignocellulosic biomass could lead to large reductions in both petroleum use and GHG emissions. While they can be introduced without major changes in fuel delivery infrastructure or vehicles, the achievable production levels are uncertain.

Battery costs are projected to drop steeply, but limited range and long recharge time are likely to limit the use of all-electric vehicles mainly to local driving. Advanced battery technologies are under development, but all face serious technical challenges.

Battery and fuel cell vehicles could become less expensive than the advanced internal combustion engine vehicles of 2050. Fuel cell vehicles are not subject to the limitations of battery vehicles, but developing a hydrogen infrastructure in concert with a growing number of fuel cell vehicles will be difficult and expensive.

The GHG benefits of all fuels will depend on their production and use without large net emissions of carbon dioxide. To the extent that fossil resources become a large source of non-carbon transportation fuels (electricity or hydrogen), then the successful implementation of carbon capture and storage will be essential.

Summary

Internal combustion engines operating on petroleum fuels have powered almost all light-duty vehicles (LDVs) for the past century. However, concerns over energy security from petroleum imports and the effect of greenhouse gas (GHG) emissions on global climate are driving interest in alternatives. LDVs account for almost half of the petroleum use in the United States, and about half of that fuel is imported (EIA, 2011). LDVs also account for about 17 percent of the total U.S. GHG emissions (EPA, 2012).

In response to a congressional mandate in the Senate's Fiscal Year 2010 Energy and Water Development Appropriations Bill (Report 111-45) for the U.S. Department of Energy, Energy Efficiency and Renewable Energy (DOE-EERE), the National Research Council (NRC) convened the Committee on Transitions to Alternative Vehicles and Fuels (see Appendix B) to assess the potential for vehicle and fuel technology options to achieve substantial reductions in petroleum use and GHG emissions by 2050 relative to 2005. This report presents the results of that analysis and suggests policies to achieve the desired reductions. The statement of task (see Appendix A) specifically asks how the on-road LDV fleet could reduce, relative to 2005,

- Petroleum use by 50 percent by 2030 and 80 percent by 2050, and
- GHG emissions by 80 percent by 2050.

SCOPE AND APPROACH

Four general pathways could contribute to attaining both goals—highly efficient internal combustion engine vehicles (ICEVs) and vehicles operating on biofuels, electricity, or hydrogen. Natural gas vehicles could contribute to the additional goal of reducing petroleum consumption by 50 percent by 2030.

This study considered the following types of LDVs:

- ICEVs that are much more efficient than those expected to be available by 2025;
- Hybrid electric vehicles (HEVs), such as the Toyota Prius;
- Plug-in hybrid electric vehicles (PHEVs), such as the Chevrolet Volt;
- Battery electric vehicles (BEVs), such as the Nissan Leaf; BEVs and PHEVs are collectively known as plug-in electric vehicles (PEVs);
- Fuel cell electric vehicles (FCEVs), such as the Mercedes F-Cell, scheduled to be introduced about 2014; and
- Compressed natural gas vehicles (CNGVs), such as the Honda Civic Natural Gas.

The non-petroleum-based fuel technologies examined in the study are hydrogen, electricity, biofuels, natural gas, and liquid fuels made from natural gas or coal. For each fuel and vehicle type, the committee determined current capability and then estimated future performance and costs, plus barriers to implementation, including safety and technology development timelines. The report also comments on key federal research and development (R&D) activities applicable to fuel and vehicle technologies.

BEVs, FCEVs, and CNGVs[1] can operate only on their specific fuel, although hydrogen and electricity can be produced from a variety of sources that might or might not involve the control of emissions of carbon dioxide, the main GHG responsible for human-induced climate change. The engines in ICEVs, HEVs, and PHEVs can use fuels produced from petroleum, biomass, natural gas, or coal.

The committee recognizes the great uncertainties regarding future vehicles and fuels, especially costs, timing of technology advances, commercialization of those advances,

---

[1] Vehicles that operate on CNG can also be designed as dual-fuel vehicles that can switch to gasoline when CNG is not available, or as hybrid electric vehicles. To keep the analysis manageable, these options are not considered in this report.

and their penetration into the market. As a result, the committee developed a range of estimates for use in this study.

For vehicle technologies, the committee used two sets of assumptions for cost and performance: (1) midrange estimates that are ambitious but reasonable goals in the committee's assessment; and (2) optimistic estimates which are potentially attainable, but will require greater successes in R&D and vehicle design. Both sets are predicated on the assumption that strong and effective policies are implemented to continually increase requirements or incentives (at least through 2050) to ensure that technology gains are focused on reducing petroleum use and GHG emissions.

Alternate assumptions were also developed for fuels to aid in assessing uncertainties. For example, several production processes were considered for hydrogen and biofuels, and both conventional generation and low-GHG-emission scenarios were considered for electricity.

In its assessment of the current state of LDV fuel and vehicle technologies and their projections to 2050, the committee built on earlier studies by the NRC and other organizations as listed in Appendix D. In addition, the committee examined publicly available literature and gathered information through presentations at open meetings. Insofar as possible, the committee assessed the fuels and vehicle technologies on a consistent and integrated basis. Its approach accounted for important effects, including the following:

- Potential projected performance characteristics of specific vehicles and fuel systems,
- Costs of the technologies including economies of scale and learning,
- Technical readiness,
- Barriers to implementation,
- Resource demands, and
- Time and capital investments required to build new fuel and vehicle technology infrastructure.

The committee also considered crosscutting technologies. For vehicles, these included weight reduction and improvements in rolling and aerodynamic resistance; for fuels, carbon capture and storage (CCS). In addition, the analysis took into account sector-wide effects such as consumer preferences and potential changes in vehicle miles traveled (VMT).

The committee then analyzed the impact of the various options. Vehicle performance was projected using a model developed by the committee and its consultants that estimates the impact of reductions in energy losses. Costs were projected for expected technologies relative to a 2010 base vehicle. These analyses and the results are described in Chapter 2. Efficiencies, costs, and performance characteristics were analyzed consistently for all vehicle classes and powertrain options, with the partial exception of travel range. Fuel technologies were analyzed individually using consistent assumptions and cost data across all fuels as shown in Chapter 3.

The vehicle and fuel data were then used to forecast future LDV fleet energy use and GHG emissions using two models described in Chapter 5. VISION was used to assess technology pathways to on-road fleets in 2050 based on inputs from the vehicle and fuel analyses developed in Chapters 2 and 3. LAVE-Trans—a spreadsheet model that takes into account consumer choices (discussed in Chapter 4), which are affected by vehicle and fuel characteristics, costs, and policy incentives—was used to compare different policy-driven scenarios. These scenarios are not intended as predictions of the future but rather to evaluate the relative potential impact on future petroleum use and GHG emissions of technological success and policy options, and the resulting costs and benefits.

By their nature, all models are simplifications and approximations of the real world and will always be constrained by computational limitations, assumptions, and knowledge gaps. All the models' estimations depend critically on assumptions about technologies, economics, and policies and should best be viewed as tools to help inform decisions rather than as machines to generate truth or make decisions. The LAVE-Trans model in particular uses the committee's assumptions about technological progress over several decades, how people behave, what things cost and what they are worth. It predicts, in a formal relational structure, how the vehicle fleet composition would then evolve and what the impact would be on petroleum use and GHG emissions. Some of the LAVE-Trans results were surprising, but the committee examined them and the model, fixed mistakes, and revised assumptions, until it was satisfied with the robustness of the outputs that resulted from the inputs. Even so, there is considerable uncertainty about the results presented here. Input assumptions are estimates that may prove inaccurate. The model's handling of market relationships may be simplistic. Nevertheless, as described in Chapter 5, the results are robust for a variety of inputs, and, as long as the results are used with an understanding of the models' strengths and weaknesses, they should be valuable assets in thinking about potential policy actions.

The major results of the committee's work are listed below; additional findings and policy options are embedded in individual chapters of the report.

MEETING THE GOALS OF REDUCING PETROLEUM USE AND GHG EMISSIONS

Finding: It will be very difficult for the nation to meet the goal of a 50 percent reduction in annual LDV petroleum use by 2030 relative to 2005, but with additional policies, it might achieve a 40 percent reduction.

Future petroleum use is likely to decline as more efficient vehicles enter the market in response to the Corporate Average Fuel Economy (CAFE) standards and GHG requirements for 2025, more than compensating for the increased number of vehicles on the road and the miles traveled. These vehicles will be mainly ICEVs, with an increasing share of HEVs. In addition, biofuels mandated by the Renewable Fuel Standard (RFS) could displace a significant amount of petroleum fuels by 2030, especially if coupled with advances in processes for producing "drop-in" cellulosic biofuels (direct substitutes for gasoline or diesel fuel).

Additional policy support may be required to promote increased sales of CNGVs, BEVs, and FCEVs. Even then the nation is unlikely to reach a 50 percent reduction in petroleum use by 2030 because very little time remains for achieving the required massive changes in the on-road LDV fleet and/or its fuel supply. Many of the vehicles on the road in 2030 will have been built by 2015, and these will lower the fuel economy of the on-road fleet.

Finding: The goal of an 80 percent reduction in LDV petroleum use by 2050 potentially could be met by several combinations of technologies that achieve at least the midrange level of estimated success. Continued improvement in vehicle efficiency, beyond that required by the 2025 CAFE standards, is an important part of each successful combination. In addition, biofuels would have to be expanded greatly or the LDV fleet would have to be composed largely of CNGVs, BEVs and/or FCEVs.

The committee considers that large reductions in LDV use of petroleum-based fuels are plausible by 2050, possibly even slightly more than the 80 percent target, but achieving reductions of this size will be difficult. A successful transition path to large reductions in petroleum use will require not only long-term rapid progress in vehicle technologies for ICEVs and HEVs, but also increased production and use of biofuels, and/or the successful introduction and large-scale deployment of CNGVs, BEVs with greatly improved batteries, or FCEVs.

Extensive new fuel infrastructure would be needed for FCEVs. CNGVs would require new supply lines in areas where natural gas is unavailable or in limited supply, and many filling stations. The infrastructure needed for BEVs would mostly be charging facilities, since electricity supply is already ubiquitous. The technology advances required do not appear to require unexpected breakthroughs and can produce dramatic advances over time, but they would have to be focused on reducing fuel use rather than allowing increases in performance such as acceleration. Thus, a rigorous policy framework would be needed, more stringent than the 2025 CAFE/GHG or RFS standards. Large capital investments would be required for both the fuel and vehicle manufacturing infrastructure. Further, alternative vehicles and some fuels will be more expensive than the current technology during the transition, so incentives to both manufacturers and consumers may be required for more than a decade to spur

purchases of the new technology. Figure S.1 shows potential petroleum use for technology-specific scenarios.

Finding: Large reductions are potentially achievable in annual LDV GHG emissions by 2050, on the order of 60 to 70 percent relative to 2005. An 80 percent reduction in LDV GHG emissions by 2050 may be technically achievable, but will be very difficult. Vehicles and fuels in the 2050 time frame would have to include at least two of the four pathways: much higher efficiency than current vehicles, and operation on biofuels, electricity, or hydrogen (all produced with low GHG emissions). All four pathways entail great uncertainties over costs and performance. If BEVs or FCEVs are to be a majority of the 2050 LDV fleet, they would have to be a substantial fraction of new car sales by 2035.

Achieving large reductions in net GHG emissions from LDVs is more difficult than achieving large reductions in petroleum use. In addition to making all LDVs highly efficient so that their fuel use per mile is greatly reduced, it will be necessary to displace almost all the remaining petroleum-based gasoline and diesel fuel with fuels with low net GHG emissions. This is a massive and expensive transition that, because LDVs emit only about 17 percent of U.S. GHGs, would have to be part of an economy-wide transition to provide major GHG reduction benefits.

The benefits of biofuels depend on how they are produced and on any direct or indirect land-use changes that could lead to GHG emissions. Several studies indicate that sufficient biomass should be available to make a large contribution to meeting the goals of this study, but the long-term costs and resource base for biofuels produced with low GHG emissions need to be demonstrated. Hydrogen and electricity must be produced with low-net-GHG emissions, and the costs of large-scale production are uncertain. Achieving the goals does not require fundamental breakthroughs in batteries, fuel cell systems, or lightweight materials, but significant continuing R&D yielding sustained progress in cost reduction and performance improvement (e.g., durability) is essential.

Overall, the committee concluded that LDV GHG emissions could be reduced by some 60 percent to somewhat more than 80 percent by 2050 as shown in Figure S.2. The cost will be greater than that for meeting the 80 percent petroleum reduction goal because options such as CNGVs, or BEVs operating on electricity produced without constraints on GHG emissions, cannot play a large role.

Finding: None of the four pathways by itself is projected to be able to achieve sufficiently high reductions in LDV GHG emissions to meet the 2050 goal. Further, the cost, potential rate of implementation of each technology, and response of consumers and

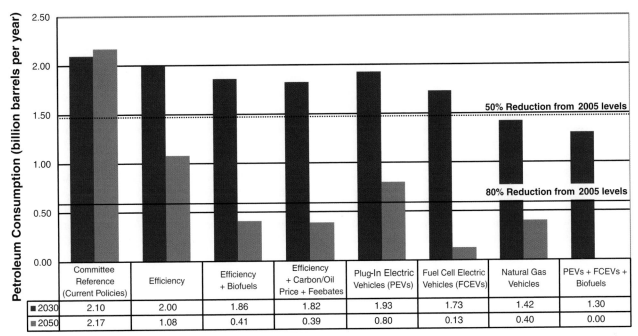

FIGURE S.1 Estimated U.S. LDV petroleum use in 2030 and 2050 under policies emphasizing specific technologies. Midrange values are the committee's best estimate of the progress of the technology if it is pursued vigorously. All scenarios except the Committee Reference Case (current policies, including the fuel economy standards for 2025) include midrange efficiency improvements. Controls for GHG emissions from hydrogen and electricity production are not assumed because the main objective is to reduce petroleum use.

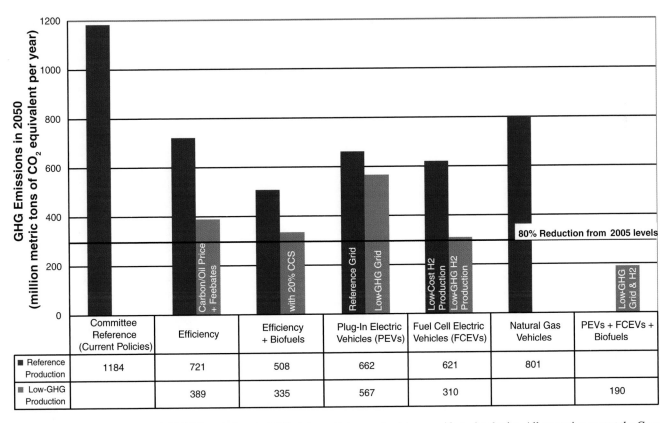

FIGURE S.2 Estimated U.S. LDV GHG emissions in 2050 under policies emphasizing specific technologies. All scenarios except the Committee Reference Case (current policies, including the fuel economy standards for 2025) include midrange efficiency improvements. Fuel production for these scenarios is assumed to be constrained by policies controlling GHG emissions (low GHG production).

manufacturers to policies are uncertain. Therefore, an adaptive framework that modifies policies as technologies develop and as conditions change is needed to efficiently move toward the long-term policy goals.

Continued improvements in vehicle efficiency, especially load reduction (e.g., through the use of light weight but strong materials), are essential to achieving high GHG reductions and are included in all scenarios as a key step in improving the feasibility of all the other pathways. In addition, some combination of biofuels, BEVs, and FCEVs (with the last two operating on low-GHG electricity or hydrogen) must play a large role. Given the uncertainties surrounding all four of these pathways, there is no single, clearly supported choice of vehicle and fuel system that will lead to 80 percent reduction in GHG emissions.

Much more efficient or alternative vehicles are currently more costly than today's ICEVs and their prices are projected to remain high until the newer technologies are more mature. Achieving an extensive transition by 2050 will thus require government action. These transition costs are in addition to those associated with bringing the technologies to readiness and providing needed infrastructure.

Displacing the incumbent ICEVs and petroleum-based fuels will be difficult. Technologies may not be as successful as anticipated, and the policies to encourage them may not be as successful as modeled by the committee. Furthermore the costs would likely be very large early on, with benefits occurring much later in time. It is essential, then, to ensure that policies, especially those that focus on investment in particular technologies, are not introduced too early (for example, before those new fuel and vehicle technologies are close to market readiness, taking into account the best available information on consumer behavior) or too late (for example, not allowing for the benefits of learning to be realized and to contribute to meeting the goals). Further, it is essential that policies are designed so that they can be adapted to changing evidence about technology and market acceptance, and to market conditions.

In pursuing these goals, costs and benefits of the intended action should both be assessed. Action should be undertaken only upon a reasoned determination that the benefits of intended proposed regulation justify its costs. Scenario analysis has identified strong tipping points for the transition to new vehicle technologies. If policies are insufficient to overcome the early cost differentials, then the transition to such technologies will not occur, and the costs will have been largely wasted.

Finding: Substantial progress toward the goals of reducing LDV petroleum use and GHG emissions is unlikely unless these goals are set and pushed on a nationwide basis through strong and effective policy intervention by the federal government.

All four transition paths are based on technology options that are currently more expensive than their ICEV equivalent, and some will require substantial infrastructure changes and possibly consumer adaptation. Thus, success will depend on consistent and sustained policies that support reduced petroleum use and GHG emissions.

Finding: Even if the nation falls short of the 2050 goals, there are likely to be environmental, economic and national security benefits resulting from the petroleum use and GHG emissions reductions that are achieved.

Finding: The CAFE standard has been effective in reducing vehicle energy intensity, and further reductions can be realized through even higher standards if combined with policies to ensure that they can be achieved.

Policy Option: The committee suggests that LDV fuel economy and GHG emission standards continue to be strengthened to play a significant role after model year 2025 as part of this country's efforts to improve LDV fuel economy and reduce GHG emissions.

Finding: "Feebates," rebates to purchasers of high-fuel-economy (i.e., miles per gallon [mpg]) vehicles balanced by a tax on low-mpg vehicles is a complementary policy that would assist manufacturers in selling the more-efficient vehicles produced to meet fuel economy standards.

Policy Option: The committee suggests that the U.S. government include "feebates" as part of a policy package to reduce LDV fuel use.

Finding: Several types of policies including a price floor for petroleum-based fuels or taxes on petroleum-based fuels could create a price signal against petroleum demand, assure producers and distributors that there is a profitable market for alternative fuels, and encourage consumers to reduce their use of petroleum-based fuels. High fuel prices, whether due to market dynamics or taxes, are effective in reducing fuel use.

The impact of increases in fuel prices, especially on low-income and rural households, could be offset by using the increased revenues from taxes or a price floor for reductions in other taxes. Alternatively, some or all of the revenue generated could be used to replace income lost to the Highway

Trust Fund as gasoline sales decline, so that transport infrastructure could continue to be supported.

Finding: Fuel cells, batteries, biofuels, low-GHG production of hydrogen, carbon capture and storage, and vehicle efficiency should all be part of the current R&D strategy. It is unclear which options may emerge as the more promising and cost-effective. At the present time, foreclosing any of the options the committee has analyzed would decrease the chances of achieving the 2050 goals.

The committee believes that hydrogen/fuel cells are at least as promising as battery electric vehicles in the long term and should be funded accordingly. Both pathways show promise and should continue to receive federal R&D support.

Policy Option: The committee supports consistent R&D to advance technology development and to reduce the costs of alternative fuels and vehicles. The best approach is to promote a portfolio of vehicle and fuel R&D, supported by both government and industry, designed to solve the critical technical challenges in each major candidate pathway. Such primary research efforts need continuing evaluation of progress against performance goals to determine which technologies, fuels, designs, and production methods are emerging as the most promising and cost-effective.

Finding: Demonstrations are needed for technologies to reduce GHG emissions at appropriate scale (for example, low-carbon hydrogen and CCS) to validate performance, readiness, and safety. Integrated demonstrations of vehicles and fueling infrastructure for alternative vehicle and fuel systems will be necessary to promote understanding of performance, safety, consumer use of these alternatives, and other important characteristics under real-world driving conditions.

Policy Option: The committee supports government involvement in limited demonstration projects at appropriate scale and at appropriate times to promote understanding of the performance and safety of alternative vehicles and fueling systems. For such projects, substantial private sector investment should complement the government investment, and the government should ensure that the demonstration incorporates well-designed data collection and analysis to inform future policy making and investment. The information collected with government funds should be made available to the public consistent with applicable rules that protect confidential data.

Finding: The commercialization of fuel and vehicle technologies is best left to the private sector in response to performance-based policies, or policies that target reductions in GHG emissions or petroleum use rather than specific technologies. Performance-based policies for deployment (e.g., CAFE standards) or technology mandates (e.g., RFS) do not require direct government expenditure for particular vehicle or fuel technologies. Additional deployment policies such as vehicle or fuel subsidies, or quantity mandates directed at specific technologies are risky but may be necessary to attain large reductions in petroleum use and GHG emissions. For alternative-vehicle and fuel systems, government involvement with industry is likely to be needed to help coordinate commercial deployment of alternative vehicles with the fueling infrastructure for those vehicles.

Policy Option: The committee suggests that an expert review process independent of the agencies implementing the deployment policies and also independent of any political or economic interest groups advocating for the technologies being evaluated be used to assess available data, and predictions of costs and performance. Such assessments could determine the readiness of technologies to benefit from policy support to help bring them into the market at a volume sufficient to promote economies of scale. If such policies are implemented, there should be specific goals and time horizons for deployment. The review process should include assessments of net reductions in petroleum use and GHG emissions, vehicle and fuel costs, potential penetration rates, and consumer responses.

TECHNOLOGY- AND POLICY-SPECIFIC FINDINGS

Vehicles (Chapter 2)

- Large increases in fuel economy are possible with incremental improvements in currently known technology for both load reduction and drivetrain improvements. The average of all conventional LDVs sold in 2050 might achieve CAFE test values of 74 mpg for the midrange case. Hybrid LDVs might reach 94 mpg by 2050. On-road fuel economy values will be lower.

- To obtain the efficiencies and costs estimated in Chapter 2, manufacturers will need incentives or regulatory standards or both to widely apply the new technologies.

- The unit cost of batteries will decline with increased production and development; in addition, the energy storage (in kilowatt-hours) required for a given vehicle range will decline with vehicle load reduction and improved electrical component efficiency. Therefore, battery pack costs in 2050 for a 100-mile real-world travel range are expected to drop by a factor of about 5. However, even these costs are unlikely to create a mass market for BEVs, because a battery large enough for a 300-mile real-world range would still present significant weight and volume penalties and probably could not be recharged in much less than 30 minutes. Therefore, BEVs may be used mainly for local travel rather than as all-purpose vehicles.

- BEVs and PHEVs are likely to use lithium-ion batteries for the foreseeable future. Several advanced battery technologies (e.g., lithium-air) are being developed that would address some of the drawbacks of lithium-ion batteries, but their potential for commercialization by 2050 is highly uncertain, and they may have their own disadvantages.

- PHEVs offer substantial amounts of electric-only driving while avoiding the range and recharge-time limitations of BEVs. However, their larger battery will always entail a significant cost premium over similar HEVs, and their incremental fuel savings will decrease as the efficiency of HEVs improves.

- The technical hurdles that must be surmounted to develop an all-purpose vehicle acceptable to consumers appear lower for FCEVs than for BEVs. However, the infrastructure and policy barriers appear larger. Well before 2050 the cost of FCEVs could actually be lower than the cost of an equivalent ICEV, and operating costs should also be lower. FCEVs are expected to be equivalent in range and refueling time to ICEVs.

- If CNGVs can be made competitive (with respect to both vehicle cost and refueling opportunities), they will offer a quick and economical way to reduce petroleum use, but as shown in Figure S.2, the reductions in GHG emissions are insufficient for CNGVs to be a large part of a fleet that meets the 2050 GHG goal.

- Although fundamental technology breakthroughs are not essential to reach the mpg, performance, and cost estimates in Chapter 2, new technology developments would substantially reduce the development cost and lead time. In particular, continued research to reduce the costs of advanced materials and battery concepts will be critical to the success of electric vehicles.

Fuels (Chapter 3)

- Meeting the GHG and petroleum reduction goals requires a massive restructuring of the fuel mix used for transportation. The use of petroleum must be greatly reduced, implying retirement of crude oil production and distribution infrastructure. Depending on the progress in drop-in biofuels versus non-liquid

fuels, refineries, pipelines, and filling stations might also become obsolete. For BEVs to operate with low GHG emissions, coal- and natural gas-fired electricity generation might have to be greatly reduced unless CCS proves cost-effective. Reliance on natural gas or hydrogen for transportation would require additional infrastructure. With currently envisioned technology, sufficient biofuels could be produced by 2050 to meet the goal of 80 percent reduction in petroleum use if the committee's vehicle efficiency estimates are attained.

- With increasing economic natural gas reserves and growing domestic natural gas production mostly from shale gas, there is enough domestic natural gas to greatly increase its use for the transportation sector without significantly affecting the traditional natural gas markets. Currently the cost of natural gas is very low ($2.5 to $3.5/million Btu) and could remain low for several decades. Environmental issues associated with shale gas extraction (fracking) must be resolved, including leakage of natural gas, itself a powerful GHG, and potential contamination of groundwater. There are several opportunities, direct and indirect, to use natural gas in LDVs, including producing electricity for PEVs and producing hydrogen for FCEVs. The fastest way to reduce petroleum use is probably by direct combustion in CNGVs coupled with efficiency improvements, but that approach is likely to interfere with achieving the GHG goal in 2050.

- Making hydrogen from fossil fuels, especially natural gas, is a low-cost option for meeting future demand from FCEVs, but such methods, by themselves, will not reduce GHG emissions enough to meet the 2050 goal. Making hydrogen with low GHG emissions is more costly (e.g., renewable electricity electrolysis) or requires new production methods (e.g., photoelectrochemical, nuclear cycles, and biological methods) or CCS to manage emissions. Continued R&D is needed on low-GHG hydrogen production methods and CCS to demonstrate that large amounts of low-cost and low-GHG hydrogen can be produced.

- Natural gas and coal conversion to liquid fuel (GTL, CTL) can be used as a direct replacement for petroleum gasoline, but the GHG emissions from these fuels are slightly greater than those from petroleum-based fuels even when CCS is employed at the production plant. Therefore, these fuels will play a small role in reducing petroleum use if GHG emissions are to be reduced simultaneously.

- Carbon capture and sequestration is a key technology for meeting the 2050 goal for GHG emissions reductions. Insofar as fossil fuels are used as a source of electricity or hydrogen to power LDVs, CCS will be essential. The only alternatives are nuclear power and renewable energy sources, including biofuels. Applying CCS to biofuel production could result in slightly negative net emissions.

Consumer Barriers (Chapter 4)

- Widespread consumer acceptance of alternative vehicles and fuels faces significant barriers, including the high initial purchase cost of the vehicles and the perception that such vehicles offer less utility and convenience than conventional ICEVs. Overcoming these barriers is likely to require significant government policy intervention that could include subsidies and vigorous public information programs aimed at improving consumers' familiarity with and understanding of the new fuels and powertrains. Consumers are used to personal vehicles that come in a wide variety of sizes, styles, and prices that can meet most needs ranging from basic transportation to significant cargo hauling. Conventional ICEVs can be rapidly refueled by a plentiful supply of retailers, effectively giving the vehicles unlimited range. Conversely, in the early years, alternative vehicles will likely be limited to a few body styles and sizes and will cost from a few hundred to many thousands of dollars more than their conventional ICEV counterparts. Some will rely on fuels that are not readily available or have limited travel range, or require bulky energy storage that will limit their cargo and passenger capacity.

Additional Findings from Policy Modeling (Chapter 5)

- Including the social costs of GHG emissions and petroleum dependence in the cost of fuels (e.g., via a carbon tax) provides important signals to the market that will promote technological development and behavioral changes. Yet these pricing strategies alone are likely to be insufficient to induce a major transition to alternative, net-low-carbon vehicle technologies and/or energy sources. Additional strong, temporary policies may be required to break the "lock-in" of conventional technology and overcome the market barriers to alternative vehicles and fuels.
- If two or more of the fuel and/or vehicle pathways identified above evolve through policy and technology development as shown in a number of the committee's scenarios, the committee's model calculations indicate benefits of making a transition to a low-petroleum, low-GHG energy system for LDVs that exceed the costs by a wide margin. Benefits include energy cost savings, improved vehicle technologies, and reductions in petroleum use and GHG emissions. Costs refer to the additional costs of the

transition over and above what the market is willing to do voluntarily. However, as noted above, modeling results should be viewed as approximations at best because there is by necessity in such predictions a great deal of uncertainty in estimates of both benefits and costs. Furthermore, the costs are likely to be very large early on with benefits occurring much later in time.
- It is essential to ensure that policies, especially those that focus on investment in particular technologies, are not introduced before those new fuel and vehicle technologies are close to market readiness and consumer behavior toward them is well understood. Forcing a technology into the market before it is ready can be costly. Conversely, neglecting a rapidly developing technology could lead to forgone significant benefits. Policies should be designed to be adaptable so that mid-course corrections can be made as knowledge is gained about the progress of vehicle and fuels technologies. Further, it is essential that policies be designed so that they can be adapted to changing evidence about technology and market acceptance, and market conditions.
- Depending on the readiness of technology and the timing of policy initiatives, subsidies or regulations for new-vehicle energy efficiency and the provision of energy infrastructure may be required, especially in the case of a transition to a new vehicle and fuel system. In such cases, policy support might be required for as long as 20 years if technological progress is slow (e.g., BEVs with lithium-ion batteries may require 20 years of subsidies to achieve a large market share).
- Advance placement of refueling infrastructure is critical to the market acceptance of FCEVs and CNGVs. It is likely to be less critical to the market acceptance of grid-connected vehicles, since many consumers will have the option of home recharging. However, the absence of an outside-the-home refueling infrastructure for grid-connected vehicles is likely to depress demand for these vehicles. Fewer infrastructure changes will be needed if the most cost-effective solution evolves in the direction of more efficient ICEVs and HEVs combined with drop-in low-carbon biofuels.
- Research is needed to better understand key factors for transitions to new vehicle fuel systems such as the costs of limited fuel availability, the disutility of vehicles with short ranges and long recharge times, the numbers of innovators and early adopters among the car-buying public, as well as their willingness to pay for novel technologies and the risk aversion of the majority, and much more. More information is also needed on the transition costs and barriers to

production of alternative drop-in fuels, especially on the type of incentives necessary for low-carbon biofuels. The models this committee and others have used to analyze the transition to alternative vehicles and/or fuels are first-generation efforts, more useful for understanding processes and their interactions than for producing definitive results.

REFERENCES

EIA (Energy Information Administration). 2011. *Annual Energy Review 2010*. Washington, D.C.: U.S. Department of Energy.

EPA (Environmental Protection Agency). 2012. *Inventory of U.S. Greenhouse Gas Emissions and Sinks:1990-2010*. Available at http://www.epa.gov/climatechange/Downloads/ghgemissions/US-GHG-Inventory-2012-Main-Text.pdf.

1

Introduction

Internal combustion engines (ICEs) operating on petroleum fuels have powered almost all light-duty vehicles (LDVs) for a century. The dominance of ICEs over steam and batteries has been due to their low cost, high power output, readily available fuel, and ability to operate for long distances in a wide range of temperatures and environmental conditions. Although ICEs can run on many fuels, gasoline and diesel have remained the fuels of choice because of their low cost and high energy density, allowing hundreds of miles of driving before refueling. Crude oil has remained the feedstock of choice for these fuels because production has kept pace with demand and world reserves have actually been expanded as a result of ongoing technological progress. The co-evolution and co-optimization of ICE and petroleum-based fuel technology, infrastructure, and markets have proven resilient to challenges from market forces such as oil price spikes in a geopolitically complex world oil market as well as environmental policies such as tailpipe pollution reduction requirements.

For nearly 40 years, energy security concerns have motivated efforts to reduce the use of petroleum-based fuels. LDVs consume about half the petroleum used in the United States, and about half is imported, tying Americans to a world oil market that is vulnerable to supply disruptions and price spikes and contributing about $300 billion to the nation's trade deficit (EIA, 2011).

More recently, concerns have been growing over emissions of carbon dioxide (CO_2), the most important of the greenhouse gases (GHGs) that threaten to cause serious problems associated with global climate change.[1] Petroleum use is the largest source of GHG emissions in the United States. Because LDVs account for the single largest share of U.S. petroleum demand and directly account for 17 percent of total U.S. GHG emissions (EPA, 2012), they have become the subject of policies for mitigating climate change.

For these reasons, U.S. policy makers seek to both improve the fuel efficiency of LDVs and promote the development and adoption of alternative fuels and vehicles (AFVs). Here "alternative fuels" refers to non-petroleum-based fuels, including plant-based fuels that are otherwise essentially identical to gasoline or diesel fuel, and to powertrains much more efficient than today's or capable of using alternative fuels, including non-liquid energy carriers such as natural gas, hydrogen, and electricity. Numerous studies have addressed these issues over the years, reflecting the interest in these goals. Substantial but uneven progress has been made on LDV efficiency, and a small but significant penetration of hybrid electric vehicles in the marketplace has contributed to this goal. Otherwise little progress has been made on AFVs in the marketplace beyond the quantities of ethanol still used almost exclusively in gasoline blends.

Since its beginnings over 100 years ago, the automotive sector has succeeded through a combination of private market forces and public policies. The energy use and GHG emissions challenges with which we now are grappling are the unintended and largely unforeseen by-products of that success.

This report is the result of a study by a committee appointed to evaluate and compare various approaches to greatly reducing the use of oil in the light-duty fleet and GHG emissions from the fleet. As specified in the statement of task (Appendix A), the Committee on Transitions to Alternative Vehicles and Fuels was charged with assessing the status of and prospects for technologies for LDVs and their fuels, and with estimating how the nation could meet one or both of two goals:

1. Reduce LDV use of petroleum-based fuels by 50 percent by 2030 and 80 percent by 2050.
2. Reduce LDV emissions of GHGs by 80 percent by 2050 relative to 2005.

---

[1]As used in this report, GHG means the total of all greenhouse gases, as converted to a common base of global warming potential, i.e., CO_2 equivalent (CO_2e). For tail pipe emissions, CO_2 is used.

The 2050 petroleum reduction goal is easier to meet than the 2050 GHG goal because more options can be employed. In fact, reducing GHGs by 80 percent is likely to require reducing petroleum use by at least 80 percent. Petroleum use by the light duty fleet was 125 billion gallons gasoline in 2005 (EIA, 2011), so the targets are 62.5 billion gallons in 2030 and 25 billion in 2050.

GHG emissions from the LDV fleet in 2005 were 1,514 million metric tons of CO_2 equivalent ($MMTCO_2e$) on a well-to-wheels basis (EPA, 2012). An 80 percent reduction from that level means that whatever fleet is on the road in 2050 can be responsible for only 303 $MMTCO_2e$/year. That is the budget within which the fleet must operate to meet the goal.

Achieving an 80 percent reduction in LDV-related emissions is only possible with a very high degree of net GHG reduction in whatever energy supply sectors are used to provide fuel for the vehicles. In short, it is not possible to greatly "de-carbonize" LDVs without greatly de-carbonizing the major energy supply sectors of the economy.

The committee determined potential costs and performance levels for the vehicle and fuel options. Because of the great uncertainty in estimating vehicle cost and performance in 2050, the committee considered two levels, midrange and optimistic. Midrange goals for cost and performance are ambitious but plausible in the committee's opinion. Meeting this level will require successful research and development and no insurmountable barriers, such as reliance on critical materials that may not be available in sufficient quantities. The more optimistic goals are stretch goals: possible without fundamental technology breakthroughs, but requiring greater R&D and vehicle design success. All the vehicle and fuel cost and performance levels are based on what is achievable for the technology.

Other factors also will be very important in determining what is actually achieved. In particular, government policy will be necessary to help some new and initially costly technologies into the market, consumer attitudes will be critical in determining what technologies are successful, and of course, the price and availability of gasoline will be important in determining the competitiveness of alternative vehicles and fuels.

1.1 APPROACH AND CONTENT

To analyze all these issues, the committee constructed and analyzed various scenarios, combining options under the midrange and optimistic cost and performance levels to see

---

BOX 1.1
Analytical Techniques Used in This Report

The committee relied on four models to help form its estimates of future vehicle characteristics, their penetration into the market, and the impact on petroleum consumption and GHG emissions. Chapter 2 and Appendix F describe two of the models. One is an ICEV model developed by a consultant that projects vehicle efficiency out to 2050 by focusing on reduction of energy losses, rather than the usual technique of adding efficiency technologies until the desired level is reached. The committee's approach avoids the highly uncertain predictions of which technologies will be employed several decades from now and ensures that efficiency projections are physically achievable and that synergies between technologies are appropriately accounted for. The second is a spreadsheet model of technology costs developed by the committee, which focused on applying consistent assumptions across all of the different powertrain types. The analytical approach for both models is fully documented and the data are available in Appendix F. The methodology and results for both of these models were intensively reviewed by the committee, the committee staff, another consultant, and experts from FEV, Inc., an engineering services company. Reviewers of this report were also selected for their ability to understand this approach, which they endorsed.

The VISION and LAVE-Trans models are described in Chapter 5 and Appendix H. VISION is a standard model for analyzing transportation scenarios for fuel use and emissions. It is freely available through the U.S. Department of Energy. The committee modified it for consistency with the committee's assumptions such as on vehicle efficiencies and usage and fuel availability. The committee carefully monitored the modifications and reviewed the results, which are consistent with other analyses.

LAVE-Trans is a new model developed by a committee member for an analysis of California's energy future and expanded to the entire nation by the committee. It is unique among models in that it explicitly addresses market responses to factors such as vehicle cost and range, aversion to new technology, and fuel availability. It analyzes the effectiveness of policies in light of these market responses. The committee and staff spent considerable time reviewing LAVE-Trans and its results. In addition to presentations and discussions at committee meetings, one committee member and the study director spent a day going over the model with the developer and his associates. Another committee member examined intermediate calculations as well as model outputs. The results were also compared to VISION results for identical inputs and assumptions. These examinations led to recalibrations and changes in model assumptions. Reviewers of this report were also selected for their ability to understand the model, and they confirmed its validity.

how the petroleum and GHG reduction goals could be met. It also explored how consumers might react to new technologies. Then the committee compared the technological and economic feasibility of meeting the goals using the available options, the environmental impacts of implementing them, and changes in behavior that might be required of drivers to accommodate new technologies. Finally, the committee examined the policies that might be necessary to implement the scenarios.

Vehicle options are explored in Chapter 2 and fuels in Chapter 3. Chapter 4 discusses factors that will affect consumer choices in considering which vehicles to purchase, and Chapter 5 describes how the scenario modeling was done and the results. Box 1.1 briefly describes the models used in Chapters 2 and 5 and how they were validated.

Chapter 6 discusses policies that could enable the various options and encourage their penetration into the market as needed to implement the scenarios. Finally, Chapter 7 discusses the committee's suggested policy options that are drawn from Chapter 6. Several current policies are encouraging actions that will reduce GHG emissions and petroleum use. The Corporate Average Fleet Economy (CAFE) standards require vehicle manufacturers to sell efficient vehicles. The Renewable Fuel Standards mandate the use of biofuels. Box 1.2 briefly describes these policies. In addition, tax credits for battery vehicles encourage consumers to buy them. Fuel taxes, carbon reduction measures such as carbon taxes, and other standards and subsidies also could be used. State and local policies may also be important, particularly in the absence of activist federal policies, but the focus of

---

BOX 1.2
U.S. Policies Directly Affecting Fuel Consumption

U.S. Corporate Average Fuel Economy (CAFE) Standards

From the mid-1970s through 2010, the United States had one set of standards that applied to passenger cars and another set that applied to light-duty trucks. These standards were administered by the National Highway Traffic Safety Administration (NHTSA) of the U.S. Department of Transportation, following requirements in legislation passed by the U.S. Congress in 1975. They first became effective in the 1978 model year. The standard for passenger cars that year was 18.0 miles per gallon (mpg). The standard increased to 27.5 mpg for the 1985 model year and varied between that level and 26.0 mpg from model year 1986 through model year 1989. In model year 1990 it was raised again to 27.5 mpg and remained at that level through model year 2010. The first combined light truck standard applied to model year 1985 vehicles and was set at 19.5 mpg. The light truck standard ranged between 20.0 and 20.7 mpg between model years 1986 and 1996, remained at 20.7 mpg for model years 1996 through 2004, and increased to 23.5 mpg by model year 2010.

More recently, the federal government implemented two new sets of standards. In 2010, complementary standards were set by the Environmental Protection Agency (EPA) based on greenhouse gas (GHG) emissions and by NHTSA based on fuel economy. NHTSA's CAFE standard for 2016 was set at 34.1 mpg for cars and light trucks. In 2012, new standards were set by EPA and NHTSA through 2025, although the NHTSA standards for 2022-2025 are proposed and not yet final, pending a midterm review. NHTSA's CAFE standard for 2025 is 48.7-49.7 mpg. If flexibilities for paying fines instead of complying, flexible fuel vehicle (FFV) credits, electric vehicle credits, and carryforward/carryback provisions are considered, NHTSA estimated that the CAFE level would be 46.2-47.4 mpg. This does not consider off-cycle credits, which could further reduce the test cycle results by up to 2-3 mpg. Thus, for comparison purposes, the committee used 46 mpg as the tailpipe mpg levels comparable to the committee's technology analyses in Figure 2.1. Also note that on-road fuel economy will be significantly lower—the committee used a discount factor of 17 percent in assessing in-use benefits in Chapter 5. The standards are discussed in more detail in Chapter 5. In particular, see Box 5.1.

Renewable Fuel Standard

The federal Renewable Fuel Standard (RFS) was created under the Energy Policy Act of 2005 because Congress recognized "the need for a diversified portfolio of substantially increased quantities of . . . transportation fuels" to enhance energy independence (P.L. 109-58). The RFS was amended by the Energy Independence and Security Act (EISA) of 2007 which created what is referred to as RFS2. RFS2 mandates volumes of four categories of renewable fuels to be consumed in U.S. transportation from 2008 to 2022. The four categories are:

- Conventional biofuels—15 billion gallons/year of ethanol derived from corn grain or other biofuels.
- Biomass-based diesel—currently 1 billion gallons/year are required.
- Advanced biofuels from cellulose or certain other feedstocks that can achieve a life-cycle GHG reduction of at least 50 percent.
- Cellulosic biofuels, which are renewable fuels derived from any cellulose, hemicellulose, or lignin from renewable biomass and that can achieve a life-cycle GHG reduction threshold of at least 60 percent. In general, cellulosic biofuels also qualify as renewable fuels and advanced biofuels.

this report is on actions the federal government can take. Chapters 6 and 7 estimate the relative effectiveness of U.S. policies in achieving the goals of this study.

The vehicle and fuel options discussed in this report generally are more expensive and/or less convenient for consumers than those that are available now. The societal benefits they provide (in particular, lower oil consumption and GHG emissions) will not, by themselves, be sufficient to ensure rapid penetration of the new technologies into the market. Therefore strong and effective policies will be necessary to meet the goals of this study. By "strong public policies," the committee means options such as steadily increasing fuel standards beyond those scheduled for 2025, measures to substantially limit the net GHG emissions associated with the production and consumption of LDV fuels, and large-scale support for electric vehicles or fuel cell vehicles to help them overcome their high initial cost and other consumer concerns. It also may be necessary to have policies that ensure that the fuels required by alternative powertrains are readily available.

Although the committee is generally skeptical of the value of the government picking winners and losers, the goal of drastically reducing oil use inherently entails a premise of picking a loser (oil) and developing (and perhaps promoting) winners among a set of vehicles and fuel resources.

In turn, implementation of such policies is likely to depend on a strong national imperative to reduce oil use and GHG emissions. The committee has not studied such an imperative but notes that, given the length of time needed to make major changes in the nation's light-duty vehicle fleet, additional policies will be needed soon to meet the goals.

1.2 REFERENCES

EIA (Energy Information Administration). 2011. *Annual Energy Review 2010.* Washington, D.C.: U.S. Department of Energy.

EPA (Environmental Protection Agency). 2012. *Inventory of U.S. Greenhouse Gas Emissions and Sinks:1990-2010.* Available at http://www.epa.gov/climatechange/Downloads/ghgemissions/US-GHG-Inventory-2012-Main-Text.pdf.

2

Alternative Vehicle Technologies: Status, Potential, and Barriers

2.1 INTRODUCTION AND OVERALL FRAMEWORK FOR ANALYSES

Virtually all light-duty vehicles on U.S. roads today have internal combustion engines (ICEs) that operate on gasoline (generally mixed with about 10 percent ethanol produced from corn) or diesel fuel. To achieve very large reductions in gasoline use and greenhouse gas emissions from the light-duty fleet, vehicles in 2050 must be far more efficient than now, and/or operate on fuels that are, on net, not based on petroleum and are much less carbon-intensive. Such fuels include some biofuels, electricity, and hydrogen. This chapter describes the vehicle technologies that could contribute to those reductions and estimates how their costs and performance may evolve over coming decades. Chapter 3 considers the production and distribution of fuels and their emissions.

Improving the efficiency of conventional vehicles, including hybrid electric vehicles (HEVs), is discussed first.[1] It is, up to a point, the most economical and easiest-to-implement approach to saving fuel and reducing emissions. It includes reductions of the loads the engine must overcome, specifically vehicle weight, aerodynamic resistance, rolling resistance, and accessories, plus improvements to the ICE powertrain and HEV electric systems However, if improved efficiency was the only way to meet the goals, then, for the expected vehicle miles traveled (VMT) in 2050, the average on-road fleet fuel economy would have to exceed 180 mpg.[2] Since that is extremely unlikely, at least with

currently identifiable technologies, additional options will be needed. Options considered by the committee include biofuels (discussed in Chapter 3), plug-in hybrid electric vehicles (PHEVs), battery-electric vehicles (BEVs [PHEVs and BEVs are collectively referred to as plug-in vehicles, PEVs]), fuel cell electric vehicles (FCEVs), and ICE vehicles (ICEVs) using compressed natural gas (CNGVs).

ICEVs and PHEVs will require little or no modification to operate on "drop-in" biofuels or synthetic gasoline derived from natural gas or coal. Vehicles that are powered by electricity or hydrogen are very different from current vehicles as described later in this chapter. CNGVs are also discussed, as they require a much larger fuel tank and other modifications. Upstream impacts of producing and providing electricity, hydrogen, and CNG are discussed in Chapter 3.

All these alternative vehicle options currently are more expensive than conventional ICEVs. The rate at which research and development (R&D) improves the performance and reduces the cost of new technologies is highly uncertain. To address this uncertainty, the analysis in this chapter considers two technology success pathways. The midrange case is the committee's best assessment of potential cost and performance should all technologies be pursued vigorously. The committee also developed a stretch case with more optimistic, but still feasible, assumptions about advances in technology and low-cost manufacturing. Details of the technology assessments are in Appendix F.

The committee's estimates are not based on detailed evaluations of all the specific technologies that might be used by 2050. It is impossible to know exactly which technologies will be used that far in the future, especially since major shifts from current technology will be necessary to meet this study's goals for reduced light-duty vehicle (LDV) petroleum

---

[1]All fuel economy (mpg) and fuel consumption numbers discussed in Chapter 2 are based on unadjusted city and highway test results or simulations, and do not include in-use efficiency adjustments.

[2]To meet the goal of 303 million metric tons of carbon dioxide equivalent (MMTCO$_2$e), 80 percent reduction from the 1514 light duty fleet emissions in 2005, with gasoline responsible for 10.85 kilograms CO$_2$e/gallon (8.92 from the tail pipe, the rest from refining and other upstream activities), at most only 28 billion gallons/year could be used (vs. 125 billion now). VMT in 2050 is expected to be about 5 trillion miles (see Chapter 5). Therefore, if the goal were to be met only with efficiency and no advanced vehicle or fuel technology, average economy would have to be 180 mpg. For this

case only, the 80 percent oil reduction goal (28 billion gallons) is identical to the GHG goal.

use and GHG emissions.[3] The optimistic and midrange estimates reflect the committee's appraisal of the overall development challenges facing the general pathways, and the promise of the various technologies that might be employed to meet the challenges. These estimates do not consider issues of market acceptance, which are addressed in Chapters 4 and 5, and are not based on specific policies to encourage market acceptance. Both estimates assume that policies are adopted that are sufficiently effective to overcome consumer and infrastructure barriers to adoption.

The committee reviewed a wide range of studies on technology potential and cost but was not able to find a study based on up-to-date technology assumptions and a consistent methodology for all types of technologies through 2050. The 2017-2025 light duty fuel economy standards were based on analyses that included major improvements in data and estimation of technology benefits and costs, but assessed technology only through 2025 (EPA and NHTSA, 2011). The 2009 MultiPath study (ANL, 2009) used a consistent methodology through 2050, but it lacked this recent data. Thus, the committee performed its own assessment of technology effectiveness and costs, as described below and in Appendix F.

In order to compare technologies, all costs discussed in this chapter assume the economies of scale from high volume production even in the early years when production is low. The modeling in Chapter 5, which estimates the actual costs of following specific trajectories, modified these costs for early and low-volume production.

Great care was taken to apply consistent assumptions to all of the technologies considered. For example, the same amount of weight reduction was applied to all vehicle types, and vehicle costs were built up from one vehicle type to the next (e.g., hybrid costs were estimated based on changes from conventional vehicles, and PEV costs were based on changes from hybrid vehicles). This approach does not reduce the large uncertainty in forecasting future benefits and costs, but it does help ensure that the relative differences in costs between different technologies are appropriately assessed and are more accurate than the absolute cost estimates.

The committee made every attempt to ensure accurate technology assumptions. Fundamental limitations for all technologies were considered for all future assessments, such as the ones discussed below for lithium-ion (Li-ion) battery chemistry and for engine losses. As these limits were approached, the rate of technology improvement was

slowed down to ensure that the estimates stayed well short of the limits.

On the other hand, learning occurs primarily because manufacturers are very good at coming up with better and more efficient incremental improvements. For example, 10 years ago technology that uses turbochargers to boost exhaust gas recirculation (EGR) was virtually unknown for gasoline engines. This new development, enabled by sophisticated computer simulations and design, has the potential to improve overall ICEV efficiency by about 5 percent. Certainly some of the currently known technologies will not pan out as planned, but it is equally certain that there will be incremental improvements beyond what we can predict now. The estimates in this chapter reflect an effort to strike a careful balance between these considerations.

Learning also applies to cost. Historically, technology costs have continuously declined due to incremental improvements. For example, 6-speed automatic transmissions, currently the most common type, are cheaper to manufacturer than 4-speed automatic transmissions, thanks to innovative power flow designs that allow additional gear combinations with fewer clutches and gearsets.

Although significant continuing R&D yielding sustained progress and cost reduction in all areas is essential, the technology estimates used for the committee's analyses do not depend on any unanticipated and fundamental scientific breakthroughs in batteries, fuel cell systems, lightweight materials, or other technologies. Therefore the estimates for improvements may be more readily attained, especially for 2050, when technology breakthroughs are quite possible. For example:

- Batteries beyond Li-ion were not considered for PEVs because the challenges facing their development make their availability highly speculative.
- Fuel cell efficiency gains were much less than theoretically possible, based on the assumption that developers will consider reducing the cost of producing a given power level to be more important.
- Reducing weight with carbon fiber materials was not included in the analyses, because the committee was uncertain if costs would be low enough by 2050 for mass market acceptance.
- The annual rate of reduction for the various vehicle energy losses was assumed to diminish after 2030, usually to about half of the historical rate of reduction or the rate projected from 2010 to 2030. This reflects reaching the limits of currently known technology and implicitly assumes that the rate of technology improvements will slow in the future, despite the current trend of accelerating technology introduction.
- Only turbocompounding was considered for waste heat recovery, even though other methods with much

---

[3]The committee did not assess GHG emissions from the production of vehicles or include such emissions in its analyses of emissions trends later in this report. Given that vehicles are expected to last about 15 years, any differences in production emissions will not make a large difference in lifetime emissions. In addition, data on emissions from the production of vehicles is poor, and estimates for advanced vehicles in several decades will be even more uncertain.

higher potential waste heat recovery rates are being researched (Ricardo, 2012).

- Radical new ICE combustion techniques with potentially higher thermal efficiency were not considered due to uncertainty about cost and durability. In fact, the assumptions for thermal energy in the committee's modeling for the 2030 optimistic and 2050 midrange cases were very similar to the efficiency levels considered achievable by Ford's next generation Eco-Boost engine with "potentially up to 40% brake thermal efficiency . . . at moderate cost" (Automotive Engineering, 2012).

2.2 VEHICLE FUEL ECONOMY AND COST ASSESSMENT METHODOLOGY

2.2.1 Fuel Economy Estimates

This committee's approach to estimating future vehicle fuel economy differs from most projections of future ICE efficiency, which have generally assessed the benefits of specific technologies that can be incorporated in vehicle designs (see Appendix F). Such assessments work well for estimates out 15 to 20 years, but their usefulness for 2050 suffers from two major problems. One is that it is impossible to know what specific technologies will be used in 2050. The traditional approaches taken to assess efficiency, such as PSAT and ADVISOR, depend on having representative engine maps, which do not exist for the engines of 2050. The second is that as vehicles approach the boundaries of ICE efficiency, the synergies, positive and negative, between different technologies become more and more important; that is, when several new technologies are combined, the total effect may be greater or less than the sum of the individual contributions.

The three-step approach used here avoids these problems. First, for ICE and HEV technologies, sophisticated computer simulations conducted by Ricardo were used to establish powertrain efficiencies and losses for the baseline and 2030 midrange cases.[4] These simulations fully accounted for synergies between technologies. Second, the efficiencies and losses of the different powertrain components and categories were determined. Using these categories to extrapolate efficiencies and losses allowed the committee to properly assess synergies through 2050. Third, the estimates of future efficiencies and losses were simultaneously combined with modeling of the energy required to propel the vehicle as loads, such as weight, aerodynamics, and rolling resistance, were reduced. This approach ensures that synergies are prop-

erly assessed and that the modeled efficiency results do not violate basic principles.

The committee estimated conventional powertrain improvements using the results of sophisticated simulation modeling conducted by Ricardo (2011). This modeling was used by the U.S. Environmental Protection Agency (EPA) to help set the proposed 2025 light-duty vehicle CO_2 standards. Ricardo conducted simulations on six different vehicles, three cars and three light trucks, which examined drivetrain efficiency (not load reduction) in the 2020-2025 timeframe. The simulations were based on both existing cutting-edge technologies and analyses of technologies at advanced stages of development.

EPA post-analyzed Ricardo's simulation runs and apportioned the losses and efficiencies to six categories—engine thermal efficiency, friction, pumping losses, transmission efficiency, torque converter losses, and accessory losses. The committee used these results as representative of potential new-vehicle fleet average values in 2025 for the optimistic case and in 2030 for the midrange case. The 2050 mid-level and 2050 optimistic vehicles were constructed by assuming that the rates of improvement in key drivetrain efficiencies and vehicle loads would continue, although at a slower rate, based on the availability of numerous developing technologies and limited by the magnitude of the remaining opportunities for improvement.

Baseline inputs for 2010 ICEVs were developed by the committee from energy audit data that corresponded with specific baseline fuel economy. The model calculates changes in mpg based on changes in input assumptions over EPA's test cycles. Additional details of the model are in Appendix F. The results were averaged to one car and one truck for analysis in the scenarios, but the analysis for all six vehicles is in Appendix F.

Starting with the results for ICEVs, the energy audit model was then applied to the other types of vehicles considered in this report for each analysis year and for the midrange and optimistic scenarios. PHEVs were assumed to have fuel economy identical to their corresponding BEVs[5] while in charge-depleting mode (that is, when energy is supplied by the battery) and to HEVs in charge-sustaining mode (when energy is supplied by gasoline or diesel). Natural gas vehicles were assumed to have the same efficiency as other gasoline fueled vehicles.

Care was taken to use consistent assumptions across the different technologies. For example, the same vehicle load reduction assumptions (weight, aero, rolling resistance) were applied to all of the drivetrain technology packages.

---

[4]The committee accepts the Ricardo results. However, it should be noted that they are based in part on input data that has not been peer reviewed because it is proprietary.

[5]The BEVs evaluated have a 100 mile range. BEVs with longer range would have substantially heavier battery packs (and supporting structures), adversely affecting vehicle efficiency. PHEVs might have higher electric efficiency than long-range BEVs.

Variables considered by the model (not all variables were used for each technology) were the following:

- Vehicle load reductions:
  —Vehicle weight,
  —Aerodynamic drag,
  —Tire rolling resistance, and
  —Accessory load;
- ICE:
  —Indicated (gross thermal) efficiency,
  —Pumping losses,
  —Engine friction losses,
  —Engine braking losses, and
  —Idle losses;
- Transmission efficiency;
- Torque converter efficiency;
- Electric drivetrain:
  —Battery storage and discharge efficiencies,
  —Electric motor and generator efficiencies, and
  —Charger efficiency (BEV and PHEV only);
- Fuel cell stack efficiency,
  —Also the FCEV battery loop share of non-regenerative tractive energy;
- Fraction of braking energy recovered; and
- Fraction of combustion waste heat energy recovered.

Details of the input assumptions for alternative technologies and of the operation of the model are described in Appendix F.

2.2.2 Vehicle Cost Calculations

Future costs are more difficult to assess than fuel consumption benefits. The committee examined existing cost assessments for consistency and validity. Fully learned out, high-volume production costs were developed as described in this chapter and in Appendix F.

The primary goal was to treat the cost of each technology type as equitably as possible. The vehicle size and utility were the same for all technology types. Range was the same for all vehicles except for BEVs, which were assumed to have a 100 mile real-world range. Care was taken to match the cost assumptions to the efficiency input assumptions. Results from the efficiency model were used to scale the size of the ICE, electric motor, battery, fuel cell, and hydrogen and CNG storage tanks (as applicable). Consistent assumptions of motor and battery costs were used for HEVs, PHEVs, BEVs, and FCVs. Costs were calculated separately for cars and light trucks.

For load reduction, the cost of lightweight materials, aerodynamic improvements, and reductions in tire rolling resistance were assumed to apply equally to all vehicles and technology types.

ICE technology includes a vast array of incremental engine, transmission, and drivetrain improvements. Past experience has shown that initial costs of new technologies can be high, but generally drop dramatically as packages of improvements are fully integrated over time. The incremental cost of other technologies was compared to future ICE costs (FEV, 2012).

For HEVs, costs specific to the hybrid system were added to ICE costs, and credits for smaller engines and components not needed were subtracted to arrive at the hybrid cost increment versus ICE. Similarly, the other vehicle costs were derived from ICEVs by adding and subtracting costs for various components as appropriate. Battery, motor, and power electronics costs were assessed separately for electric drive vehicles.

2.3 LOAD REDUCTION (NON-DRIVETRAIN) TECHNOLOGIES

Many opportunities exist to reduce fuel consumption and CO_2 emissions by reducing vehicle loads, as shown in Table 2.1. The load reduction portion of improved efficiency will benefit all the propulsion options by improving their fuel efficiency, reducing their energy storage requirements, and reducing the power and size of the propulsion system. This is especially important for hydrogen- and electricity-fueled vehicles because battery, fuel cell, and hydrogen storage costs are quite expensive and scale more directly with power or energy requirements than do internal combustion powertrain costs. In particular, load reduction allows a significant reduction in the size and cost of electric vehicle battery packs.

TABLE 2.1 Non-drivetrain Opportunities for Reducing Vehicle Fuel Consumption

Light weighting	Structural materials
	Component materials
	Smart design
Rolling resistance	Tire materials and design
	Tire pressure maintenance
	Low-drag brakes
Aerodynamics	C_d (drag coefficient) reduction
	Frontal area reduction
Accessory efficiency	Air conditioning
	Efficient alternator
	Efficient lighting
	Electric power steering
	Intelligent cooling system

2.3.1 Light Weighting

Reducing vehicle weight is an important means of reducing fuel consumption. The historical engineering rule of thumb, assuming appropriate engine resizing is applied and vehicle performance is held constant, is that a 10 percent weight reduction results in a 6 to 7 percent fuel consumption savings (NHTSA/EPA/CARB, 2010). The committee specifically modeled the impact of weight reduction for each technology type, as this rule of thumb was derived for conventional drivetrain vehicles and other technologies may differ in their response to weight reduction.

A variety of recent studies (see Appendix) have evaluated the weight reduction potential and cost impact for light duty vehicles through material substitution and extensive vehicle redesign. The long-term goal of the U.S. DRIVE Partnership sponsored by the U.S. Department of Energy DOE) is a 50 percent reduction in weight (DOE-EERE, 2012).[6] Lotus Engineering projects a 2020 potential for about a 20 percent weight reduction at zero cost and 40 percent weight reduction potential at a cost of about 3 percent of total vehicle cost, from an aluminum/magnesium intensive design (Lotus Engineering, 2010).

2.3.1.1 Factors That May Affect Mass Reduction Potential

Towing Capacity Mass reduction potential for some light trucks will be constrained by the need to maintain towing capacity, which limits the potential for engine downsizing and requires high structural rigidity. Towing capacity is the only advantage of body-on-frame over unibody construction, thus it was assumed that the historical trend for conversion of minivans and sport utility vehicles (SUVs) from body-on-frame to unibody construction would continue and all vehicles that did not need significant towing capacity would convert to unibody construction. The committee accounted for towing capacity by reducing the weight of body-on-frame trucks (pickups and some SUVs) by only 80 percent of the mass reduction of passenger cars and unibody trucks (minivans and most SUVs). In other words, if a car in 2050 is estimated to be 40 percent lighter, a corresponding mass reduction for a body-on-frame truck would be limited to 32 percent.

Mass Increases Due to Safety Standards Weight associated with increased safety measures is likely to be lower than in the past. The preliminary regulatory impact analysis for the 2025 Corporate Average Fuel Economy (CAFE) standards looked at weight increases for a variety of safety regulations, including proposed rules that would affect vehicles through 2025 and estimated a potential weight increase of 100-120 pounds (EPA and NHTSA, 2011). That is about a 3 percent mass increase, which was factored into the committee's assessment of weight reduction potential.

Mass Increases for Additional Comfort and Accessories Vehicle weight decreased rapidly in the late 1970s and early 1980s because of high fuel prices and implementation of the initial CAFE standards, then increased significantly during the period from the mid 1980s to the mid 2000s when fuel prices fell and fuel economy standards were kept constant (EPA, 2012). Thus, projecting weight trends into the future is very uncertain.[7] Continued weight increases are inconsistent with the assumptions driving this study, i.e., a future that emphasizes improved vehicle efficiency, increased fuel costs, and strong policies to reduce fuel consumption. Not only will manufacturers have strong incentive to reduce weight, but the historical increase in comfort and convenience features is likely to slow and historical increases in weight associated with emission control technology should not continue.[8] The committee estimated that weight increases associated with additional comfort and accessories for the midrange scenarios would be roughly half of the historical annual weight increase during a period of fixed fuel economy standards, or 5 percent by 2030 and 10 percent by 2050. This adjustment was applied after the weight reductions considered here for lightweight materials. The optimistic cases did not include weight increases for additional comfort and accessories.

Mass Reductions Related to Smart Car Technology In the 2050 timeframe, a significant portion of LDVs may include crash avoidance technology and other features of smart car technology. Although it is possible that such features might lead to weight reduction, that is speculative and was not considered. The committee also did not consider driverless (or

---

[6]U.S. DRIVE is a government-industry partnership focused on advanced automotive and related energy infrastructure technology R&D. The partnership facilitates pre-competitive technical information to accelerate technical progress on technologies that will benefit the nation. Further information can be found at http://www1.eere.energy.gov/vehiclesandfuels/pdfs/program/us_drive_partnership_plan_may2012.pdf.

[7]In addition to weight increases, improvement in powertrain efficiency has been used to increase performance instead of improving fuel economy in the past. The committee concluded that, as for weight discussed above, power is unlikely to grow significantly under the conditions postulated for this study. Past performance increases occurred primarily during periods of little regulatory pressure, and this study assumes that strong regulations or high gasoline prices will be required to reach the levels of fuel economy discussed here. In addition, the average performance level of U.S. vehicles already is high, and many drivers aren't interested in faster acceleration. Finally, the advanced vehicles expected in the future are likely to operate at high efficiency over a broader range than current engines, so high power engines will detract less from fuel economy. Hence the committee decided that performance increases may not happen to a great degree and, if they did, would likely not have a significant impact on fuel economy in the future.

[8]Future emission reductions will be accomplished largely with improved catalysts and better air/fuel ratio control—neither of which will add weight to the vehicle.

autonomous) vehicles because it is not clear what the impact on fuel use may be. While they may lead to smaller cars and mass reduction because of improved safety, and driving a given route may be more efficient with computer controlled acceleration and braking and continuous information on congestion, people may be encouraged to live further away from their workplaces and other destinations because they can use the time in their vehicles more productively. More information on the potential impact of autonomous vehicles is in Appendix F.

2.3.1.2 Safety Implications

Any effects of fleet-wide weight reduction on safety will depend on how the reductions are obtained and on the distribution of weight reduction over different size classes and vehicle types. However, the footprint-based standards implemented in 2005 for light trucks and 2011 for cars eliminate any regulatory incentive to produce smaller vehicles, and there are few indications that substantial weight reduction through the use of lightweight materials and design optimization will have significant adverse net effects on safety (DOT, 2006). Advanced designs that emphasize dispersing crash forces and optimizing crush stroke and energy management can allow weight reduction while maintaining or even improving safety. Advanced materials such as high strength steel, aluminum and polymer-matrix composites (PMC) have significant safety advantages in terms of strength versus weight. The high strength-to-weight ratio of advanced materials allows a vehicle to maintain or even increase the size and strength of critical front and back crumple zones and maintain a manageable deceleration profile without increasing vehicle weight. Finally, given that all light duty vehicles likely will be down-weighted, vehicle to vehicle crash forces should also be mitigated, and vehicle handling may improve because lighter vehicles are more agile, helping to avoid crashes in the first place.

2.3.1.3 Weight Reduction Amount and Cost

Table 2.2 summarizes the weight reductions and costs per pound saved that are used in the committee's scenarios. The

table also includes carbon fiber in 2050 for context, even though the committee considers it unlikely that costs will drop sufficiently for widespread use in vehicles and it was not used in the vehicle benefit and cost analyses. As noted above, the midrange case includes some weight growth from additional consumer features.

The costs of weight reduction are ameliorated by the cost *savings* associated with the corresponding secondary weight savings from downsizing chassis, suspension and engine and transmission to account for the reduced structural requirements and reduced drivetrain loads from the reduced mass. Although estimates of the secondary savings vary, they may approach an additional 30 percent of the initial reduction (NRC, 2011).

2.3.2 Reduced Rolling Resistance

Rolling resistance, and the energy required to overcome it, is directly proportional to vehicle mass. The tire rolling resistance coefficient depends on tire design (shape, tread design, and materials) and inflation pressure. Reductions in rolling resistance can occur without adversely affecting wear and traction (Pike Research and ICCT, 2011). The fuel consumption reduction from a 10 percent reduction in rolling resistance for a specific vehicle is about 1 to 2 percent. If in addition the engine is downsized to maintain equal performance, historically fuel consumption was reduced 2.3 percent (NRC, 2006).

In 2005, measured rolling resistance coefficients ranged from 0.00615 to 0.01328 with a mean of 0.0102. The best is 40 percent lower than the mean, equivalent to a fuel consumption reduction of 4 to 8 percent (8 to 12 percent with engine downsizing). Some tire companies have reduced their rolling resistance coefficient by about 2 percent per year for at least 30 years. Vehicle manufacturers have an incentive to provide their cars with low rolling resistance tires to maximize fuel economy during certification. The failure of owners to maintain proper tire pressures and to buy low rolling resistance replacement tires increases in-use fuel consumption.

For this study, scenario projections of reductions in light-duty new-vehicle-fleet rolling resistance for the midrange case average about 16 percent by 2030, resulting in about a

TABLE 2.2 Summary of Weight Reduction and Costs Relative to Base Year 2010

| Year | Cars and Unibody Light Trucks | | | Body-on-Frame Light Trucks | | |
	Weight Reduction (%)	Cost ($/lb)	Reduction with Weight Growth (%)	Weight Reduction (%)	Cost ($/lb)	Reduction with Weight Growth (%)
2030	25	1.08	Midrange 20 Optimistic 25	20	0.86	Midrange 15 Optimistic 20
2050	40	1.73	Midrange 30 Optimistic 40	32	1.38	Midrange 22 Optimistic 32
2050 carbon fiber	50	6.0	Optimistic 50	40	6.0	Optimistic 40

4 percent decrease in fuel consumption, and about 30 percent in 2050, for about a 7 percent fuel consumption decrease. For the optimistic case, rolling resistance reductions were projected to be about 25 percent in 2030 and 38 percent in 2050.

2.3.3 Improved Aerodynamics

The fraction of the energy delivered by the drive-train to the wheels that goes to overcoming aerodynamic resistance depends strongly on vehicle speed. Unlike rolling resistance, the energy to overcome drag does not depend on vehicle mass. It does depend on the size of the vehicle, as represented by the frontal area, and on how "slippery" the vehicle is designed to be, as represented by the coefficient of drag. For low speed driving, e.g., the EPA city driving cycle, about one-fourth of the energy delivered by the drivetrain goes to overcoming aerodynamic drag; for high speed driving, one-half or more of the energy goes to overcoming drag. Under average driving conditions, a 10 percent reduction in drag resistance will reduce fuel consumption by about 2 percent. Vehicle drag coefficients vary considerably, from 0.195 for the General Motors EV1 to 0.57 for the Hummer 2. The Mercedes E350 Coupe has a drag coefficient of 0.24, the lowest for any current production vehicle (Autobloggreen, 2009). Vehicle drag can be reduced by measures such as more aerodynamic vehicle shapes, smoothing the underbody, wheel covers, active cooling aperture control (radiator shutters), and active ride height reduction.

For this study's scenarios, reduction in new-vehicle-fleet aerodynamic drag resistance for the midrange case is estimated to average about 21 percent (4 percent reduction in fuel consumption) in 2030 and 35 percent (7 percent reduction in fuel consumption) in 2050. For the optimistic case, the aerodynamic drag reductions are estimated to average about 28 percent in 2030 and 41 percent in 2050.

2.3.4 Improved Accessory Efficiency

Accessories currently require about 0.5 horsepower from the engine for most vehicles on the EPA city/highway test cycle. While small, this is a continual load that affects fuel economy. Accessory load reductions were assessed using Ricardo simulation results and the EPA Energy Audit data, as described above. Overall, test cycle accessory loads were reduced about 21-25 percent by 2030 and 25-35 percent by 2050.

2.4 DRIVETRAIN TECHNOLOGIES FOR REDUCING FUEL CONSUMPTION

Currently, conventional gasoline-fueled ICE drivetrains generally convert about 20 percent of the energy in the gasoline into power at the wheels. The engine cannot operate at peak efficiency most of the time. Within the engine, energy is lost as heat to the exhaust or transferred to the cooling system. Moving parts create frictional losses, intake air is throttled (called "pumping" losses), accessories are powered, and the engine remains in operation at idle and during deceleration. In the transmission, multiple moving parts create friction, and pumps and torque converters create hydraulic losses. Also, when the vehicle brakes, much of the potential energy built up during acceleration is lost as heat in the friction brakes. Many or most of these losses and limitations can be reduced substantially by a variety of technological improvements. The technologies discussed below are just a few of the options. More information can be found in Appendix F. Note that biomass-fueled vehicles are being treated as conventionally powered vehicles in this study.

2.4.1 Conventional Internal Combustion Engine Vehicles

2.4.1.1 Gasoline Engine Drivetrains

Engines will improve efficiency in the future by increasing the maximum thermal efficiency and reducing friction and pumping losses. There are multiple technology paths for accomplishing these improvements.

Although the dominant technology used to control fuel flow in gasoline engines currently is port fuel injection, engines with direct injection of fuel into the cylinders have been rapidly entering the U.S. fleet. Gasoline direct injection (GDI, or just DI) systems provide better fuel vaporization, flexibility as to when the fuel is injected (including multiple injections), more stable combustion, and allow higher compression ratios due to intake air charge cooling. Direct injection reduces fuel consumption across the range of engine operations, including high load conditions, and increases low-rpm torque by allowing the intake valve to be open longer. Future GDI systems using spray-guided injection can deliver a stratified charge allowing a lean air/fuel mixture (i.e., excess air) for greater efficiency.

One approach that is rapidly penetrating the market is to combine direct injection with down-sized turbocharged engines. Turbocharging increases the amount of fuel that can be burned in the cylinders, increasing torque and power output and allowing engine downsizing. The degree of turbocharging is enhanced by GDI because of its cooling effect on the intake (air) charge and reduction of early fuel detonation. Further efficiency improvements are available with more sophisticated turbocharging techniques (e.g., dual-stage turbochargers) and combining turbocharging with some combination of variable valve timing, lean-burn, Atkinson cycle, and cooled and boosted EGR.

Ricardo developed engine maps specifically for an EGR DI turbo system, which uses the turbocharger to boost EGR in addition to intake air. This recirculates additional cooled exhaust gas into the cylinder to reduce intake throttling (and pumping losses), increase compression ratio, enable higher boost and further engine downsizing, and reduce combustion temperatures and early fuel detonation (Ricardo, 2011). This engine is projected to have a fuel economy benefit of 20 to 25 percent, compared to the baseline port fuel injected, naturally aspirated engine, by 2020-2025.

Turbocharging with GDI engines is likely to become very common by 2030 because the costs are modest and the fuel economy improvement significant.

Engine friction is an important source of energy losses. Friction reduction can be achieved by both redesign of key engine parts and improvement in lubrication. The major sources of friction in modern engines are the pistons and piston rings, valve train components, crankshaft and crankshaft seals, and the oil pump. Key friction reduction measures include the following (EEA, 2007):

- Low mass pistons and valves,
- Reduced piston ring tension,
- Reduced valve spring tension,
- Surface coatings on the cylinder wall and piston skirt,
- Improved bore/piston diameter tolerances in manufacturing,
- Offset crankshaft for inline engines, and
- Higher-efficiency gear drive oil pumps.

Over the past two and one half decades, engine friction has been reduced by about 1 percent per year (EEA, 2007). Continuing this trend would yield about an 18 percent reduction by 2030, but considerably greater reduction than this should be possible, especially with continued aggressive vehicle efficiency requirements. For example, surface technologies such as diamond-like carbon and nanocomposite coatings can reduce total engine friction by 10 to 50 percent. Laser texturing can etch a microtopography on material surfaces to guide lubricant flow, and combining this texturing with ionic liquids (made up of charged molecules that repel each other) can yield 50 percent or more reductions in friction.

There will also be improvements to transmission efficiency and reductions in torque converter losses. The primary advanced transmissions over the next few decades are expected to be advanced versions of current automatic transmissions, with more efficient launch-assist devices and more gear ratios; and dual-clutch automated manual transmissions (DCTs). Transmissions with 8 and 9 speeds have been introduced into luxury models and some mass market vehicles, replacing baseline 6-speed transmissions. The overdrive ratios in the 8- and 9-speed transmissions allow lower engine revolutions per minute (rpm) at highway speeds, and the higher number of gears allows the engine to operate at higher efficiency across the driving cycle. A 20 to 33 percent reduction in internal losses in automatic transmissions is also possible by 2020-2025 from a combination of advances, including improved finishing and coating of components, better lubrication, improvements in seals and bearings, and better overall design (Ricardo, 2011). Dual clutch transmissions, currently in significant use in Europe,

will also improve with the perfection of dry clutches and other improvements, with an additional reduction in internal losses (beyond advanced automatic transmissions) of about 20 percent. Their cost should also be lower than advanced automatic transmissions.

2.4.1.2 Estimation of Future Internal Combustion Engine and Powertrain Efficiency Improvements

As discussed earlier in Chapter 2, the committee estimated conventional powertrain improvements using the results of sophisticated simulation modeling on six different vehicles conducted by Ricardo for baseline (2010) and future (2025) vehicles. EPA post-analyzed Ricardo's simulation runs and apportioned the losses and efficiencies to six categories—engine thermal efficiency, friction, pumping losses, transmission efficiency, torque converter losses, and accessory losses.

The committee directly used EPA's 2025 results for the 2030 midrange case to ensure adequate time for the technologies to fully penetrate the entire fleet. These results were also extrapolated to 2050 by assuming that the percent annual improvements in each of the six categories after 2030 would be at most half the percent annual improvement calculated for 2010 to 2030. Optimistic estimates were calculated the same way, except that the Ricardo runs were used for 2025 instead of delaying the results until 2030. The total reductions for the various vehicles and losses are shown in Tables 2.9, 2.10, and 2.11, and in Appendix F.

2.4.1.3 Diesel Engines

This report has not explicitly considered diesel engines. Today's diesels are about 15-20 percent more efficient than gasoline engines, which would seem to mandate their inclusion in a study of greatly improved fuel economy. The committee ultimately decided, however, that a diesel case would not add significant value to the results of the study, primarily because the efficiency advantage of the diesel will be much smaller in the future as gasoline engines improve. Current diesels have a much higher level of technology than gasoline engines in order to address diesel drivability, noise, smell, and emission concerns, such as direct fuel injection, sophisticated turbocharging systems using variable geometry or dual turbochargers, and cooled EGR systems. As this same level of technology is added to the gasoline engine, the efficiency advantage of the diesel will be much smaller. Another consideration is that combustion technology by 2050 may blur, if not completely eliminate, the distinction between diesel and gasoline engine combustion. For example, diesel engines are reducing compression ratio in order to increase turbocharger boost and reduce emissions, while gasoline engines are increasing compression ratio due to improvements in combustion chamber design, increasing use of variable valve timing, and better control of EGR. Another example is development of homogenous charge compression ignition engines, which combine features of both gasoline and diesel engines.

2.4.2 Conventional Hybrid Electric Vehicles

HEVs combine an ICE, electric motor(s), and a battery or ultracapacitor. All the energy comes from the fuel for the ICE. HEVs reduce fuel consumption by:

- Turning off the engine during idling, deceleration, and coasting;
- Capturing a percentage of the energy that is normally lost to friction braking (i.e., regenerative braking);
- Engine downsizing (because the electric motor provides a portion of the maximum tractive power required);
- Allowing easier electrification of accessories such as power steering;
- Allowing the engine to operate more efficiently. By using the electric motor to drive the wheels at low load, or by operating the engine at a higher power (and higher efficiency) during low loads and capturing excess energy in the battery; and
- By allowing the use of efficient engine cycles, e.g. Atkinson cycle, that are impractical for conventional drivetrains.

The simplest HEV configuration has a "stop-start" system which shuts off the engine when idling and restarts it rapidly when the accelerator is depressed. These "micro-hybrids" need a higher capacity battery and starter motor than ICEVs. Stop-start systems are rapidly growing and are likely to be universal by 2030 because they are a relatively inexpensive way to achieve substantial fuel economy improvements. The benefits of stop-start systems are included in the committee's calculations for future ICEV efficiency. The hybrid vehicle projections assess the incremental efficiency above that of the stop-start system.

More complex systems that allow electric drive and substantial amounts of regenerative braking include parallel hybrid systems with a clutch between the engine and the motor, commonly referred to as P2 parallel hybrids (e.g., Hyundai Sonata hybrid). They have an electric motor inserted between the transmission and wheels, with clutches allowing the motor to drive the wheels by itself or in combination with the engine, or allowing the engine to drive the wheels without motor input. Powersplit hybrids (e.g., Prius) are another approach, with two electric machines connected via a planetary gearset to the engine and the powertrain. The committee determined that there is more opportunity for cost reduction on P2 hybrid systems in the future and used P2

systems for the future hybrid efficiency and cost assessments (see Appendix F).

About 60 percent of the fuel energy in an ICE is rejected as heat, roughly evenly divided between the engine cooling system (through the radiator) and the exhaust. Some of this heat can be recovered and used to reduce fuel consumption, especially from the exhaust, which is at a high temperature. Turbines, such as used for turbo-chargers, can generate electric power or transfer power to the crankshaft. Alternatively, thermoelectric couples can generate electric power directly, reducing fuel consumption by about 2 to 5 percent. HEVs would likely benefit more than ICEVs from waste heat recovery, as generated electric power could be used in their hybrid propulsion systems or to recharge the battery. This analysis assumes waste heat recovery systems will be applied starting in 2035, and only to HEVs. More efficient forms of waste heat recovery, such as Rankine cycle devices, were not included in the analyses.

There is some uncertainty about the fuel consumption benefit of advanced hybrid systems in the future. While hybrid systems will improve (more efficient components, improved designs and control strategies), advanced engines will reduce some of the same losses that hybrids are designed to attack (e.g., advanced engines will have reduced idle and braking fuel consumption, yielding less benefit from stopping the engine during braking and idling). In addition, even as hybrid drivetrains improve, conventional ICE fuel consumption will shrink, and the actual volume of fuel saved will go down. As done for ICEVs, the committee used the Ricardo simulations of 2025 hybrid vehicles to directly estimate losses and efficiency for the optimistic case in 2025 and for the midrange case in 2030. Unfortunately, Ricardo did not conduct simulations of baseline hybrid systems, so the annual rate of improvement from 2010 to 2025/2030 was assessed using Ricardo's ICE baseline simulations and differences in the 2025 simulations for ICE and hybrid vehicles to establish baseline hybrid energy losses. The committee's estimates are shown in Table 2.3.

TABLE 2.3 Estimated Future Average Fuel Economy and Fuel Consumption

	Cars				Trucks			
	Midrange		Optimistic		Midrange		Optimistic	
	ICE	HEV	ICE	HEV	ICE	HEV	ICE	HEV
Average Fuel Economy (miles per gallon)								
2010	31	43	31	43	24	32	24	32
2030	65	78	74	92	46	54	52	64
2050	87	112	110	145	61	77	77	100
Average Fuel Consumption (gallons per 100 miles)								
2010	3.20	2.34	3.20	2.34	4.24	3.10	4.24	3.10
2030	1.55	1.28	1.36	1.09	2.19	1.84	1.91	1.56
2050	1.15	0.89	0.91	0.69	1.64	1.30	1.30	1.00

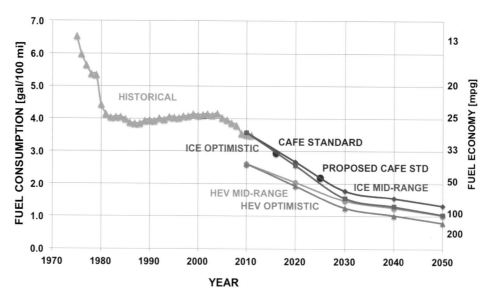

FIGURE 2.1 Historical and projected light-duty vehicle fuel economy.
NOTE: All data is new fleet only using unadjusted test values, not in-use fuel consumption.

While the gains projected by the committee are clearly ambitious, the rate of improvement for conventional vehicles (including use of stop-start systems and advanced alternators) is about 3 percent/year from 2010-2050. Light-duty trucks are expected to improve almost as much. Figure 2.1 compares these rates of improvement to past experience and the 2016 and 2025 CAFE standards. All of the vehicle modeling was assessed as percentage improvements over baseline vehicles. These results were adjusted by the ratio of the baseline used for the modeling in Chapter 5 to the average efficiency of the baseline vehicles used in Chapter 2.

The committee estimated HEV costs by adding the cost of the battery pack, electric motor, and other hybrid system components to the cost previously estimated for conventional vehicles. Credits were also applied for engine downsizing and deletion of the torque converter and original equipment alternator, with the exception that engine size was not reduced

on body-on-frame light trucks in order to maintain towing capacity. Weight and other load reductions were incorporated into calculations of the size of the engine, motor, and battery pack for each of the six vehicles. Credits associated with engine downsizing and eliminating the torque converter were subtracted. Except for the battery pack, hybrid system costs were based on detailed and transparent tear-down cost assessments conducted by FEV, Inc., on current production HEV vehicles, with learning factors and suitable design improvements applied to future HEV vehicles (FEV, 2012). Batteries are discussed in Section 2.5, below.

Currently, an HEV costs about $4,000 to $5,000 more than an equivalent ICEV, mostly for the battery, electric motor, and electronic controls. The committee's total direct manufacturing cost increments for hybrids, compared with 2010 reference vehicles, are shown in Table 2.4. Details on projected costs for hybrid systems are in Appendix F. Retail

TABLE 2.4 Efficiency Cost Increment Over Baseline 2010 Vehicle

| | Cars | | | | Trucks | | | |
| | Midrange | | Optimistic | | Midrange | | Optimistic | |
	ICE	HEV	ICE	HEV	ICE	HEV	ICE	HEV
2010	$0	$4,020	$0	$4,020	$0	$4,935	$0	$4,935
2015	$435	$3,510	$376	$3,006	$460	$4,228	$400	$3,601
2020	$986	$2,989	$867	$2,485	$1,059	$3,516	$939	$2,890
2025	$1,652	$3,017	$1,473	$2,590	$1,798	$3,446	$1,618	$2,942
2030	$2,433	$3,280	$2,195	$2,765	$2,676	$3,711	$2,436	$3,160
2035	$2,675	$3,357	$2,432	$2,973	$2,978	$3,834	$2,734	$3,408
2040	$2,960	$3,638	$2,713	$3,267	$3,332	$4,171	$3,085	$3,770
2045	$3,288	$3,949	$3,036	$3,577	$3,738	$4,540	$3,487	$4,142
2050	$3,659	$4,347	$3,403	$3,960	$4,196	$5,022	$3,941	$4,611

price markups are discussed in Chapter 5. Additional information on how the committee arrived at its estimates of fuel economy improvements and direct manufacturing costs are in Appendix F.

2.5 PLUG-IN ELECTRIC VEHICLES

Three distinctly different configurations that utilize battery power for propulsion are in production: HEVs, discussed in the previous section; PHEVs; and BEVs. Each has a rechargeable battery designed for a specific service. The Chevrolet Volt is the first mass-produced PHEV,[9] and Nissan's Leaf the first mass produced BEV[10] introduced into the U.S. market. Other manufacturers are introducing electric vehicles of both types over the next several years. Improvements in battery technology will be critical to the success of electric vehicles.

Plug-in hybrids are conceptually similar to HEVs. The same set of improvements in fuel economy that will benefit HEVs will also benefit PHEVs. PHEV batteries have about 4-20 kilowatt-hours (kWh) of stored energy that can be charged from the grid. PHEVs can travel 10 to 40 miles on electricity before the engine is needed. Thus a driver who does not exceed the electric range and charges the vehicle before using it again will use little or no gasoline. However, when driven beyond the charge depletion mode of the first 10 to 40 miles, the vehicles operate as conventional hybrid vehicles (in a charge sustaining mode), eliminating the range anxiety associated with BEVs. PHEV efficiency was assumed to be the same as BEV efficiency when operating on the battery pack and the same as HEV when the engine is running.

A BEV has no engine, a significant cost savings relative to PHEVs, but currently the battery pack for even a small, short-range vehicle is likely to be at least 20 kWh, and a large SUV might require 100 kWh for a range of 200 miles. The Nissan Leaf has a battery of 24 kWh. Battery cost will thus be a key determinant for the success of PHEVs and BEVs. Based on the energy modeling described earlier in this chapter, a car that today gets 30 mpg would, if built as a BEV, require about 26 kWh/100 miles. For a range of 300 miles, the battery would need at least 78 kWh of available energy.[11] With current technology and costs, this would be prohibitively expensive, heavy, and bulky for most applica-

tions and would take prohibitively long to charge. At \$450/kWh, the current battery pack cost estimate (see Section 2.5.3 below), a 78 kWh battery costs \$35,000. Prospects for reducing the cost are discussed below.

Other considerations for plug-in vehicles include the range that can be achieved in an affordable vehicle and the time required for recharging. As vehicle weight, aerodynamic resistance, and rolling resistance are improved, range can be improved for the same battery size, or a smaller, less expensive battery may be used for the same range. Many PHEVs and BEVs can be plugged in at home overnight on regular 110 or 220 volt lines. Gradual charging is generally best for the batteries, and night-time charging is best for the power supplier, as power demand is lower than during the day and excess generating capacity is available (see Chapter 3). Fast charging is more challenging for batteries, requires more expensive infrastructure, and is likely to use peak-load electricity with higher cost, lower efficiency, and higher GHG emissions.

2.5.1 Batteries for Plug-In Electric Vehicles

There is general agreement that the Li-ion battery will be the battery of choice for electric vehicles for the foreseeable future. It was developed for the portable electronics industry 20 years ago because of its light weight, superior energy storage capability, and long cycle life, attributes, which also are important for electric vehicles. Cell performance has increased steadily by improvements in the internal electrode structure and cell design and manufacturing processes, as well as the introduction of higher performance anode and cathode materials.

There are several Li-ion chemistries that are being investigated for use in vehicles, but none offers an ideal combination of energy density, power capability, durability, safety, and cost. HEVs are also shifting to Li-ion from the original nickel-metal-hydride chemistry. HEV batteries, which are optimized for high power, may differ from those for PHEVs and BEVs, which will be optimized for high energy and low cost.

Development of the cylindrical 18650 Li-ion cell for the portable electronics industry is representative of how automotive batteries may develop. In 1991, the cost of the 18650 was \$3.17/Wh. Twenty years later, the same cell costs \$0.20/Wh, while the charge capacity of the cell went from 1 Amp-hour (Ah) to over 3 Ah in the same volume (see Figure 2.2). These improvements resulted from the introduction of new, high-performance materials, improvements to the cell and electrode structure design, and high volume production processes with reduced wastage. As a rule of thumb for highly automated cell production, cell materials account for about 60 to 80 percent of the cell cost in volume production.[12]

---

[9]The Volt's all-electric range is certified by EPA as 38 miles. General Motors refers to the Volt as an extended range electric vehicle because all power to the wheels is delivered by the electric motor, unlike, say, Toyota's Prius PHEV. However, both are hybrids in that they have two fuel sources.

[10]The EPA certified range is 73 miles, but estimates vary widely; also, range is extremely sensitive to weather, driving conditions, and driver behavior.

[11]Available energy is typically less than nameplate battery pack capacity because batteries may not completely discharge to avoid damage to battery life and loss of power. In addition, available energy could effectively be reduced by energy required to offset the loss of vehicle efficiency caused by the additional weight of a larger battery for longer range.

---

[12]As used here, "materials" means processed materials ready for cell manufacture. It does not mean raw materials, which may be much cheaper.

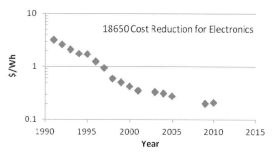

FIGURE 2.2 Cost of the 18650 portable electronics Li-ion cell (current dollars).
SOURCE: H. Takeshita, Tutorials, Florida International Battery Seminars, 1974-2010.

Cells for vehicles are likely to be prismatic (flat plate) or pouch-type rather than cylindrical, because these are easier to cool and arrange in stacks. The production process for flat plate vehicle cells differs from that for cylindrical cells, but it is anticipated that the cost will follow a similar learning pattern as the 18650 cell. Both the Volt and Leaf use a manganese spinel cathode and a graphite anode in a flat-plate configuration with a $LiPF_6$ electrolyte for long cycle life and relatively low cost.

Global R&D activity in Li-ion battery technology is funded at a level of several billion dollars annually. It explores all aspects of the technology and aims to improve energy-storage capacity per unit weight and volume, durability, safety characteristics, operating temperature range, manufacturing processes, and of course cost. Technologies that will offer improved performance without negatively affecting safety, durability, and cost, or, alternatively, improved cost without negatively affecting durability and safety are the only ones likely to find high-volume commercial application. In the next five years or so, optimization of the use of existing materials, engineering optimization of cell and component design, manufacturing process improvement, and economy of scale will support moderate improvements in performance and steady reduction in cost. In the longer term (8 to 15 years), introduction of materials with higher energy density could provide enhanced performance. Further out, probably beyond 2030, new chemistry may be developed but at this point in time no chemistry other than Li-ion is promising enough to be included in this analysis.

2.5.2 Automotive Battery Packs

A battery pack for vehicles consists of an assembly of cells, electrical components, structural components, a cooling system, module management electronics, and battery management system (BMS). A typical pack consists of 30 to several hundred cells configured in a series/parallel arrangement. The series arrangement includes 30 to 100 "virtual"

cells in strings that provide a battery voltage of 100 to 400 volts. The virtual cells include a single cell or several cells in parallel to provide the desired Ah capacity. In other combinations, several strings could be put in parallel to provide the total energy capacity required. Cells typically represent 50 to 60 percent of the cost of a battery in HEV applications, 60 to 70 percent of the cost of the pack in PHEV applications and 70 to 80 percent of the cost of pack in BEV applications. The BMS, structural components, electrical components, cooling systems, and assembly account for the balance. While the non-cell portion of the pack grows in complexity and cost from HEVs to PHEVs to BEVs, the number and cost of the cells increases faster.

The BMS is designed to maximize battery life, to minimize the risk of safety incidents, and to communicate to the vehicle controller the state of charge and state of health of the battery. The BMS monitors individual cell voltages, battery current, and battery temperature (measured in several places in the pack). When abnormal cell voltages, temperatures or current are measured, the BMS "takes action" to minimize damage to the battery or risk of safety events.

2.5.3 Battery Cost Estimates

Estimates of future vehicle battery costs vary widely and depend greatly on assumed production levels as well as technology development. Even current costs are uncertain because of proprietary information, and battery companies may sell batteries below costs in order to gain market share in the early stages of growth. The committee assumed that future costs of Li-ion cells for vehicles are likely to follow a similar (but dropping somewhat more gradually) trajectory as that for the 18650 cell shown in Figure 2.2. Those costs fell in a regular manner for 10 years and then began to level off as production processes matured and improved in reliability. Costs of the battery pack (in addition to the cells) also should decline at about the same rate as cells as manufacturers and suppliers improve designs and production techniques.

The starting point for the committee's projected costs for BEV battery packs in Figure 2.3 is $450/kWh for high rates of production.[13] Midrange BEV pack costs for 2030 are estimated at $250/kWh and $160/kWh in 2050. Optimistically, pack costs might reach $200/kWh in 2030 and $150/kWh in 2050.

The battery packs used in PHEVs, FCVs, and HEVs are smaller and must still provide high levels of power. This requires the use of batteries with higher power densities, which increases the cost per kWh of energy storage. PHEV pack costs are likely to be $60-70/kWh higher than BEV pack costs. HEV costs are highest because they are much smaller and require different characteristics. Batteries for

---

The processing of these materials is subject to considerable cost reduction, as is the cell manufacture.

[13]Actual costs for the Leaf and Volt battery packs in 2012 are estimated at about $500/kWh, which reflect lower production volumes. However, note that the Leaf battery does not have a liquid cooling system, and the packs may deteriorate faster. Hence that cost may not be typical.

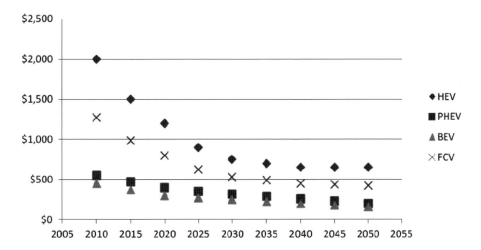

FIGURE 2.3 Estimated battery pack costs to 2050 ($/kWh).

fuel cell vehicles are between HEVs and PHEVs, as discussed later. Details on the committee's assessment of batteries are in Appendix F.

Using costs in Figure 2.3, the committee's estimate for a 30 mile range, as shown in Appendix F, is $4,000 (optimistic) to $4,600 (midrange). In comparison, DOE's 2015 goal for a battery pack for a PHEV with a 10-mile all-electric range in 2015 is $1,700 and $3,400 for a 40-mile range (Howell and Elder, 2012). PHEV battery costs depend on assumptions such as available energy (state of charge range) as well as how deterioration is handled and the vehicle that is to be propelled, but in general, the committee's assessment is less optimistic than DOE's targets.

A battery recycling effort will be needed when large numbers of battery packs reach the end of their useful lifetimes, and that will help to control costs. Recycling already works well for lead acid batteries, almost all of which are returned and the components reused in construction of new batteries.

2.5.4 Battery Technology for Future Applications

Li-ion battery technology for automotive applications may be limited to about 250 to 300 Wh/kg and $175 to $200/kWh (all at the pack level), although this report estimates that costs could get down to $150/kg by 2050. Research work around the world is examining other potential technologies that can yield higher energy density and/or lower cost per unit of energy. As noted before, none of the more futuristic systems has achieved enough maturity to be considered in this evaluation. Lithium sulfur chemistry utilizes a lithium metal anode and a cathode based on sulfur compounds. That system could theoretically double the specific energy of Li-ion batteries and offer competitive cost, but to date the cycling of both electrodes is quite problematic. Even more attention is given to the Li-Air chemistry. This chemistry

utilizes lithium-metal anodes and an air electrode so that the cathodic active material (oxygen) is taken from the air and at the charged state does not add to the weight of the battery (the battery gains weight as it discharges). This chemistry can theoretically provide a battery system with a specific energy of several kWh/kg. However, there are multiple independent technical challenges including the cyclability of the lithium electrode, cyclability of the air electrode, charge and discharge rate capability of the air electrode, finding suitable electrolyte, and finding a durable membrane permeable to the electrolyte but impermeable to water and CO_2. Several independent breakthroughs would have to occur to make the technology viable, and overall its chance of success is low.

2.5.5 Electric Motors

Almost all HEVs and PEVs use rare-earth-based interior permanent magnet (IPM) motors. IPM motors are by far the most popular choice for hybrids and EVs because of their high power density, specific power, efficiency, and constant power-to-speed ratio. Performance of these motors is optimized when the strongest possible magnets (NdFeB) are used. Cost and power density (power density equates to torque and acceleration) are emerging as the two most important properties of motors for traction drives in hybrid and EVs, although high efficiency is essential as well.

China currently has a near monopoly on the production of rare-earth materials, and since 2008 it has steadily raised the price of rare-earth magnet materials to as high as $60/kg. An automotive traction motor uses 1 to 1.5 kg of rare-earth magnet materials, which influences the cost of motors for electric vehicles.

The potential for a future shortage of rare-earth materials has led DOE to search for technologies that either eliminate or reduce the amount of rare-earth magnets in motors. The

TABLE 2.5 Motor Cost Estimates

	HEV/PHEV Costs		BEV/FCEV Costs	
	Fixed	$/kW	Fixed	$/kW
Midrange				
2010	$668	$11.6	$668	$11.6
2030	$393	$6.3	$425	$7.3
2050	$322	$5.2	$347	$6.0
Optimistic				
2010	$668	$11.6	$668	$11.6
2030	$349	$5.5	$381	$6.5
2050	$286	$4.5	$311	$5.3

DOE strategy continues ongoing cost-reduction efforts for rare-earth-based motors while also searching for new permanent magnet materials that do not use rare earths and motor designs that do not use permanent magnets.

Recently Toyota announced that it has developed a new material with equivalent or superior capability as rare-earth materials for the electric motors in its line of electric vehicles (Reuters, 2012). Toyota has also developed an induction motor that it claims is lighter and more efficient than the magnet-type motor now used in the Prius and does not use rare-earth materials.

In addition, U.S. production of rare-earths is resuming. Therefore, rare-earth materials are not likely to cause major increases in motor costs in the future. Overall, motor costs are likely to decline from about $2,000 now to less than $1,000 in 2050 for a typical electric car. This decline will result from better design and manufacturing and from the smaller size that will be needed to power more efficient future vehicles.

Table 2.5 presents the committee's motor cost estimates. These are based upon detailed tear-down cost estimates by FEV and include the cost of the motor, case, launch clutch, oil pump and filter, sensors, connectors, switches, cooling system, motor clutch, power distribution, and electronic control module. Some costs are independent of the size of the motor within the range considered here (fixed), and others are directly dependent (variable). Future cost projections included learning and incorporation of the electric motor into the transmission for HEV and PHEV applications. Further details on electric motors are in Appendix F.

2.5.6 Barriers to the Widespread Adoption of Electric Vehicles

2.5.6.1 Battery Cost

Cost is a key issue for the success of the electric vehicle. Lower cost electrode materials will be an important step. Cathode, separator and electrolyte are the main contributors to the cell cost. Most of the new cathode materials are composed of high cost nickel and cobalt materials. However,

lower cost, lower performance materials such as lithium iron phosphate and manganese spinel for cathodes and graphite for anodes can be made for about $10/kg or less in large volume. Battery pack costs per kWh are expected to decline by as much as two-thirds by 2050, as noted above, and pack size will also decline as vehicles become more efficient.

2.5.6.2 BEV Range and Recharge Time

Even with expected cost reductions, batteries will still be expensive and bulky, limiting the size that can be installed in most vehicles. BEVs must have reasonable range at reasonable cost if they are to widely replace ICEVs. The average conventional vehicle has a range of at least 300 miles on a tank of gasoline, but more range in a BEV requires a bigger battery, and that raises costs significantly as discussed above. Very few affordable BEVs will greatly exceed 100 miles for the next several years and possibly much longer. An even larger problem is recharge time. Unless batteries can be developed that can be recharged in 10 minutes or less, BEVs will be limited largely to local travel in an urban or suburban environment.

Battery swapping is being tested as a solution to the range and recharge time problems. A vehicle with a nearly discharged battery pack would drive into a station where a large machine would extract the pack and replace it with a fresh one. While battery swapping would, if widely available, solve the recharging and range problems, it also faces significant problems: (1) vehicles and battery packs would have to be standardized; (2) the swapping station would have to keep a large and very expensive inventory of different types and sizes of battery packs; (3) swapping stations are likely to start charging the incoming batteries right away in order to have them available for the next vehicle, possibly aggravating grid peaking problems; (4) batteries deteriorate over time, and customers may object to getting older batteries, not knowing how far they will be able to drive on them; and (5) most battery swapping will occur only when drivers make long trips, thus seasonal peaks in long-distance travel, e.g., during holidays, are likely to aggravate inventory problems. Although Israel has begun development of a battery swapping network and other countries appear to be considering it, the committee considers it unlikely that battery swapping will become an important recharging mechanism in this country.

2.5.6.3 Durability and Longevity

Battery life expectancy is a function of battery design and manufacturing precision as well as battery operating and charging behavior. Rapid charging and discharging can shorten the lifetime of the cell. This is particularly important because the goal of 10 to 15 years service for automotive applications is far longer than for use in electronic devices. Current automotive batteries are not expected to last for 15 years, the average lifetime of a car. Replacing the battery would be a very expensive repair, even as costs decline. Thus improved longevity is an important goal.

2.5.6.4 Safety

Battery safety is a critical issue. There are three major components that characterize the safety of a battery pack: the failure rate of an individual cell, the probability of propagation of a single cell fault to the pack, and the failure rate of the electronics. Li-ion batteries are high-energy-density systems that utilize a flammable electrolyte and highly reactive cathode and anode materials separated by a thin micro-porous separator. The potential thermal energy in the cell is much larger than the electro-chemical energy because the electrolyte is flammable in air and most anodes are metastable compounds that require kinetic protection at the surface. Li-ion cells contain sufficient energy to heat the cell to over 500°C if this energy is released rapidly inside the cell. That could cause neighboring cells to also fail, leading to a catastrophic event. Ensuring safe operation of vehicles that utilize large Li-ion batteries is a significant engineering task that includes the following:

a. Protection from overcharge;
b. Protecting the battery cells from deforming during crash;
c. Reducing the likelihood of an internal short that could develop due to poor cell design or to a manu-facturing defect (BMS should remove the cell from the circuit);
d. Designing a cell in such a way that even if an internal short does occur, it does not lead to thermal runaway of the cell;
e. Designing the BMS in such a way that even if a single cell experiences thermal runaway, the process does not propagate to neighboring cells and to the pack; and
f. Avoiding external shorts of the whole battery or sec-tions of it during installation, servicing, or normal usage.

Cell, battery, and vehicle engineers have developed mul-tiple tests to assess the ability of the cell, battery, and vehicle to operate without endangering human life. In most tests, bat-tery failure is allowed but fire or explosions are unacceptable.

The failure rate for Li-ion 18650 cells equates to a reli-ability rate of about 1 out of 10,000. This level of reliability is not satisfactory for electric vehicle batteries, where 1 out of a million is the minimum required (Takeshita, 2011). There-fore, it is essential to essentially eliminate cell construction defects in the individual cells, as well as defects in the battery pack electronics, in order to virtually eliminate the chances of a catastrophic event. Since increasing the energy density of the cell is associated with an increase of the thermal energy available per unit weight and volume, insuring safety while increasing energy density is particularly challenging.

2.6 HYDROGEN FUEL CELL ELECTRIC VEHICLES

The hydrogen FCEV is an all-electric vehicle similar to a BEV except that the electric power comes from a fuel cell system with on-board hydrogen storage. FCEVs are com-monly configured as hybrids in that they use a battery for cap-turing regenerative braking energy and for supplementing the fuel cell output as needed. Power electronics manage the flow of energy between the fuel cell, battery and electric motor.

The fuel cell system consists of a fuel cell stack and sup-porting hardware known as the balance of plant (BOP). The fuel cell stack operates like a battery pack with the anodes fueled by hydrogen gas and the cathodes fueled by air. The BOP consists of equipment and electrical controls that man-age the supply of hydrogen and air to the fuel cell stack and provide its thermal management. The vehicle is fueled with hydrogen at a fueling station much like gasoline fueling, and hydrogen is stored on the vehicle as a compressed gas or cryogenic liquid in a storage tank.

The key advantages of FCEVs include the following:

- High energy efficiency;
- No tailpipe emissions—neither GHG nor criteria pollutants—other than water;
- Quiet operation;
- Hydrogen fuel can be produced from multiple sources, thereby enabling diversity in energy sources (including low carbon and renewable energy sources) away from near-total reliance on petroleum;
- Full vehicle functionality for safe on-road driving, including 300-mile driving range;
- Rapid refueling; and
- Source of portable electrical power generation for off-vehicle use.

The key challenges of FCEVs are the following:

- Demonstration of on-road durability for 15-year service life;
- Maturation of the technology for cost reduction, greater durability, and higher efficiency; and
- Availability of fuel while few FCEVs are on the road and the eventual production and distribution of hydrogen at competitive costs (discussed in Chap-ter 3).

Several companies (e.g., Hyundai, Daimler, Honda, and Toyota) have announced plans to introduce FCEVs commer-cially by 2015, but mainly in Europe, Asia, California, and Hawaii where governments are coordinating efforts to build hydrogen infrastructures.

2.6.1 Current Technology for Hydrogen Fuel Cell Electric Vehicles

2.6.1.1 Fuel Cell Powertrain

The power electronics and electric motor/transmission are similar in efficiency and cost as for PHEVs and BEVs. Future improvements in the performance and cost of those systems will apply to FCEVs as well.

The battery in FCEVs has comparable power but greater energy content than that in current HEVs because it must power driving for 2 to 5 miles while the fuel cell warms up in cold weather. The fuel cell must be sized for nominal driving requirements and efficient operation. The battery will recharge from the fuel cell directly and through regenerative braking.

Over the past decade, FCEVs used in demonstration fleets have shown significant technology advances toward commercial readiness in the areas of performance and cost. For example, the cost of automotive fuel cell systems has been reduced from $275/kW in 2002 to $51/kW in 2010 (based on projections of high-volume manufacturing costs), and vehicle range has increased to at least 300 miles (James et al., 2010). Vehicles have demonstrated the capability to meet all urban and freeway driving demands. A remaining development challenge is proving the capability for high load driving at high ambient temperatures.

2.6.1.2 Fuel Cell Systems

Fuel cell stacks currently used in automotive applications are based on the polymer-electrolyte membrane/proton-exchange membrane (PEM). PEMs operate at moderate temperatures that can be achieved quickly so they are suitable for the infrequent and transient usage of on-road automotive service. Catalysts using precious metals (primarily platinum) are needed to promote the hydrogen/oxygen reaction that generates electricity in the fuel cell stack. Improvements in stack durability, specific power and cost have resulted from methods to improve the stability of the active catalytic surface area, and from new membrane materials and structures. For example, stack lifetimes of 2,500 operating hours (equivalent to approximately 75,000 mile range) have been demonstrated in on-road vehicles, and laboratory tests with newer membrane technologies have demonstrated (using accelerated test protocols) over 7,000 hours.

The BOP consists primarily of mature technologies for flow management of fluids and heat. Significant improvements in efficiency and cost result from continuing simplifications in BOP design, as illustrated in Figure 2.4.

Further reductions in the cost of fuel cell systems are expected to result from down-sizing associated with improved stack efficiency and improved response to load transients. Significant additional cost reductions will result if vehicle loads (weight, rolling resistance, and aerodynam-

ics) are reduced because that will allow the use of smaller hydrogen tanks and fuel cells with lower total power.

2.6.1.3 Fuel Cell System Efficiency

Fuel cell system efficiency measured for representative FCEVs driven on chassis dynamometers at several steady-state points of operation has shown a range of first-generation net system efficiencies from 51 to 58 percent. Second-generation vehicle systems have shown 53 to 59 percent efficiency at one-quarter rated power. System efficiency has improved slightly while the major design changes have focused instead on improving durability, freeze performance, and cost (Wipke, 2010a,b). With current fuel cell system efficiencies, fuel storage capacity and vehicle attributes (weight, aerodynamics, rolling resistance), FCEVs are currently capable of 200 to 300 miles of real-world driving range, and fuel efficiency over twice that of the comparable conventional ICEV.[14]

2.6.1.4 Fuel Cell System Cost

Projected costs for high volume production of fuel cells have dropped steeply with improved technology, dropping to $51/kW in 2010 for the fuel cell system, as shown in Figure 2.5. The fuel cell stack generally accounts for 50 to 60 percent of the system cost. Costs are very sensitive to production volume as shown in Figure 2.6.

2.6.1.5 Onboard Hydrogen Storage

Hydrogen storage costs are a significant element in the overall costs of a FCEV. Compressed gas at 5,000 psi (35 MPa) or 10,000 psi (70 MPa) has emerged as the primary technology path for the introduction of FCEVs because it is a proven technology that can meet the needs of the fuel cell (Jorgensen, 2011). Other possible future means of hydrogen storage (cryogenic or solid state) that have not been deployed in FCEV fleets were not considered by the committee.

The compressed gas storage capacity, and hence the vehicle driving range, is limited by the volume and cost of tanks that can be packaged in vehicles. Driving ranges over 300 miles are expected to be achieved, and a 300 mile real-world range, plus a 10 percent reserve, was used by the committee to calculate the size and cost of the storage tank.

Carbon-fiber reinforced composite (CFRC) tanks have been employed to achieve sufficient strength at manageable weight. Detailed cost analyses in Appendix F show total costs for representative 5.6 kg usable hydrogen systems are $2,900 for 35 MPa and $3,500 for 70 MPa (Hua et al., 2011). Car-

---

[14] 2011 Honda Clarity: ICEV fuel economy = 27 mpg, FCEV fuel economy > 60 mpg, with both mpg values based on (adjusted) fuel economy label values; ICEV fuel economy based on EPA, 2012; FCEV fuel economy from DOE, 2012a.

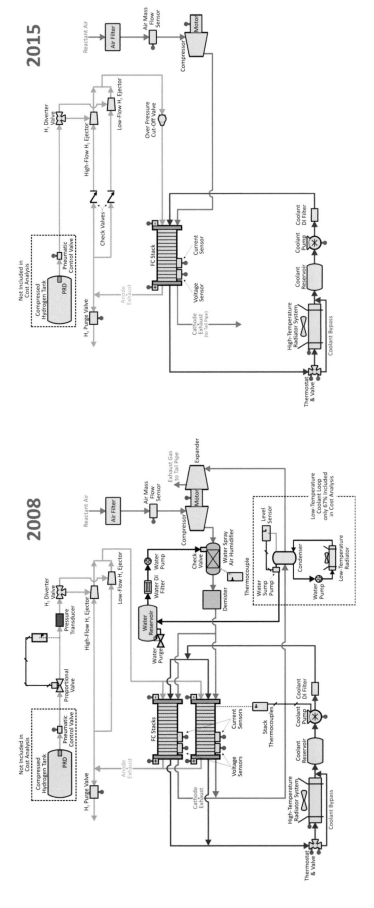

FIGURE 2.4 Continuing system simplification contributes to cost reduction.
SOURCE: James et al. (2010).

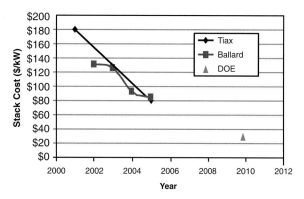

FIGURE 2.5 Historical progression of high-volume fuel-cell stack cost projections.
SOURCES: Kromer and Heywood (2007), NRC (2005, 2008), and Carlson et al. (2005).

bon fiber, priced at roughly $30/kg of the hydrogen stored, accounts for most of the cost of the CFRC wrapped layers that provide the structural strength of the storage system. The remaining costs are primarily attributed to flow-regulating hardware.

2.6.1.6 Vehicle Safety

The two primary features that distinguish FCEVs from ICEVs with respect to safety are high-voltage electric power and hydrogen fuel. The safety of high voltage electric power is managed on FCEVs similarly to HEVs, where safety requirements have resulted in on-road safety comparable to that of ICEVs. Experience from decades of safe and extensive use of hydrogen in the agriculture and oil refining industries has been applied to vehicle safety, and verified in vehicle maintenance and on-road demonstration programs.

Fire risk is mitigated because hydrogen dissipates much faster than do gasoline fumes and by regulatory provisions for fuel system monitoring. The safety of high-pressure onboard gaseous fuel storage has been demonstrated worldwide in decades of use in natural gas vehicles. Comparable safety criteria and engineering standards, as applied to ICEVs, HEVs, and CNGVs, have been applied to FCEVs with adaptation of safety provisions for differences between properties of natural gas and hydrogen. The United Nations has drafted a Global Technical Regulation for hydrogen-fueled vehicles to provide the basis for globally harmonized vehicle safety regulations for adoption by member nations (UNECE, 2012). Codes and standards will also be required for hydrogen fueling stations, as discussed in Chapter 3, but DOE has greatly reduced its work in developing them.

2.6.2 FCEV Cost and Efficiency Projections

Detailed analyses of current fuel cell costs and near-term improvements yield an estimated fuel cell system cost estimate of $39/kW for a high volume FCEV commercial introduction in 2015 (James 2010). This estimate reflects recent advances in technology and material costs, especially sharp reductions in the loading of precious metal in fuel cell electrodes. The platinum (Pt) loading in an earlier-generation 100 kW stack with ~80 g Pt at $32/g (2005 Pt price) would cost ~$2,500. For the 2010 loading of only 10 g Pt in a higher-technology alloyed-Pt 100 kW stack, the cost would be only ~$600 even at the higher 2011 Pt price of $58/g.

The committee estimates a midrange fuel cell system cost of $40/kW in 2020, and an optimistic cost of $36/kW, assuming additional cost benefit from potential near term technology developments. All cost estimates assume commercial introduction of FCEVs at annual production volumes over

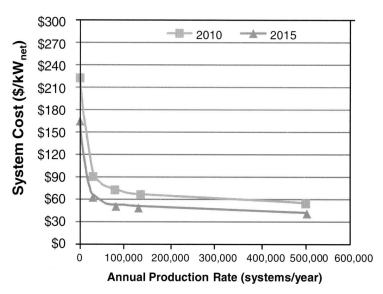

FIGURE 2.6 Progression of fuel cell system costs with production volume.
SOURCE: James et al. (2010).

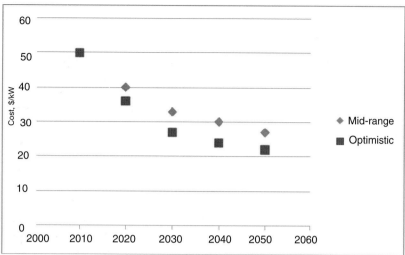

FIGURE 2.7 Fuel cell system estimated costs.

200,000 units, with the primary economy of scale occurring at 50,000 units (James, 2010).

Costs are likely to drop more rapidly in the earlier years of deployment because automotive fuel cell systems are in an early stage of development. Historically, reductions in weight, volume and cost and improvements in efficiency between successive early generations of a new technology are much more substantial than between more mature generations. Reductions of 2.3 percent per year in high volume cost in early generations of a technology, and 1 percent per year in later generations have commonly been observed. Therefore, for purposes of this report, technology-driven cost reductions from 2020 to 2030 of 2 percent per year were used for the midrange case and 3 percent per year for the optimistic case. This report assumes that improved technology will reduce costs by 2030 to $33/kW for the midrange and $27/kW for the optimistic scenarios.

Because of the major focus of fuel cell research and development on cost reduction prior to 2030, the committee expects that subsequent cost reduction rates will be slower, at 1 percent per year. By 2050, the midrange cost estimate is $27/kW and the optimistic is $22/kW. Cost estimates are shown in Figure 2.7. The supporting analysis is in Appendix F.

An evaluation of potential world Pt supply to support FCEVs as 50 percent of the on-road light-duty vehicle sales by 2050 assumed the conservative achievement of 15 g Pt per FCEV by 2050. Key documented findings are that (1) there are sufficient Pt resources in the ground to meet long-term projected Pt demand; (2) the Pt industry has the potential for expansion to meet demand for 50 percent market penetration of FCEVs (15 g Pt/vehicle) by 2050; and (3) the price of Pt may experience a short-term rise in response to increasing FCEV penetration, but is expected to return to its long-term mean once supply adjusts to demand (TIAX LLC, 2003). Scaled to 10 g Pt per FCEV (already achieved by 2010),

TABLE 2.6 Fuel Cell Efficiency Projections

	2010	2020	2030	2050
Midrange	53%	53%	55%	60%
Optimistic	53%	55%	57%	62%

the same conclusions apply to 80 percent penetration of the light-duty sales by 2050.

For the foreseeable future, technology developments for fuel cell systems are expected to prioritize reducing the cost of producing a given level of power (kW), rather than efficiency improvements. Therefore, even though significant gains in fuel cell efficiency are theoretically possible, this report assumes only modest improvements from the 2010 level of 53 percent as shown in Table 2.6.[15]

The cost of a CFRC hydrogen storage tank varies with the pressure and volume capacity. In addition, there is a fixed cost, independent of size, from equipment such as valves, pressure regulators and sensors. Reduction in the cost of CFRC tanks can be expected from two sources: new manufacturing/design techniques and the decreasing size of tanks as demand for fuel is reduced with improved vehicle efficiency.

Significant cost reduction from technology advancement is not expected by 2020, but several improvements in processing techniques are expected to reduce the cost of carbon fiber used in CFRC by 25 percent by 2030. The fixed cost fraction, which is associated with flow-control equipment, is expected to have modest potential for cost reduction because the technologies are mature. Therefore, a 1 percent per year cost reduction is applied to the fixed cost fraction, resulting

---

[15]The efficiency improvements in Table 2.6 were included in assessing the size and cost of the fuel cell stack.

TABLE 2.7 Illustrative Hydrogen Storage System Cost Projections

	2010	2020	2030	2050
Midrange				
Capacity (kg)	5.5	4.6	3.8	2.8
Cost ($)	3,453	3,031	2,402	1,618
$/kg-H$_2$	628	659	632	578
$/kWh	19	20	19	17
Optimistic				
Capacity (kg)	5.5	4.4	3.3	2.4
Cost ($)	3,453	2,938	2,055	1,326
$/kg-H$_2$	628	668	623	553
$/kWh	19	20	19	16

in a 10 percent cost reduction in the fixed cost fraction over the 2020-2030 period.

The midrange estimate for 2050 hydrogen storage cost results from continuation of the technology-driven 1 percent per year cost improvement over the 2030-2050 period in recognition of research into improvements in CRFC winding patterns and expectation of further improvements in manufacturing costs from added experience with high-volume production using new techniques (Warren, 2009). Hence, improved technology is estimated to reduce costs by 26 percent between 2020 and 2050. Research on cost reduction of structural CFRC is expected to accelerate with the new market driver of its broadened application to airplane fuselages, and other forms of hydrogen storage could become commercially viable.

Due to the difficulty in confirming promise among early stage research possibilities for manufacturing carbon fibers derived from polyacrylonitrile (PAN), or replacing it as the precursor for carbon fiber, the committee did not assume dramatic cost reductions for CFRC even by 2050. However, it is noted that a reduction in storage cost associated with achievement of a targeted <$10/kg carbon fiber and pressure shift to 50 MPa would be consistent with a cost reduction of 35 to 40 percent, the optimistic technology-driven projection in Table 2.7.

In addition to these technology-related cost projections, additional reductions can be expected when the storage system is down-sized. The volume of hydrogen that needs to be stored for full vehicle range declines as vehicle efficiency increases. This reduction in the variable fraction of the storage cost is directly proportional to the reduced vehicle load.

Promising areas for research and future technology development for improved energy efficiency, performance and cost of fuel cell systems and hydrogen storage are listed in Appendix F.

2.7 COMPRESSED NATURAL GAS VEHICLES

Increasing the use of natural gas in U.S. LDVs would displace petroleum with a domestic fuel, reduce fuel costs,

and reduce tailpipe GHG emissions.[16] A key driver of recent interest in natural gas vehicles is the potential from shale-based resources using hydraulic fracturing ("fracking"), and the likelihood that natural gas prices will remain well below gasoline prices for the foreseeable future. The supply of natural gas, and its potential for conversion into liquid fuels, electricity, or hydrogen, are discussed in Chapter 3. This section considers its direct use as a fuel in CNGVs with conventional ICE engines.

Adding a compressed gas storage tank is a larger problem for ICE vehicles than for fuel cell vehicles. This is because vehicle interior space is highly optimized and the large CNG tank compromises the interior space and utility. In contrast, FCEVs eliminate the internal combustion engine and drivetrain, plus the fuel cell stack can be configured in many different ways to optimize interior space. This allows additional room and flexibility for hydrogen storage tanks.

Some vehicles have been converted to burn CNG, but until recently the only dedicated CNG light-duty vehicle sold new in the United States was the Honda Civic Natural Gas vehicle (formerly called the GX). Chrysler has just introduced a CNG pickup, and Ford and General Motors are expected to follow soon. CNGVs have been much more popular in other countries, especially Italy, although sales recently plummeted in Italy after the end of incentives.

2.7.1 Fuel Storage

The key issue is the vehicle storage tank. In order to store enough natural gas for a reasonable driving range, it must be compressed to high pressure. CNGVs can be fast-filled at fueling stations that have natural gas storage facilities and large compressors, or they could be filled overnight, typically at a rate of 1 gallon of gasoline equivalent per hour (gge/hr where gge is the amount of energy equivalent to a gallon of gasoline) at home, tapping into the residential natural gas service and employing smaller compressors.[17]

At 3,600 psi and 70°F, a CNG tank is about 3.8 times larger than a gasoline tank with the same energy content. CNG tanks also are heavier in order to manage the high pressure. The cheapest solid steel (type 1) cylinders weigh 4 to 5 times as much as the same capacity gasoline tank; advanced (Type 3) cylinders with thin metal liners wrapped with composite weigh about half as much as Type 1 tanks, though at higher cost. Tanks with polymer liners weigh even less, but at higher cost. The tank on the 2012 Honda Civic NG vehicle holds about 8.0 gge of CNG at 3,600 psi, giving the vehicle a range of 192 miles (EPA city) to 304 miles

---

[16]A CNGV emits about 25 percent less CO$_2$ than a comparable vehicle operating on gasoline. Upstream emissions of methane, including leakage, are discussed in Chapter 3.

[17]The natural gas must be of sufficiently high quality; Honda does not recommend home refueling at this time because of concern over moisture in the fuel in some parts of the country.

(EPA highway), while taking up half of the vehicle's trunk space. Higher pressure tanks (up to 10,000 psi) can reduce fuel storage space, though at added cost and increased energy required to compress the gas.

In the future, it may be possible to store CNG at 500 psi (within the 200-1500 psi range of the pressure of gas in natural gas transmission pipelines) in adsorbed natural gas (ANG) tanks using various sponge-like materials, such as activated carbon. This technology, which is still under development, could allow vehicles to be refueled from the natural gas network without extra gas compression, reducing cost and energy use and allowing the fuel tanks to be lighter. Also, at lower pressure, the shape of the tank can be adjusted as needed to fit the space available, thus minimizing the impact on cargo space. The committee did not include ANG tanks in its modeling.

2.7.2 Safety

When used as an automobile fuel, CNG is stored onboard vehicles in tanks that meet stringent safety requirements. Natural gas fuel systems are "sealed," which prevents spills or evaporative losses. Even if a leak were to occur in a fuel system, the natural gas would dissipate quickly up into the atmosphere as it is lighter than air—unlike gasoline, which in the event of a leak or accident pools on the ground and creates a cloud of evaporated fuel that is easily ignited. Natural gas has a high ignition temperature, about 1,200° F, compared with about 600° F for gasoline. While fires or even explosions could occur, overall the safety of CNGVs should be no worse than gasoline vehicles and is likely to be better.

2.7.3 Emissions

Compared with vehicles fueled with conventional diesel and gasoline, natural gas vehicles can produce significantly lower amounts of harmful emissions such as particulate matter and hydrocarbons. Natural gas has a higher ratio of hydrogen to carbon than gasoline, reducing CO_2 emissions for the same amount of fuel consumed. However, methane is a potent greenhouse gas, so it is important to prevent methane leakage throughout the well-to-wheels life cycle if the greenhouse gas benefits of natural gas are to be realized, as discussed in Chapter 3.

2.7.4 Vehicle Costs and Characteristics

Other than the tank, CNGVs do not require significant re-engineering from their gasoline counterparts, although the cylinder head and pistons must be redesigned for a higher compression ratio and the ignition system modified. These design costs are significant for low volume production, but should be almost zero at high-volume. The lower density of the fuel means that CNG engines have lower output than gasoline engines of the same size, though this is mitigated to

TABLE 2.8 Comparison of the Honda Civic NG with Similar Vehicles

	Civic NG	Civic LX	Civic Hybrid
MSRP[a]	$26,805	$18,505	$24,200
mpg	27/38	28/39	44/44
Fuel cost	$1,050	$1,800	$1,300
Power	110 HP	140 HP	110 HP
Cargo (cubic feet)	6.1	12.5	10.7
Weight (pounds)	2848	2705	2853
CO_2 (grams/mile)	227	278	202

[a]Manufacturer's suggested retail price.
SOURCE: American Honda Motor Company; available at http://www.honda.com/.

some extent by the higher compression ratios possible with the high octane of the fuel. For the analysis in this report, CNGVs are assumed to operate with the same efficiency as gasoline-powered vehicles, including future efficiency improvements. CNG engines were assumed to be 10 percent larger than other ICE engines for the purpose of calculating engine cost at the same power output.

CNGV vehicles currently are sold in very low volumes and, partly due to that, cost significantly more than their gasoline-powered counterparts. For example, the base price of the 2012 Honda Civic NG vehicle is about $8,000 more than a similarly equipped Civic LX. Table 2.8 compares the 2012 Honda Civic NG with the LX and the Civic Hybrid.

The CNGV has higher up-front vehicle costs mainly because its high-pressure storage tanks are bulky and expensive. Currently, a CNGV might require nearly ten years to recover the higher purchase price, but these costs should come down significantly as production volume increases. The large fuel tank also reduces vehicle interior space, especially in the trunk. CNGVs could also be built as hybrids with the same incremental cost and benefits as gasoline HEVs.

2.8 SUMMARY OF RESULTS

The previous sections present a variety of options for reducing oil use and GHG emissions in LDVs and a methodology for estimating how much might be accomplished by 2050. This section summarizes those results. An example of how one vehicle might evolve illustrates how the benefits and costs were determined. This is followed by a series of tables showing the technology results that were input into the energy audit model, the results of those analyses, and the data that was input to the scenario models discussed in Chapter 5. Detailed results can be found in Appendix F.

2.8.1 Potential Evolution of a Midsize Car Through 2050

As an illustration of how a vehicle might evolve with increasing fuel economy technology, this section examines a midsize car, one of the six vehicles the committee analyzed.

Both a conventional drivetrain and a hybrid electric drivetrain are traced from a baseline 2007 vehicle to a 2050 advanced vehicle. Similar information for a BEV and FCEV and for the other vehicle types is shown in Appendix F. This evolution assumes that there is continuous pressure (from either or both regulatory pressure and/or market forces) to improve fuel economy and reduce emissions of greenhouse gases.

Table 2.9 shows details of the evolution of the vehicle with a conventional drivetrain. As can be seen in the table, the combination of shifting to a downsized turbocharged direct injection engine with high EGR and an advanced 8-speed automatic transmission drastically reduces pumping losses within the engine and, to a lesser extent, reduces friction losses and increases indicated thermal efficiency. The combination of idle-off and an advanced alternator allow fuel use during idling to be virtually eliminated. In addition, engine efficiency at low loads can be improved by increasing the charging rate of the alternator to the battery, thereby storing the energy for later use and allowing the engine to operate at more efficient load levels. In addition, smart alternators can improve the capture of regenerative braking energy. There are also improvements in transmission and torque converter efficiency and reductions in accessory loads.

The overall result in both the 2030 mid-level and optimistic case is nearly a 50 percent increase over the EPA 2-cycle tests in overall brake thermal efficiency, and a similar increase in fuel economy (50.5 mpg for the mid-level case) with no changes in vehicle loads. With load reduction, 2030 fuel economy levels of nearly 66 mpg (mid) and 75 mpg (optimistic) are possible without full hybridization. The added benefits of the vehicle load reduction—in particular, the weight reduction, which pays back about 6 to 7 percent fuel economy improvement for every 10 percent reduction in weight—are quite powerful.

By 2050, strong additional benefits can be gained by further vehicle load reductions and, within the drivetrain, primarily by continued improvements in indicated efficiency and reductions in friction losses. Improvements in the transmission and torque converter are minor because most of the possible improvements have been done by 2030, but some further reduction in pumping losses and improvements in accessories is possible. Successful achievements of these improvements can yield startling levels of fuel economy—88.5 mpg for the mid-level case, and 111.6 mpg for the optimistic case. Note that these estimates are for the EPA test cycle, and on-road results will be significantly lower.

Table 2.10 tracks the evolution of the benefits of adding a hybrid drivetrain to the technologies already onboard the advanced conventional vehicles. Note that part of the "standard" benefits of hybrid drivetrains are already captured by the combination of stop-start and advanced alternators in the conventional vehicles. While the hybrid system allows elimi-

TABLE 2.9 Details of the Potential Evolution of a Midsize Car, 2007-2050

Conventional Drivetrain	Baseline	2030 Midrange	2030 Optimistic	2050 Midrange	2050 Optimistic
Engine type	Baseline	EGR DI turbo	EGR DI turbo	EGR DI turbo	EGR DI turbo
Engine power, kW	118	90	84	78	68
Transmission type	6-sp auto	8-sp auto	8-sp auto	8-sp auto	8-sp auto
Drivetrain improvements					
Brake energy recovered through alternator, %	—[a]	14.1	14.1	14.1	14.1
Reduction in transmission losses, %	n/a	26	30	37	43
Transmission efficiency, %	87.6	91	91	92	93
Reduction in torque converter losses, %	n/a	69	75	63	88
Torque converter efficiency, %	93.2	98	99	99	99
Reduction in pumping losses, %	n/a	74	76	80	83
Reduction in friction losses, %	n/a	39	44	53	60
Reduction in accessory losses, %	n/a	21	25	30	36
% increase in indicated efficiency	n/a	5.6	6.5	10.6	15.6
Indicated efficiency, %	36.3	38.4	38.7	40.2	42
Brake thermal efficiency, %	20.9	29.6	30.3	32.5	34.9
Load changes					
% reduction in CdA	n/a	15	24	29	37
CdA (m^2)	7.43	6.31	5.64	5.29	4.68
% reduction in Crr	n/a	23	31	37	43
Crr	0.0082	0.0063	0.0057	0.0052	0.0047
% reduction in curb weight	n/a	20	25	30	40
Curb weight, lb	3325	2660	2494	2328	1995
Fuel economy, test mpg	32.1	65.6[b]	74.9	88.5	111.6

NOTE: All conventional drivetrains have stop-start systems and advanced alternators that can capture energy to drive accessories.
[a]Ricardo assumed stop start and smart alternator, with 14.1 percent of braking energy recovered, resulting in fuel economy = 34.9 mpg.
[b]Fuel economy with drivetrain changes only = 50.5 mpg.

TABLE 2.10 Details of the Potential Evolution of a Midsize Car Hybrid, 2007-2050

Hybrid Drivetrain—P2 hybrid with DCT8 transmission	2030 mid	2030 opt	2050 mid	2050 opt
Engine power, kW	88	82	77	68
Drivetrain improvements				
% additional pumping loss reduction[a]	80	80	80	80
% additional friction loss reduction[a]	30	30	30	30
% tractive energy provided by regen	20	22	24	26
Brake thermal efficiency, %	33.7	34.3	36.3	38.5
% of waste heat recovered	0	0	1	2
Fuel economy, test mpg	81.7[b]	95.1	115.8	150.9
Hybrid benefit over conventional, %	25	27	31	35

[a]Additional from conventional drivetrain in that year.

[b]Fuel economy with drivetrain changes only = 62.6 mpg.

nation of most pumping losses, the actual overall efficiency improvement is modest as pumping losses were already reduced to low levels in the conventional ICE case. Most of the incremental efficiency gains from the hybrid system are due to the tractive energy provided by capture of regenerative braking energy.

As shown in Table 2.10, the overall hybrid fuel economy benefit over the corresponding conventional drivetrain vehicle increases from 25 to 27 percent in 2030 to 31 to 35 percent in 2050; however, the hybrid benefit in terms of actual fuel consumption actually declines in the future—from about 0.30 gallons per 100 miles for the 2030 mid-level case to 0.23 gallons per 100 miles in the 2050 optimistic case. In other words, as non-hybrid ICEVs grow more efficient, the actual fuel savings and monetary benefit of hybridization may decline even as hybrid systems improve. For example, as vehicle mass decreases, the potential energy savings from regenerative braking also decreases.

An interesting aspect of the evolution of hybrids is the improvement in the efficiency of electric components, not shown in the table but included in the fuel economy calculations. For example, the benefits of hybridization will increase with improvements in electric motor/generator efficiency, battery in/out efficiency, and improving control strategies as onboard computer power increases over time. Note that these benefits also apply to BEVs and FCVs. Also, the 2050 hybrids benefit from waste heat recovery.

2.8.2 Technology Results, Performance, and Costs

Tables 2.11 and 2.12 and Figures 2.8 through 2.11 summarize the results from the committee's vehicle analyses. Note that fuel consumption was directly assessed only for 2030 and 2050. Between 2010 and 2030 and between 2030 and 2050, fuel consumption was assumed to have a constant multiplicative reduction each year.

Table 2.11 presents the load reductions assessed by the committee. These reductions were applied consistently to the calculations of costs and benefit for all of the technology types. Note that "Trucks" in this table is the sales-weighted average of unibody and body-on-frame light trucks from Table 2.2.

Table 2.12 presents the overall fuel economy calculated by the committee for the average car and light truck of each type. It is presented in miles per gallon because that is the metric usually used in the United States. There are three caveats with the numbers in Table 2.12. First, at very high mpg levels, large changes in mpg are needed to have much impact on fuel consumption (see Figure 2.1 for an illustration of this effect). Second, Table 2.12 shows the mpg results of the test cycles which do not include the adjustment for real-world fuel consumption. Third, the BEV and FCEV numbers are for the vehicle and do not account for the energy needed to produce the electricity or hydrogen. This is especially important for BEVs, where there are substantial losses in electricity generation. Chapter 3 adds assessments of upstream energy losses.

Figures 2.8 through 2.11 present the incremental cost calculated for each of the technology types. Note that these costs are all incremental to a baseline 2010 conventional vehicle. They are also direct manufacturing costs to the manufacturer.

TABLE 2.11 Load Reduction, Percent Relative to 2010

	Rolling Resistance		Aerodynamic Drag		Mass	
	Cars	Trucks	Cars	Trucks	Cars	Trucks
2030 Midrange	26%	15%	18%	15%	20%	18%
2030 Optimistic	40%	30%	31%	29%	30%	27%
2050 Midrange	33%	23%	26%	24%	25%	23%
2050 Optimistic	46%	37%	39%	37%	40%	37%

TABLE 2.12 Estimated Miles per Gallon Gasoline Equivalent (mpgge) on EPA 2 Cycle Tests

	ICEV		HEV		BEV		FCEV	
	Cars	LT	Cars	LT	Cars	LT	Cars	LT
2010 Baseline (mpgge)	31	24	43	32	144	106	89	65
2030 Midrange (mpgge)	64	46	78	54	190	133	122	86
2050 Midrange (mpgge)	87	61	112	77	243	169	166	115
2030 Optimistic (mpgge)	74	52	92	64	219	154	145	102
2050 Optimistic (mpgge)	110	77	146	100	296	205	206	143

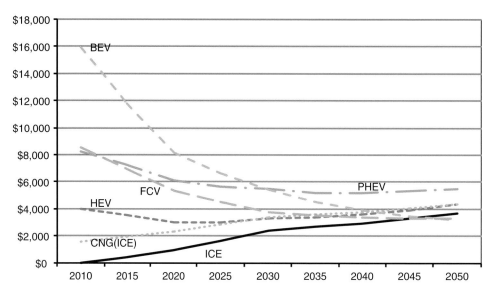

FIGURE 2.8 Car incremental cost versus 2010 baseline ($26,341 retail price)—Midrange case.

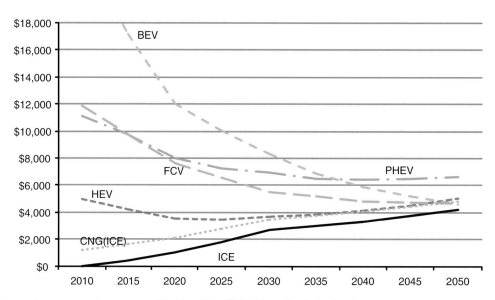

FIGURE 2.9 Light truck incremental cost versus 2010 baseline ($32,413 retail price)—Midrange case.

Markups for retail prices are evaluated in Chapter 5. Finally, the cost estimates assume that high volume production has already been realized. While this is not realistic for BEV, PHEV, and FCEV production in the near term, it allows all technologies to be evaluated on a consistent basis. Cost increases for near term, lower volume production are incorporated into the modeling in Chapter 5.

2.9 COMPARISON OF FCEVs WITH BEVs

FCEVs and BEVs are electric vehicles having no tailpipe GHG emissions. Both are "fueled" by an energy carrier (electricity or hydrogen) that can be produced from a myriad of traditional and renewable energy sources (biofuels, natural gas, coal, wind, solar, hydroelectric, and nuclear). Three primary considerations differentiate their prospects for introduction and acceptance as LDVs: vehicle attributes, rate of technology development, and infrastructure:

- *Vehicle attributes.* FCEVs provide the full utility of current on-road vehicles. BEVs, however, require time consuming "refueling" (recharging) and only offer limited driving range between "refuelings."
- *Rate of technology development.* A key requirement for realization of projected technology advances for battery and fuel cell systems is the continued dedication of research and development resources. Because demand for improved battery technologies is driven

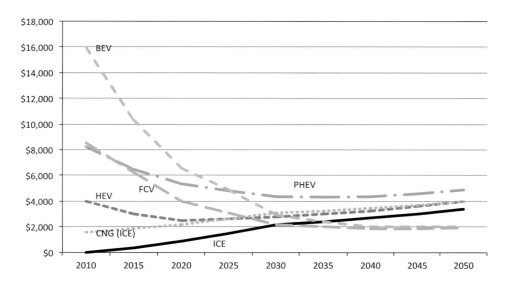

FIGURE 2.10 Car Incremental cost versus 2010 baseline ($26,341 retail price)—Optimistic case.

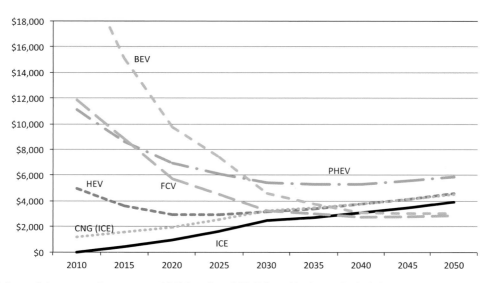

FIGURE 2.11 Light truck incremental cost versus 2010 baseline ($32,413 retail price)—Optimistic case.

by their established application in portable communication/computer devices, prospects for short-term return on R&D investments are substantial.

- *Infrastructure* is discussed in Chapter 3, but it should be noted that the barriers facing hydrogen are more formidable than those facing electricity. A brand new infrastructure for producing and distributing hydrogen would have to be built in concert with FCEV manufacturing. Neither is viable without the other, and the investments required both for manufacturing vehicles and hydrogen are extremely large. Both industries would require guarantees that the other will produce as promised, and that probably will entail a government role.

2.10 FINDINGS

- Large increases in fuel economy are possible with incremental technology that is known now for both load reduction and drivetrain improvements. The average of all conventional LDVs sold in 2050 might achieve EPA test values of 74 mpg for the midrange case and 94 mpg for the optimistic case. Hybrid LDVs might reach 94 mpg for the midrange case and 124 mpg for the optimistic case by 2050. On-road fuel economy values will be significantly lower.
- To obtain the efficiencies and costs estimated in this chapter, manufacturers will need incentives or regu-

latory standards, or both, to widely apply the new technologies.

- The unit cost of batteries will decline with increased production and development; additionally, the energy storage (in kWh) required for a given vehicle range will decline with vehicle load reduction and improved electrical component efficiency. Therefore battery pack costs in 2050 for a 100-mile real-world range are expected to drop by a factor of about 5 for the midrange case and at least 6 for the optimistic case. However, even these costs are unlikely to allow a mass-market vehicle with a 300-mile real-world range. In addition to the weight and volume requirements of these batteries, they are unlikely to be able to be recharged in much less than 30 minutes. Therefore BEVs may be used mainly for local travel rather than as all-purpose vehicles.

- BEVs and PHEVs are likely to use Li-ion batteries for the foreseeable future. Several advanced battery technologies (e.g., lithium-air) are being developed that would address some of the drawbacks of Li-ion batteries, but their potential for commercialization by 2050 is highly uncertain and they may have their own disadvantages.

- PHEVs offer substantial amounts of electric-only driving while avoiding the range and recharge time limitations of BEVs. However, their larger battery will always entail a significant cost premium over the cost of HEVs, and their incremental fuel savings will decrease as the efficiency of HEVs improves.

- The technical hurdles that must be surmounted to develop an all-purpose vehicle acceptable to consumers appear lower for FCEVs than for BEVs. However, the infrastructure and policy barriers appear larger. Well before 2050, the cost of FCEVs could actually be lower than the cost of an equivalent ICEV, and operating costs should also be lower. FCEVs are expected to be equivalent in range and refueling time to ICEVs.

- Making CNG vehicles fully competitive will require building large numbers of CNG fueling stations, moving to more innovative tanks to extend vehicle range and reduce the impacts on interior space, and developing manufacturing techniques to reduce the cost of CNG storage tanks.

- If CNGVs can be made competitive (both vehicle cost and refueling opportunities), they offer a quick way to reduce petroleum consumption, but the GHG benefits are not great.

- Codes and standards need to be developed for the vehicle-fueling interface.

- International harmonization of vehicle safety requirements is needed.

- While fundamental research is not essential to reach the targets calculated in this chapter, new technology

developments would substantially reduce the cost and lead time to meet these targets. In addition, continued research on advanced materials and battery concepts will be critical to the success of electric drive vehicles. The committee recommends the following research areas as having the greatest impact:

—Low-cost, conductive, chemically stable plate materials: fuel cell stack;

—New durable, low-cost membrane materials: fuel cell stack and batteries;

—New catalyst structures that increase and maintain the effective surface area of chemically active materials and reduce the use of precious metals: fuel cell stack and batteries;

—New processing techniques for catalyst substrates, impregnation and integration with layered materials: fuel cell stack and batteries;

—Energy storage beyond Li-ion: PHEVs and BEVs;

—Reduced cost of carbon fiber and alternatives to PAN as feedstock;

—Replacements for rare earths in motors;

—Waste heat recovery: ICEVs, HEVs, and PHEVs; and

—Smart car technology.

2.11 REFERENCES

ANL (Argonne National Laboratory). 2009. Multi-Path Transportation Future Study: Vehicle Characterization and Scenario Analyses. Chicago, Ill.: Argonne National Laboratory.

Autobloggreen. 2009. New Mercedes E-class coupe couples low drag coefficient to efficient engines Available at http://green.autoblog.com/2009/02/17/new-mercedes-e-class-coupe-couples-low-drag-coefficient-to-effic/. Accessed August 1, 2012.

Automotive Engineering. 2012. Ford's next-gen EcoBoost aims for 25% vehicle fuel economy improvements. Available at http://www.sae.org/mags/AEI/11043. Accessed August 1, 2012.

Carlson, E.J., P. Kopf, J. Sinha, and S. Sriramulu. 2005. *Cost Analysis of PEM Fuel Cell Systems for Transportation.* National Renewable Energy Laboratory. Subcontract Report NREL/SR-560-39104. September 30, 2005.

DOE (U.S. Department of Energy). 2012a. Fuel Economy. Available at http://www.fueleconomy.gov/feg/fcv_sbs.shtml.

DOE-EERE (U.S. Department of Energy, Energy Efficiency and Renewable Energy). 2012. U.S. DRIVE. Available at http://www1.eere.energy.gov/vehiclesandfuels/about/partnerships/usdrive.html. Accessed June 30, 2012.

DOT (U.S. Department of Transportation). 2006. 49 CFR Parts 523, 533 and 537 [Docket No. 2006-24306] RIN 2127-AJ61 Average Fuel Economy Standards for Light Trucks Model Years 2008-2011. Washington, D.C.: National Highway Traffic Safety Administration.

EEA (Energy and Environmental Analysis, Inc.). 2007. Update for Advanced Technologies to Improve Fuel Economy of Light Duty Vehicles. Prepared for U.S. Department of Energy. August. Arlington, Va.

EPA (U.S. Environmental Protection Agency). 2012. Light-Duty Automotive Technology, Carbon Dioxide Emissions, and Fuel Economy Trends: 1975 Through 2011. Washington, D.C.: U.S. Environmental Protection Agency.

EPA and NHTSA (U.S. Environmental Protection Agency and National Highway Traffic Safety Administration). 2011. 2017 and later model year light-duty vehicle greenhouse gas emissions and Corporate Average Fuel Economy Standards. *Federal Register* 76(213):74854-75420.

FEV (FEV, Inc.). 2012. FEV, Light-Duty Vehicle Technology Cost Analysis—European Vehicle Market (Phase 1). Prepared for the International Council on Clean Transportation. Analysis Report BAV 10-449-001.

Howell, D., and R. Elder. 2012. Electrochemical Energy Storage Technical Team (EESTT) Presentation to National Research Council Committee on Review of the U.S. DRIVE Partnership. January 26.

Hua, T.Q., R.K. Ahluwalia, J.K. Peng, M. Kromer, S. Lasher, K. McKenney, K. Law, and J. Sinha. 2011. Technical assessment of compressed hydrogen storage tank systems for automotive applications. *International Journal of Hydrogen Energy* 36(4):3037-3049.

James, B.D., J.A. Kalinoski, and K.N. Baum. 2010. Mass Production Cost Estimation for Direct H2 PEM Fuel Cell Systems for Automotive Applications: 2010 Update. Arlington, Va.: Directed Technologies, Inc. Available at http://www1.eere.energy.gov/hydrogenandfuelcells/pdfs/dti_80kwW_fc_system_cost_analysis_report_2010.pdf.

Jorgensen, S. 2011. Hydrogen storage tanks for vehicles: Recent progress and current status. *Current Opinion in Solid State and Materials Science* 15:39-43.

Kromer, M.A., and J.B. Heywood. 2007. Electric Powertrains: Opportunities and Challenges in the U.S. Light Duty Vehicle Fleet. MIT publication LFEE 2007-02 RP. Available at http://web.mit.edu/sloan-auto-lab/research/beforeh2/files/kromer_electric_powertrains.pdf.

Lotus Engineering. 2010. An Assessment of Mass Reduction Opportunities for a 2017-2020 Model Year Vehicle Program. Washington, D.C.: The International Council on Clean Transportation.

NHTSA/EPA/CARB (2010). *Interim Joint Technical Assessment Report: Light-Duty Vehicle Greenhouse Gas Emission Standards and Corporate Average Fuel Economy Standards for Model Years 2017-2025*.

NRC (National Research Council). 2005. *Review of the Research Program of the FreedomCar and Fuel Partnership*. Washington, D.C.: The National Academies Press.

———. 2006. *Tires and Passenger Vehicle Fuel Economy. Informing Consumers, Improving Performance*. Washington, D.C.: The National Academies Press.

———. 2008. *Transitions to Alternative Transportation Technologies—A Focus on Hydrogen*. Washington, D.C.: The National Academies Press.

———. 2011. *Assessment of Fuel Economy Technologies for Light-Duty Vehicles*. Washington, D.C.: The National Academies Press.

Pike Research and ICCT (International Council on Clean Transportation). 2011. Opportunities to Improve Tire Energy Efficiency. Available at http://theicct.org/tire-energy-efficiency. Accessed August 1, 2012.

Reuters. 2012. Toyota Finds Way to Avoid Using Rare Earth: Report. Available at http://www.reuters.com/article/2012/01/23/us-toyota-rare-earth-idUSTRE80M0JK20120123. Accessed August 1, 2012.

Ricardo. 2012. War on waste. *Ricardo Quarterly Magazine*, Q1.

———. 2011. Project Report: Computer Simulation of Light-Duty Vehicle Technologies for Greenhouse Gas Emission Reduction in the 2020-2025 Timeframe. Prepared for Office of Transportation and Air Qualtiy, U.S. Environmental Protection Agency. November 29.

Takeshita, H. 2011. Market Update on NiMH, Li Ion and Polymer Batteries. Proceedings of the International Battery Seminar. Boca Raton, Fla.

TIAX LLC. 2003. Platinum Availability and Economics for PEMFC Commercialization. Washington, D.C.: U.S. Department of Energy. Available at http://www1.eere.energy.gov/hydrogenandfuelcells/pdfs/tiax_platinum.pdf.

UNECE (United Nations Economic Commission for Europe. 2012. Working Party on Passive Safety (GSRP) Working Party 29 (WP29) Informal Safety Subgroup on Hydrogen Fuel Cell Vehicles (HFCV-SGS). May. Available at http://www.unece.org/trans/main/wp29/wp29wgs/wp29grsp/grspinf50.html.

Warren, C.D. 2009. ORNL Carbon Fiber Technology Development. Available at http://www.ms.ornl.gov/PMC/carbon_fiber09/pdfs/C_David_Warren_ORNL.pdf. Accessed August 1, 2012.

Wipke, K., S. Sprik, J. Kurtz, and T. Ramsden. 2010a. *2010 Annual Progress Report for NREL's Controlled Hydrogen Fleet and Infrastructure Analysis Project, Section VII.1*. Available at http://www.nrel.gov/hydrogen/pdfs/viii_controlled_fleet.pdf.

———. 2010b. *Controlled Hydrogen Fleet and Infrastructure Demonstration and Validation Project, Fall 2010*. NREL Report No. TP-5600-49642. Available at http://www.nrel.gov/hydrogen/pdfs/49642.pdf.

3

Alternative Fuels

This chapter discusses the fuel production and use associated with striving to meet the overall study goals of a 50 percent reduction in petroleum use by 2030 and an 80 percent reduction in petroleum use and in greenhouse gas (GHG) emissions from the light-duty vehicle (LDV) fleet by 2050 compared to the corresponding values in 2005. It addresses the primary sources of energy for making alternative fuels, the costs of alternative fuels, and the investment needs and the net GHG emissions of the fuels delivered to the LDV fleet over time. Alternative fuels are transportation fuels that are not derived from petroleum, and they include ethanol, electricity (used in plug-in electric vehicles [PEVs] such as plug-in hybrid electric vehicles [PHEVs] or battery electric vehicles [BEVs]), hydrogen, compressed or liquid natural gas, and gasoline and diesel derived from coal, natural gas, or biomass. Petroleum-based fuels are liquid fuels derived from crude oil or unconventional oils.

The chapter opens with a summary discussion of the study goals, fuel pathways, trends in the fuels market, fuel costs, investment costs, and GHG emissions for an LDV in 2030 using each fuel, and it includes a summary table for each of the last three categories, as well as some cross-cutting findings. More detailed discussions of each fuel follow the summary discussion, with a section devoted to each fuel. Also discussed are carbon capture and storage, and resource needs and limitations.

3.1 SUMMARY DISCUSSION

3.1.1 The Scope of Change Required

The study goals are aggressive and require significant improvements to the vehicle and the fuel system to meet the desired goals. The number of LDVs and the vehicle miles traveled (VMT) are expected to nearly double from 2005 to 2050, adding challenges to meeting the goals.[1] To reach the

goals with twice as many LDVs on the road in 2050 means that each LDV would consume on average only 10 percent of the petroleum consumed compared to 2005 and emit only 10 percent of the net GHG emissions. Gasoline and diesel made from petroleum would be nearly eliminated from the fuel mix to reach the petroleum reduction goal. The 80 percent net GHG emissions reduction goals can be met by various combinations of lower fuel consumption rate (inverse of fuel economy) and lower fuel net GHG emission (Table 3.1). The higher the reductions in LDV fuel consumption rate, the lower the reductions in fuel net GHG emissions would need to be to reach the GHG reduction goal. As discussed in Chapter 2, LDV fleet economy improvements of 3 to 5 times may be technically feasible by 2050, meaning that the average net GHG emissions of the fuel used in the entire LDV

TABLE 3.1 LDV Fuel Economy Improvement and Fuel GHG Impact Combinations Needed to Reach an 80 Percent Reduction in Net GHG Emissions Compared to 2005 Assuming a Doubling in Vehicle Miles Traveled (VMT)

LDV Fuel Economy Increase versus 2005	LDV Fuel Consumption Rate Relative to 2005 (percent)[a]	Requisite Reduction in Net Fuel System GHG Impact versus 2005 (percent)[b]
2×	50	80
3×	33	70
4×	25	60
5×	20	50
6×	17	40

[a]The vehicle fuel consumption rate (e.g., gal/100 mi) corresponding to a given increase in fuel economy (e.g., miles per gallon) relative to the base year level. For example, a quadrupling (4×) of fuel economy simply means that the fuel consumption rate is 25 percent of the base level.

[b]The net reduction of system-wide GHG emissions from fuel supply sectors needed to meet an LDV sector-wide 80 percent GHG reduction goal for a given fuel economy gain when assuming a fixed doubling of VMT, that is, without accounting for induced effects such as VMT rebound due to higher fuel economy.

---

[1]The EIA *Annual Energy Outlook 2011* (EIA, 2011a) is the basis for these projections.

fleet would have to be reduced by 50 to 70 percent per gallon of gasoline equivalent (gge) by that time.

Finding: Meeting the study goals requires a massive restructuring of the fuel mix used for transportation. Petroleum-based fuels must be largely eliminated from the fuel mix. Other alternative fuels must be introduced such that the average GHG emissions from a gallon equivalent of fuel are only about 40 percent of today's level.

3.1.2 Fuel Pathways

Many different alternative fuel pathways have been proposed, and this study selected seven different fuel pathways to analyze: conventional petroleum-based gasoline, biofuels (including ethanol and "drop-in"[2] biofuels), electricity, hydrogen, compressed natural gas (CNG), gas to liquids (GTL), and coal to liquids (CTL). These were selected because of their potential to reduce petroleum use, to be produced in large quantities from domestic resources, and to be technically and commercially ready for deployment within the study period. Most fuels selected have lower net GHG emissions than petroleum-based fuels. Other alternative-fuel pathways were discussed but not included for detailed analysis because they did not meet the first three criteria. For example, methanol is discussed in Appendix G.8 but was not included for detailed analysis because of environmental and health concerns that inhibit fuel distribution and retail companies from broadly offering methanol as a fuel.

The fuel costs, net GHG emissions, investment needs, and resource requirements were analyzed on a consistent basis for the different fuels to facilitate comparisons among fuels. Future technology and cost improvements for the selected fuels are considered and compared on a consistent basis, even though the extent of improvement for different fuels is likely to vary.

3.1.3 Developing Trends in the Fuels Market

Several developments in the energy markets over the past few years will have large impacts on long-term LDV fuel-use patterns. First, the fuel economy of the LDV fleet will increase rapidly over the next decade because of higher Corporate Average Fuel Economy (CAFE) standards effective through 2016 and proposed through 2025. The CAFE standards increase requirements from 23.5 mpg in 2010 to 34.1 mpg in 2016 to 49.7 mpg in 2025. Alternative fuels and new LDV technologies would compete with future gasoline or diesel LDVs that use much less petroleum and have lower net GHG emissions. From a consumer viewpoint, the decreasing volume of gasoline needed to travel a mile

reduces the economic motivation to switch from gasoline to an alternative fuel.

Second, biofuel production is expected to increase as a result of the Renewable Fuel Standard 2 (RFS2) passed as part of the 2007 Energy Independence and Security Act (EISA). This legislation mandated the consumption of 35 billion gallons of ethanol-equivalent[3] biofuel and 1 billion gallons of biodiesel (about 24.3 billion gge/yr based on energy content) by 2022. The detailed requirements of RFS2 are discussed in Appendix G.1. Based on the 2010 gasoline use of 136 billion gge/yr (8.88 million bbl/d), this mandate increases biofuel use from 9.9 percent (0.87 million bbl/d) to 18 percent (1.59 million bbl/d) of the gasoline mix by volume (EIA, 2011b). Although the mandated volume for cellulosic biofuel is not expected to be met by 2022, any additional biofuel volume in the conventional gasoline mix reduces the need for gasoline from petroleum and the volume of other alternative fuels needed to reach the study goals. See Section 3.2, "Biofuels," in this chapter for a detailed discussion.

Third, the volume of economic natural gas from shale deposits within the United States has been increasing rapidly. In its June 18, 2009, report the Potential Gas Committee upgraded by 39 percent the estimated U.S. potential natural gas reserves (defined as being potentially economically extractable by the use of available technology at current economic conditions) compared with its previous biannual estimate (Potential Gas Committee, 2009). Based on the new estimates, the probable natural gas reserves would provide about 86 years of consumption if the consumption rate stays at the current level. In 2011, the Potential Gas Committee increased its estimates such that 90 years of probable reserves exist based on 2010 consumption. Many previous studies on alternative fuels did not include natural gas as a possible source for LDV fuel because of limited domestic supply, and the likely price increase in electricity and residential heating costs associated with high natural gas use in the transportation market. With increasing domestic production, natural gas now is a viable option for providing transportation fuels through multiple pathways including electricity, hydrogen, GTL, and CNG. See Section 3.5, "Natural Gas," in this chapter and Appendix G.7 for a detailed discussion.

3.1.4 Study Methods Used in the Analysis

This study considers conventional and alternative fuels for the 2010-2050 period, and this committee undertook a number of tasks to generate possible fuel scenarios and data for use in the modeling efforts described in Chapter 5. The primary sources for the data are different for each fuel and are explained in the sections that provide details on each fuel below in this chapter. The committee made efforts to standardize input data and definitions between the primary

---

[2] Drop-in fuel refers to nonpetroleum fuel that is compatible with existing infrastructure for petroleum-based fuels and with LDV ICEs.

[3] A gallon of ethanol has about 77,000 Btu, compared with 116,000 Btu in 1 gallon of gasoline equivalent.

information sources. The tasks the committee performed include:

- Assessed the current state of the technology readiness for each fuel using information gathered from presentations made to this committee and published literature.
- Estimated future improvements to these technologies that could be broadly deployed in the study period.[4]
- Estimated the range of costs based on future technology for each fuel delivered to the LDV at a fueling station in a similar way for each fuel. The reference price basis in the Energy Information Administration's (EIA's) *Annual Energy Outlook 2011* (EIA, 2011a) is used for all primary fuel prices. Investment costs are expressed in 2009 dollars.
- Estimated the initial investment costs needed to build the infrastructure for each fuel pathway.[5]
- Estimated the net GHG emissions per gallon of gasoline-equivalent for each fuel based on the methods selected for producing the fuel. An upstream GHG component, a conversion component, and a combustion component were included in the estimate of net GHG emissions.

3.1.5 Costs of Alternative Fuels

The costs of alternative fuels through 2035 are estimated based on the energy raw material prices in the reference case of the *Annual Energy Outlook 2011* (AEO; EIA, 2011a), and the basis and assumptions for the estimates are explained in the individual fuel sections. Fuel prices beyond 2035 were estimated by the committee. Table 3.2 summarizes the expected alternative fuel costs for 2030 on a $/gge or $/kWh basis for some of the fuel pathways and shows the consumer's annual fuel costs for a new vehicle of that type based on 2030 estimated vehicle mileage.

While the values in Table 3.2 are useful guideposts for this analysis, there are a few factors to keep in mind. First, the fuel costs shown in Table 3.2 are untaxed—current or future taxes are not included and could alter the actual annual cost that consumers pay. Second, the per-gallon of gasoline-equivalent fuel cost estimates in 2030 are a snapshot in time and will likely change as technology develops and world energy prices change. Third, the untaxed fuel-purchase costs to consumers each year appear similar for most fuels except for CNG and the BEV, which are significantly lower than others. Given the small separation for the other options in 2030, untaxed fuel costs are not expected to be a significant driv-

TABLE 3.2 2030 Annual Fuel Cost per LDV, Untaxed Unless Noted

Fuel	Fuel Cost ($/gge or kWh)	Annual Consumer Use (gge or kWh)	Annual Consumer Fuel Cost ($/yr)
Gasoline (taxed)	3.64/gge	325 gge	1,183
Biofuel (drop in)	3.39/gge	325 gge	1,102
Gasoline (untaxed)	3.16/gge	325 gge	1,027
PHEV10[a]	3.16/gge	260 gge	913
	0.141/kWh	650 kWh	
CTL with CCS	2.75/gge	325 gge	894
GTL	2.75/gge	325 gge	894
PHEV40[b]	3.16/gge	130 gge	752
	0.175/kWh	1,950 kWh	
Hydrogen—CCS case	4.10/gge	165 gge	676
Natural gas—CNG	1.80/gge	325 gge	585
BEV	0.143/kWh	3,250 kWh	465

NOTE: All fuel costs are based on the 2011 AEO (EIA, 2011a) for 2030. The assumed fuel economies are representative of on-road LDV averages for 2030 described in the scenarios in Chapter 5. The following assumptions were made: 13,000 mi/yr traveled and 40 mpgge for liquid and CNG vehicles, 80 mpgge for hydrogen and 4.0 mi/kWh for electric vehicles. PHEV10 gets 20 percent of miles on electric, PHEV40 gets 60 percent. All costs are untaxed unless noted. Electricity cost includes the retail price plus amortization of the cost of a home charger.

[a]PHEV10 is a plug-in hybrid vehicle designed to travel about 10 miles primarily on battery power only before switching to charge-sustaining operation.

[b]PHEV 40 is a plug-in hybrid vehicle designed to travel about 40 miles primarily on battery power only before switching to charge-sustaining operation.

ing force for consumers to switch from gasoline to alternate vehicle technologies in this timeframe. Untaxed fuel cost differences of only several hundred dollars per year will not cover the additional vehicle costs described in Chapter 2.[6]

Finding: As the LDV fleet fuel economy improves over time, the annual fuel cost for an LDV owner decreases. With high fleet fuel economy, the differences in annual fuel cost between alternative fuels and petroleum-based gasoline decreases and the annual costs become similar to one another. Therefore, over time fuel-cost savings will become less important in driving the switch from petroleum-based fuels to other fuels.

3.1.6 Investment Costs for Alternative Fuels

The investment costs to build the fuel infrastructure are sizable for all of the alternative fuel and vehicle pathways. In fact, these costs remain among the most important barriers

---

[4]Some future technologies that might be developed during the study period are not included for detailed analysis because future efficiencies and costs are not well understood. Examples of this include photoelectrochemical hydrogen production and biofuels from algae.

[5]Investment costs are explained in Appendix G.2, "Infrastructure Initial Investment Cost."

---

[6]As pointed out in Chapters 4 and 5, consumers tend to value about 3 years worth of fuel savings when making decisions on initial vehicle purchases. Using the numbers in Table 3.2, 3 years of untaxed hydrogen saves only $1,501 compared with taxed gasoline during 2030. The cost saved is not enough to cover the higher cost of a fuel cell electric vehicle (FCEV).

to rapid and widespread adoption of alternatives. Table 3.3 shows the investment costs on a \$/gge per day basis and on a \$/LDV basis. This calculation includes only the investment in building a new form of infrastructure needed to make and deliver the fuel to the customer. It does not include investment to expand an already large and functioning infrastructure associated with producing more of the basic resource. For instance, for hydrogen made from natural gas, the investment cost includes the cost of converting natural gas to hydrogen, pipelines to deliver the hydrogen, and the full cost of a hydrogen station, but it does not include investments to produce natural gas or deliver it to a plant. A complete list of which costs are included or excluded is shown in Appendix G.2 "Infrastructure Initial Investment Cost." Details for these investment costs are found in the individual fuel sections below in this chapter.

The investment cost for a new petroleum refinery is included in Table 3.3 for perspective. However, with increasing fuel economy for the LDV fleet, no new refinery capacity will be needed during the study period. So in effect the initial investment cost for gasoline is near zero. The alternative-fuel-producing industry, in 2030, must make a \$1,000 to \$3,000 investment for each new alternative-fuel LDV, whereas almost none is needed for new petroleum gasoline LDVs. This cost differential is a major barrier to large-scale deployment of alternative fuels.

The scale, pace, and modularity of the infrastructure investments vary for the different vehicles and fuels. These differences are noted in the right-most column of Table 3.3. Two basic categories are used to describe the infrastructure requirements: centralized and distributed. Centralized infrastructure investments are those that are borne by a select number of decision makers. For example, the infrastructure for CTL, GTL, or gasoline requires large-scale plants (which cost billions of dollars each) that individual companies would pay for. Biofuels require large-scale investments for biore-

fineries. Hydrogen requires hydrogen production plants plus smaller-scale distributed investments by retailers to install new storage tanks and fuel pumps. The investment costs for BEVs and PHEVs in Table 3.3 include only the costs for home, workplace, and public chargers. The centralized infrastructure for CNG has already been built, and so the incremental CNG infrastructure costs include home fueling systems (paid for by car owners), or new filling stations (paid for by retailers). Thus, the infrastructure requirements vary from a few very large, multibillion-dollar investments (e.g., for biorefineries) made by a few decision makers in industry, to millions of small multithousand-dollar investments made by millions of decision makers such as consumers, ratepayers, and retailers.

Finding: The investment cost for a new fuel infrastructure using electricity, biofuels, or hydrogen is in the range of \$2,000 to \$3,000 per LDV. This is a significant barrier to large-scale deployment when compared with an infrastructure cost for using petroleum of only about \$530 per LDV.

3.1.7 GHG Emissions from the Production and Use of Alternative Fuels

Operational and infrastructure costs (as noted in Tables 3.2 and 3.3) are critical factors to consider for deployment. However, the net GHG emissions for the different vehicle and fuel options need to be examined to determine how the goal of 80 percent GHG reduction could be met. The estimates of annual GHG emissions in 2030 for different vehicle and fuel options are shown in Table 3.4.

Each vehicle and fuel option has a range of net annual GHG emissions because GHG emissions depend on how the fuels are produced. The range of net GHG emissions for biofuels is large because the net GHG emissions depend

TABLE 3.3　2030 Fuel Infrastructure Initial Investment Costs per Vehicle

Alternative Fuel	2030 Investment Cost	LDV Fuel Use per Day	Infrastructure Investment Cost (\$/vehicle)	Cost Burden
Electricity BEV	\$330/kWh per day	8.9 kWh	2,930	Distributed (car owners, ratepayers)
Electricity (PHEV40)	\$530/kWh per day	5.4 kWh	2,880	Distributed (car owners, ratepayers)
Biofuel (thermochemical)	\$3,100/gge per day	0.89 gge	2,760	Centralized (industry)
CTL (with CCS)	\$2,500/gge per day	0.89 gge	2,220	Centralized (industry)
Hydrogen (with CCS)	\$3,890/gge per day	0.45 gge	1,750	Centralized (industry) and distributed (retailers)
GTL	\$1,900/gge per day	0.89 gge	1,690	Centralized (industry)
Natural gas—CNG	\$910/gge per day	0.89 gge	810	Distributed (retailers and car owners)
Electricity (PHEV10)	\$370/kWh per day	1.75 kWh	650	Distributed (car owners, ratepayers)
Gasoline (new refinery—if needed)	\$595/gge per day	0.89 gge	530	Centralized (industry)

NOTE: Basis: 13,000 mi/yr and 40 mpgge for liquid and natural gas vehicles, 80 mpgge for hydrogen, and 4.0 mi/kWh for electric vehicles. PHEV10 gets 20 percent of miles on electric; PHEV40 gets 60 percent. Investment costs are explained in the individual fuel sections.

TABLE 3.4 Estimates of 2030 Annual Net GHG Emissions per Light-Duty Vehicle Used in the Modeling in Later Chapters

Fuel	Net GHG Emissions (kg CO_2e)	Annual Use		Annual GHGs Emissions per LDV (kg CO_2e)
CTL with CCS	12.29/gge	325 gge		4,000
GTL	11.47/gge	325 gge		3,730
Gasoline	11.17/gge	325 gge		3,630
PHEV10	0.590/kWh	650 kWh	380	3,290
	11.17/gge	260 gge		2,910
Natural gas	9.20/gge	325 gge		2,990
PHEV40	0.590/kWh	1,950 kWh	1,146	2,600
	11.1/gge	130 gge	1,454	
Hydrogen—low cost	12.2/gge	165 gge		2,010
BEV—reference grid	0.590/kWh	3,250 kWh		1,920
Biofuel—with ILUC[a]	5.0/gge	325 gge		1,620
BEV—low-GHG grid	0.317/kWh	3,250 kWh		1,030
Biofuel—without ILUC	3.2/gge	325 gge		1,040
Hydrogen—with CCS	5.1/gge	165 gge		840
Hydrogen—low-GHG case	2.6/gge	165 gge		430
Biofuel—with ILUC,CCS	−9.0/gge	325 gge		−2925

[a]Indirect land-use changes (ILUC) can have large impacts on net GHG emissions but can vary considerably.

Basis: 13,000 mi/yr and 40 mpgge for liquid and NGVs, 80 mpgge for hydrogen and 4.0 miles/kWh for electric vehicles. PHEV10 gets 20 percent of miles on electric; PHEV40 gets 60 percent. GHG estimates are explained in the individual fuel sections.

on many factors, including the type of feedstock used,[7] the management practices used to grow biomass (e.g., overuse of nitrogen fertilizer could increase N_2O flux), any land-use changes associated with feedstock production,[8] and the use of carbon capture and storage (CCS) with biofuel production. The range of differences for a BEV is determined by the average GHG emissions of the grid and over time may be quite different than shown in Table 3.4. Hydrogen has a large range of possible GHGs determined by the several different choices of production method.

The net GHG emissions from the three typical alternative fuels—biofuels, hydrogen, and electricity—can be either high or low depending on technology choices, carbon costs, regulations, and other factors. Choices driven by technology, economics, and policy determine the GHG emissions for future alternative fuels.

Finding: The GHG emissions from producing biofuels, electricity, and hydrogen can vary depending on the basic resource type and conversion methods used. Making these fuels with methods involving very low GHG emissions increases the technical and cost hurdles, especially during the introductory period. Actions to encourage the use of these more challenging methods should be timed to coincide with large-scale deployment and not be a burden during the introductory period for the fuel. Needed policy actions for each fuel pathway are listed in Appendix G.3.

3.2 BIOFUELS

3.2.1 Current Status

Biofuel is a generic term that refers to any liquid fuel produced from a biomass source. A number of different biofuel products (e.g., biobutanol and drop-in biofuels[9]) derived from different feedstocks (e.g., lignocellulosic[10] biomass and algae) have been proposed, but only corn-grain ethanol and biodiesel were produced in commercially relevant quantities in the United States as of the drafting of this report. Ethanol and biodiesel have been of interest because they can be easily synthesized using well-known processes from commercially available agricultural products (such as corn and soybeans in the United States, sugar cane in Brazil, and other oil seeds elsewhere). However, neither ethanol nor biodiesel is fully fungible with the current infrastructure and LDV fleet designed for petroleum-based fuels.

Ethanol and biodiesel are usually shipped separately and blended into the fuel at the final distribution point. Ethanol can be blended into gasoline in various proportions but has only about two-thirds of the volumetric energy content of petroleum-based gasoline. As of 2011, ethanol supplied almost 10 percent by volume of the U.S. gasoline demand (Figure 3.1). Biodiesel, produced via the transesterification of various vegetable oils or animal fats, supplied less than 1 percent of U.S. transportation fuel demand in 2011 (see Figure 3.1). U.S. biodiesel production capacity was about 2.7 billion gal/yr in 2010 (NBB, 2010), but actual production is significantly lower. Biomass can also be used to synthesize drop-in fuels, that is, synthetic hydrocarbons that would be fully fungible with existing infrastructure and vehicles.

The EISA included an amendment to the Renewable Fuel Standard in the Energy Policy Act (EPAct) of 2005. RFS2 mandated an increase of over 200 percent in the use of biofuels between 2009 and 2022. (See Box 1.1 in Chapter 1.) Biofuels, including corn-grain ethanol and biodiesel, currently require government subsidies or mandates to compete economically with petroleum-based fuels. Increases in ethanol consumption can also be limited by the "blend wall"

---

[7]Corn-grain ethanol is likely to have different net GHG emissions than cellulosic biofuel.

[8]Uncertainties in GHG emissions from land-use changes are a key contributor to the wide range of estimates for net GHG emissions from biofuels. Some biofuel feedstock such as corn stover would not contribute much to GHG emissions from land-use changes.

[9]Biofuels that are compatible with existing infrastructure and internal combustion engine vehicles (ICEVs) for petroleum-based fuels.

[10]Plant biomass composed primarily of cellulose, hemicellulose, and lignin.

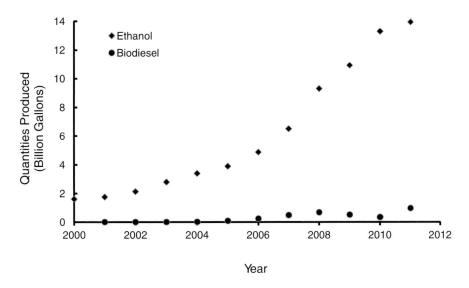

FIGURE 3.1 Amount of fuel ethanol produced in the United States.
SOURCE: Data from EIA (2012b,c).

(NRC, 2011). In 2010, the U.S. Environmental Protection Agency (EPA) approved the use of E15 in internal combustion engine vehicles (ICEVs) of model year 2001 or newer in response to a waiver request by Growth Energy and 54 ethanol manufacturers. Although EPA approved the use of E15 in 2010, its sale just began in July 2012 (Wald, 2012). In April 2012, EPA approved 20 companies for the manufacture of E15 (EPA, 2012a).[11] Without an approved method for eliminating misfueling of older cars,[12] increased ethanol use is likely to be constrained in the near term. In addition, auto manufacturers do not recommend using E15 in any vehicles that were initially designed to use E10 because of concerns that E15 might damage older engines (McAllister, 2012).

Flex-fuel vehicles (FFVs) can use higher concentrations of ethanol (up to 85 percent), and many auto manufacturers produce flex-fuel vehicles because of the CAFE credit[13] they receive (DOE-EERE, 2012c). However, the number of E85 fueling stations is limited (about 2,500 stations across the United States) and varies by state (DOE-EERE, 2012a). The price of E85 has always been higher than petroleum-based gasoline on an equivalent energy content basis.

Although the use of corn-grain ethanol can reduce petroleum imports, its effects on GHG emissions are ambiguous. Life-cycle assessments by various authors have estimated a 0 to 20 percent reduction in GHG emissions from corn-grain ethanol, relative to gasoline (Farrell et al., 2006; Hill et al., 2006; Hertel et al., 2010; Mullins et al., 2010).

The EISA requires the use of additional advanced and cellulosic biofuels that will reduce petroleum imports, lower CO_2e emissions, and be produced predominantly from lignocellulosic biomass. (See Appendix G.1 for definitions of biofuels in EISA.) To qualify as an advanced biofuel, a biofuel would have to reduce life-cycle GHG emissions by at least 50 percent compared with petroleum-based fuels.[14] To qualify as a cellulosic biofuel, a biofuel would have to be produced from cellulose, hemicellulose, or lignin and reduce life-cycle GHG emissions by at least 60 percent compared with petroleum-based fuels. Although RFS2 specified life-cycle GHG reduction thresholds for each type of fuel and EPA makes regulatory determinations accordingly, the actual life-cycle GHG emissions of biofuels could span a wide range (NRC, 2011). Biofuels facilities that began construction after 2007 would have to be individually certified for both biomass source and production pathway to qualify for renewable identification numbers (RINs).[15]

The U.S. government and private investors have invested billions of dollars to develop cellulosic biofuels (see Tables

---

[11]When the U.S. Environmental Protection Agency (EPA) approves a new fuel or fuel component, EPA only evaluates the fuel's impact on the emission control system and its ability to meet the evaporative and tailpipe emission standards. EPA does not evaluate the impact of the new fuel on any other aspect of vehicle performance, including degradation of vehicle components and performance that are not associated with the emission control system.

[12]The Renewable Fuels Association submitted a Model E15 Misfueling Mitigation Plan to EPA for review and approval on March 2012. The plan includes fuel labeling to inform customers, a product transfer documentation requirement, and outreach to public and stakeholders. However, those measures will not eliminate the possibility of accidental misfueling.

[13]CAFE credits were used to incentivize vehicle manufacturers to sell large numbers of vehicles that run on natural gas or alcohol fuels. See Chapter 6 for details.

[14]In its *Renewable Fuel Standard Program (RFS2) Regulatory Impact Analysis* (EPA, 2010b), EPA determined the life-cycle GHG emissions to be 19,200 g CO_2e/million Btu for petroleum-based gasoline and 17,998 g CO_2e/million Btu for petroleum-based diesel.

[15]The Renewable Identification Number (RIN) system was created by EPA to facilitate tracking of compliance with RFS. A RIN is a 38-character numeric code that corresponds to a volume of renewable fuel produced in or imported into the United States.

2.3 and 2.4 in NRC, 2011); however, no commercially viable processes are operational as of the drafting of this report. Initial research focused on cellulosic ethanol; however, the difficulties associated with integrating ethanol into the existing fuel distribution system and the inability to increase ethanol yields to the desired levels have resulted in a shift in research emphasis away from the biochemical conversion processes to the thermochemical or hybrid conversion processes. Conversion processes of lignocellulosic biomass to fuels are discussed below in this chapter.

3.2.2 Capabilities

The production potential of cellulosic biofuels is determined by the ability to grow and harvest biomass and the conversion efficiency of the processes for converting the biomass into a liquid fuel. Many studies have been published, and they show that the currently demonstrated conversion potential is about 46-64 gge/ton of dry biomass feedstock (as summarized in NRC, 2011). This represents an energy-conversion efficiency to liquid fuel of 25 to 50 percent based on the ratio of the lower heating value of the fuel product to that of the biomass feedstock. Much of the balance of the biomass-energy content is used to produce electricity and to power the conversion processes.

3.2.3 Biomass Availability

Multiple potential sources of lignocellulosic biomass can be used to produce biofuels. They include crop residues such as corn stover and wheat straw, fast-growing perennial grasses such as switchgrass and *Miscanthus*, whole trees and wood waste, municipal solid waste, and algae. Each potential source has a production limit. The consumptive water use and other environmental effects of producing biomass for fuels are discussed in detail in *Renewable Fuel Standard: Potential Economic and Environmental Effects of U.S. Biofuel Policy* (NRC, 2011).

Several studies have been published on the estimated amount of biomass that can be sustainably produced in the United States (NAS-NAE-NRC, 2009b; DOE, 2011; NRC, 2011, and references cited therein). All of the studies focused on meeting particular production goals and none of them projected biomass availability beyond 2030; they are discussed in Appendix G.4. The studies had different target production dates ranging from 2020 to 2030. The most recent study (DOE, 2011) projected that 767 million tons of additional biomass (above that currently consumed) could be available in 2030 at a farm gate price of less than $60/ton. This estimate was based on an annual yield growth of 1 percent and would require a shift of 22 million acres of cropland (or 5 percent of 2011 cropland) and 41 million acres of pastureland (or 7 percent of 2011 pastureland) into energy crop production. That amount was assumed to be available in 2050 in this report.

Finding: Sufficient biomass could be produced in 2050, when converted with current biofuel technology and consumed in vehicles with improved efficiencies consistent with those developed by the committee in Chapter 2 (about a factor-of-four reduction in fuel consumption per mile by 2050), that the goal of an 80 percent reduction in annual petroleum use could be met.[16]

3.2.4 Conversion Processes

Several technologies can be used to process biomass into liquid transportation fuels for the existing LDV fleet. Converting corn starch to ethanol and converting vegetable and animal fats to biodiesel or renewable (green) diesel are well-established commercial technologies. As of 2012, the collective capacity of corn-grain ethanol and biodiesel refineries in the United States is sufficient to essentially meet the 2022 RFS2 consumption mandates for conventional biofuels and biomass-based diesel.

There are a number of potential processes for converting cellulosic biomass into liquid transportation fuels. Demonstration facilities have been built for some of the various technologies. Much of the focus on cellulosic biofuel has switched away from ethanol to producing a biofuel that is a drop-in fuel.

Three main pathways are being developed to produce cellulosic biofuels: biochemical, thermochemical, and a hybrid of thermochemical and biochemical pathways. The pathways are discussed in detail in the report *Liquid Transportation Fuels from Coal and Biomass: Technological Status, Costs, and Environmental Impacts* (NAS-NAE-NRC, 2009b). Briefly, biochemical processes use biological agents at relatively low temperatures and pressures to convert the cellulosic material to biofuels—primarily ethanol and higher alcohols.

Thermochemical conversion uses heat, pressure, and chemicals to break the chemical bonds of the biomass and transform the biomass into many different products. Three main pathways are being considered for thermochemical conversion: gasification followed by Fischer-Tropsch (FT) catalytic processing to make naphtha and diesel, gasification followed by conversion of the syngas into methanol and subsequent conversion into gasoline via the methanol-to-gasoline (MTG) process, and pyrolysis (either high-temperature or lower-temperature hydropyrolysis) followed by hydroprocessing of the pyrolysis oil to produce gasoline and diesel. Other thermochemical pathways are also under development. Thermochemical and biochemical processes can be combined—for example, gasification of the biomass followed by fermentation of the syngas to produce ethanol or other alcohols.

---

[16]See Chapter 5 modeling results for further detail.

3.2.5 Costs

The economics of biofuel production have been discussed in a number of studies. Both NAS-NAE-NRC (2009b) and NRC (2011) compared recent information to develop comparative economics. The report *Renewable Fuel Standard: Potential Economic and Environmental Effects of U.S. Biofuel Policy* (NRC, 2011) and the references cited therein form the bases for the discussion of economics in this chapter.

Conversion of cellulosic biomass to drop-in biofuels is a relatively new and evolving suite of technologies. Predicting the future developments that can lower the cost of biofuel production is difficult. The cost of production is primarily a function of the cost of biomass, the yield of biofuels, and the capital investment required to build the biofuel conversion facility. Current conversion efficiencies are 46-64 gge/ton of dry biomass (which gives an average value of 55 gge per dry ton with a range of ±9 gge per dry ton).

Current capital costs to build a cellulosic biorefinery vary between 10 and 15 $/gge per year for all of the technologies discussed above. Thus, a biorefinery that would produce 36 million gge/yr consumes about 2,000 dry tons of biomass per day. The biorefinery would cost between $360 million and $540 million to build. An average capital cost would be 12.5 ± 2.5 $/gge per year. Because biorefining is a developing and evolving technology, it is reasonable to assume that yields will increase and that the capital costs will decrease as the technology matures. Yields will increase because of improvements in the catalysts used and in the process configurations. The capital costs are expected to decline primarily because

of economies of scale and improvements in the process configurations. Biorefineries that are bigger and more efficient than the first-mover facilities will be built as engineering and construction techniques are refined over time. The analysis is this chapter assumes that yields will increase from a baseline of 55 gge per dry ton in 2012 at a rate of 0.5 percent per year to a yield of 64 gge per dry ton by 2028. The capital costs are assumed to decrease by 1 percent per year through 2050 for an overall reduction in capital cost of 31 percent compared to the present cost. The capital costs given in this report are for fully engineered facilities for a relatively new technology. Others (Wright et al., 2010) have estimated a 60 percent decrease in capital costs as the technology evolves. Figure 3.2 shows the current and future costs to produce cellulosic biofuels based on these assumptions and the assumption that bioenergy feedstock is $75 or $133 per dry ton. Current estimates are for a biomass cost of $75 per ton, but a sensitivity to a higher cost is also included (see Figure 3.2).

Table 3.5 is a summary of projections of cellulosic biofuels that could be available, in addition to the 2012 ethanol and biodiesel production of 14 to 15 billion gal/yr, using different investment rates for new plant capacity. This committee estimated that about 45 billion gge of biofuel would be required to meet the target of 80 percent reduction in petroleum use for the LDV fleet in 2050 and would require about 703 million dry tons per year of biomass feedstock. A uniform annual construction rate of about $10 billion per year can easily produce the projected biofuel needs in 2050. The fuel availabilities are based on the projections discussed

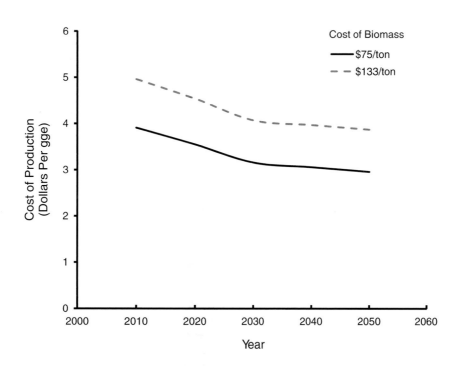

FIGURE 3.2 Sensitivity of biofuel cost to biomass cost.

TABLE 3.5 Estimates of Future Biofuel Availability

	Annual Plant Investment Rate (billion dollars per year)			
	1	4	7.2	10.4
Biofuel production (billion gge per year) by				
2022	0.9	3.7	6.7	9.7
2030	1.8	7.4	13.3	19.2
2050	4.3	17.3	31.2	45.0
Biomass required in 2050 (million dry tons per year)	68	270	488	703
Estimated land-use change (million acres)	5.5	22.2	40.1	57.8
Total investment to 2050 (billion dollars)	38	152	275	396
Average number of biorefineries built per year	2.7	10.8	19.5	28.2

above. Land requirements are scaled from the *U.S. Billion-Ton Update* previously discussed (DOE, 2011).

Worldwide expenditures on exploration and production of petroleum are high (Milhench and Kurahone, 2011). For example, ExxonMobil alone invested over $32 billion globally in capital and exploration projects in 2010. The November 7, 2011, issue of the *Oil and Gas Journal* (2011) reported that the National Oil Companies of the Middle East and North Africa planned to invest a total of $140 billion in oil and natural gas projects in 2012, with even more investments to follow in coming years.

If the biofuels industry grows as projected, many U.S. petroleum refineries will close or be converted to biorefineries. Conversion of a petroleum refinery to a biorefinery will be significantly less costly and labor-intensive than the construction of a "grass-roots" biorefinery.

In all future years, the amount of biofuels that can be produced will most likely be limited not by biomass availability, but rather by the availability of capital to build the biorefineries. However, a potential investor will not start construction without secure contracts for biomass supply and a guaranteed market for the product.[17]

3.2.6 Infrastructure Needs

A large number of biomass conversion facilities would have to be built along with specialized harvesting equipment and a truck fleet to transport the biomass from the fields to the conversion facilities. Economic studies have shown that the conversion facilities need to be near where the crops are grown. Therefore, additional product pipelines would be

---

[17]Factors that can affect actual supply of biomass for fuels are discussed in the report *Renewable Fuel Standard: Potential Economic and Environmental Effects of U.S. Biofuel Policy* (NRC, 2011).

needed to transport the biofuels from the conversion facilities to the existing petroleum product distribution system. Although drop-in biofuels can use the existing petroleum-product distribution system, feeder lines will most likely be required between the biorefineries and the major petroleum pipelines. However, adding feeder lines will require a relatively small incremental investment.

3.2.7 Regional or Local Effects

Biomass can be grown only in certain parts of the country, and so the conversion facilities will also be located nearby. If drop-in fuels are produced, then the fuels can be shipped via the existing system of petroleum-product pipelines. This system efficiently transports large volumes of petroleum products. Initially, the biofuel refineries will be sited near the locations where the lowest-cost biomass is grown or harvested. Many of these locations are in the Southeast and Midwest United States. The major petroleum pipelines between the Gulf Coast and the Northeast and North Central United States bisect these regions. Tie-ins to these pipeline systems would be relatively short.

3.2.8 Safety

The chemical properties of drop-in cellulosic biofuels will be similar to those of existing, petroleum-based LDV fuels, with no additional fuel-related safety hazards. Truck traffic in rural areas is expected to increase, which could increase traffic accidents in these areas.

3.2.9 Barriers

The primary barrier to displacing petroleum with biofuels is economic. At present, biofuels are more expensive to produce than petroleum-based fuels. The corn-grain ethanol industry had many years of government subsidies and is currently supported by the RFS2 consumption mandate. Subsidies or mandates are projected to be required to support cellulosic biofuel unless the price of oil is close to $190/bbl or conversion costs decline as projected.

As discussed above and in detail in other reports (NAS-NAE-NRC, 2009b; NRC, 2011), ethanol involves definite infrastructure issues. Pure ethanol cannot be used in conventional ICEs because of cold-start problems. It has to be blended with petroleum-based gasoline. The highest content allowed in the United States is 85 percent ethanol by volume (E85). Although E85 could contain up to 85 percent ethanol, its ethanol content typically averages only 75 percent or even less in the winter.

As of 2012, the fuel industry was close to reaching the maximum amount of ethanol that can be consumed by blending into E10. Total U.S. gasoline consumption in 2010 was just over 138 billion gallons. Blending all of this as E10 would consume only 13.8 billion gallons of ethanol, which is less than the 15 billion gallons of conventional ethanol mandated by RFS2. Fewer than 0.1 billion gallons of E85 were sold in 2009. As the fuel economy of vehicles improves and gasoline sales decline, even less gasoline will be available to be blended with the volume of ethanol mandated. Drop-in biofuels do not have this limitation.

3.2.10 GHG Reduction Potential

There is ongoing debate regarding the GHG emissions from the production of biofuels, including the time profile of the emissions. The uncertainties and variability associated with the GHG reduction potential of biofuels are discussed in detail in NRC (2011). The values for GHG emissions used in this study were a modified version of those developed by EPA for the RFS2 final regulations. The difference was the treatment of emissions attributable to indirect land-use change (ILUC). The EPA analysis distributes the GHG emissions from ILUC over a 30-year period. For the analysis in this report, all emissions contributed by ILUC were attributed to the first year's operation of the biofuel conversion facility rather than spread over 30 years. This alternate ILUC treatment and its impact on annual biofuel GHG emissions are discussed in detail in Appendix G.5. These predicted GHG emissions do not include the use of CCS in the production facility to reduce overall well-to-wheels GHG emissions. Applying CCS to a biofuel production facility can potentially provide slightly negative well-to-wheels GHG emissions (NAS-NAE-NRC, 2009a).

3.3 ELECTRICITY AS A FUEL FOR LIGHT-DUTY VEHICLES

3.3.1 Current Status

In the United States, electricity is widely available, plentiful, and relatively inexpensive. It already is used as fuel for some LDVs available on the general market, including PHEVs (e.g., the Chevrolet Volt) and BEVs (e.g., the Nissan Leaf). Further, electric-power vehicles are in wide use in commercial applications such as in warehouses and factories.

3.3.2 Capabilities

Table 3.6 shows the 2010 capability of the U.S. electricity system (EIA, 2011a). The capacity factor measures the ability of a power source to produce power and reflects both availability to produce power and whether or not the plant is dispatched. Capacity factor is estimated as the annual electricity production for each source divided by the power production it would have achieved when operating at its net summer capacity 24 hours per day for the entire year. Power dispatch is affected by the price of the source relative to other competing sources because lower-priced sources are dispatched preferentially.

TABLE 3.6 Capability of the U.S. Electricity System in 2010

Source	Net Summer Capacity (GW)	Electricity Production (thousand GWh)	Capacity Factor
Coal	318.1	1,879.9	0.67
Oil and natural gas steam	113.5	123.9	0.13
Natural gas combined cycle	198.2	733.8	0.42
Diesel/conventional combustion turbine	138.6	51.0	0.11
Nuclear	101.1	802.9	0.90
Pumped storage	21.8	–0.2	–0.001
Renewables	123.0	371.6	0.35
Total	1,014.4	3,962.8	0.45

The average U.S. retail price for electricity is about \$0.10/kWh with substantial variation across the country because of the time of use, local generation mix, and various incentives or taxes. In general, electricity produced by hydro power costs the least, followed closely by coal, nuclear, and natural gas. Electricity generation from natural gas is expanding rapidly for the following reasons:

- The cost of natural gas generation strongly depends on the cost of fuel. Currently the cost of natural gas is low (\$2.5 to \$3.5/million Btu) and could remain low for a decade or more.
- CO_2 emissions per unit of power generated by natural gas are about half of the CO_2 emissions per unit of power generated by coal.
- Emissions of sulfur oxides (SO_x), nitrogen oxides (NO_x) and other toxic air pollutants from natural gas are much lower than the emissions from coal.

Gas turbines are well suited to provide backup power for intermittent renewable energy generation sources, such as wind and solar, because they can be ramped up relatively quickly. Because of this characteristic, the share of electricity generation from natural gas tends to increase as renewable energy increases. The generation of electricity produces GHG emissions, mainly CO_2. In 2010, total GHG emissions from electric power as reported in the AEO 2011 were 2.3 billion metric tons CO_2e (EIA, 2011a). There are additional emissions further upstream in the process, for example, in mining coal, producing natural gas, transporting fuels to the power plant, and building solar panels, wind turbines, and power plants. These upstream emissions can be added to the combustion emissions to estimate the total life-cycle emission of any process, including electricity generation. Life-cycle emissions are considered in this report's analyses of GHG emissions.

The capability (and demand) for electricity generation in the United States is expected to grow slowly from the present to 2050. For the purposes of this study, two cases in the

AEO 2011 (EIA, 2011a) were examined: the 2011 reference case and the GHG price case (hereafter referred to as the low-GHG case). The low-GHG case is based on a steadily escalating carbon tax beginning at \$25/metric ton of CO_2e in 2013 and escalating at 5 percent per year, reaching \$152/metric ton in 2050. The National Energy Modeling System (NEMS) is used by EIA to produce the AEO projections up to 2035. Therefore, the reference and low-GHG cases had to be extrapolated to 2050. For the low-GHG case, the total GHG emissions, power output, and cost data were extrapolated to 2050 using the years 2031 to 2035 to better capture the accelerating effects of the carbon tax increase in shifting the mix of generation sources. For the reference case, data from the period 2020 to 2035 were used because the mix of generation sources does not change much.

The low-GHG case shows that the annual GHG emissions in 2050 are reduced from the reference-case emissions by more than the desired 80 percent; however, this result does not account for the life-cycle emission effects in the electricity-generating sector because in the AEO analyses some of the emissions are attributed to other sectors. To compare fuels used in transportation on a consistent basis, the additional upstream generation of GHG emissions for combusted fuels will have to be included to account for the life-cycle emissions for non-combusted fuels, for example, renewables and nuclear.

For coal and natural gas, the upstream emission factors in the Greenhouse Gases, Regulated Emissions, and Energy Use in Transportation Model (GREET model; Argonne National Laboratory) were used to calculate the total life-cycle emissions.

The AEO 2011 estimated GHG emissions from coal combustion to be 0.9552 kg CO_2e/kWh.[18] For coal, the upstream emissions embedded in the GREET model are 3.74 kg CO_2e/GJ. Using a conversion factor of 1.055 GJ per million Btu and assuming a heat rate of 10,000 Btu/kWh for the conversion of coal to electricity, the upstream emissions are 0.04 kg CO_2e/kWh. Accounting for transmission line losses of 7 percent, the correction from both upstream and transmission line losses is an additional 0.042 kg CO_2e/kWh, making the total emissions for coal-fired electricity 1.0 kg CO_2e/kWh.

The existing value for natural gas combustion emissions in the AEO model is 0.433 kg CO_2e/kWh.[19] The upstream GHG emissions for natural gas in the GREET model are 13.4 kg CO_2e/GJ. The heat rate used in AEO 2011 for converting natural gas to electricity is 8,160 Btu/kWh. Using this as a conversion factor, the upstream emissions of natural gas are 0.115 kg CO_2e/kWh. Correcting for transmission line losses of 7 percent makes the total correction 0.123 kg CO_2e/kWh, and the total GHG emissions for natural gas are 0.556 kg CO_2e/kWh.

There are no GHG emissions assumed in the AEO cases for nuclear and renewable electricity. The life-cycle emissions for nuclear and renewable energy sources were assumed to be 0.02 kg CO_2e/kWh, based on the values used in the NRC report *America's Energy Future. Technology and Transformation* (NAS-NAE-NRC, 2009a). Table 3.7 summarizes the results for GHG emissions from fuels.

In addition to extending beyond the AEO's 2035 projections, the current study had to verify that the low-GHG case still gives the desired result of about an 80 percent reduction in GHG emissions by 2050 after all emissions in the life cycle are accounted for. The fraction of electricity generated by each fuel was estimated by extrapolating the 2035 AEO results to 2050. Because the changes in the fuel mix were accelerating in the latter period of the EIA case, 2031-2035, the rate in that period was used as a reasonable basis from which to extrapolate. The result is shown in Table 3.8, which indicates that the GHG emissions are still reduced by more than 80 percent in 2050.

TABLE 3.7 2010 Electricity-Generation GHG Emissions by Source

Source	Combustion Emissions (kg CO_2e/kWh)	Upstream Emissions (kg CO_2e/kWh)	Life-Cycle Emissions (kg CO_2e/kWh)
Coal	0.9552	0.042	1.0
Natural gas	0.433	0.123	0.556
Nuclear	0	0.02	0.02
Hydro	0		
Renewables	0	0.02	0.02

SOURCE: EIA (2011a).

TABLE 3.8 Key Parameters of the AEO Base Case and Low-GHG Case

Parameter	2010	2020	2035	2050
AEO base-case cost (\$/kWh)	9.6	8.8	9.2	9.4
AEO low-GHG case cost (\$/kWh)	9.6	11.2	12.7	14.8
Carbon tax (\$/metric ton CO_2e)	0	35	73	152
AEO base-case output (billions kWh)	3,963	4,158	4,633	5,140
AEO low-GHG case output (billions kWh)	3,963	3,823	3,976	4,190
AEO base-case GHG emissions (kg CO_2e/kWh)	0.586	0.535	0.545	0.541
AEO low-case GHG emissions (kg CO_2e/kWh)	0.586	0.412	0.256	0.111

---

[18]See http://205.254.135.24/oiaf/1605/coefficients.html.
[19]See http://205.254.135.24/oiaf/1605/coefficients.html.

3.3.3 Grid Impact of Plug-in Electric Vehicles

Neither of the AEO grid models account for the additional load if a large number of electric-powered vehicles are added. To assess the importance of this effect, the energy demand in 2020, 2035, and 2050 was estimated (Table 3.9).

The electricity generation projection in the low-GHG case is the comparison standard because the grid capacity is lower than that in the reference case. The result of this comparison shows that the additional load from PEVs in 2020 and 2035 is a small fraction of the projected total electricity usage and probably well within the uncertainty in the projections. Between 2035 and 2050, the power demand for PEVs is assumed to rise quickly. By 2050, it is assumed to reach 7 percent of the projected power usage and has a growth rate of about 0.5 percent per year. This load increase is well within the historic growth of the grid, which has been as high as 7 percent per year in the mid-1980s, and even the growth rate of 1 to 2 percent per year that has been true over the past 10 years in the United States. However, the low-GHG case projects load growth of less than 0.1 percent a year in the absence of BEV demand. Further, adding plants to the grid is a time-consuming process, and construction of a new plant can take a few years to a decade or more. Therefore, if the low-GHG case is an accurate projection of electricity usage, additional capacity has to be planned, permitted, funded, and constructed at a more rapid pace than projected for the next 20 years as large numbers of PEVs come into service (Table 3.10). If these additional plants cannot be brought online quickly enough, then the growth of PEV use may be restrained or the low GHG emissions may not be achieved as older plants with higher emissions may be required to be kept in service. New plant demand can be reduced to the degree that load shifting to off peak can be used. The amount of this reduction is not well defined.

There are also temporal and local effects on power demand from PEV charging. If owners charge their PEVs during times that the grid is highly used (e.g., during peak

TABLE 3.9 Electric Vehicle Energy Demand Compared to Low-GHG Case

	2030	2035	2050
AEO low-GHG output (billion kWh)	3,823	3,976	4,190
Electric vehicle energy demand (billion kWh)	3.4	72	286
Electric vehicle energy demand (percent of output)	0.1	1.8	6.8

NOTE: The demand for electric vehicles was estimated assuming 13,000 miles as the base. The number of miles driven for each vehicle was taken from Elgowainy et al. (2009). The assumed number and mix of vehicles used to estimate the charging load are shown in Table 3.10. The number of vehicles, number of miles, and fraction of the fleet are not predictions by the committee, but were selected to be conservative (high) to illustrate the impact of the charging demand on the grid. For all vehicles the energy consumption is 0.286 kWh/mi.

TABLE 3.10 Assumed Number of Electric Vehicles in Fleet

	2020	2035	2050
Total electric vehicles	2 million	30 million	100 million
Fraction PHEV10	0.4	0.1	0
Fraction PHEV40	0.4	0.5	0.3
Fraction BEVs	0.2	0.4	0.7

NOTE: BEV, battery electric vehicle; PHEV, plug-in hybrid electric vehicle.

load periods), there could be problems with supplying enough electricity. For instance, if most PEVs are returned to their home base late in the afternoon with depleted batteries and are plugged in to charge, this load will be superimposed on the grid at a time when the daily load is already highest. This is especially true in the summer and winter seasons because of air conditioning and heating demands. It also may be desirable to move the load off peak to reduce GHG emissions because when peak loads are high, the oldest and likely dirtiest sources of power will be forced into service. They would not be used when power demands are well below the peak. Based on the estimates above, the peak loading issue until 2035 is unlikely to be a problem overall. But as the LDV charging load on the grid grows, the peak loading becomes of greater concern. However, studies have shown that practical, effective means are available to move the load to alternate charging times (e.g., late at night when other loads are low). One method that utilities are considering using to change consumer behavior is time-of-use (TOU) pricing, which would charge consumers lower rates during off-peak hours (generally between 11 p.m. and 5 a.m.). However, studies show that more comprehensive, integrated, and intrusive load management approaches based on the wide use of smart grid technology can be even more effective than incentives such as TOU pricing in reducing the peak load.

The present power grid has an estimated capability to handle a large fraction of the nationwide LDV fleet simply by taking advantage of the excess capacity in off-peak hours at night (PNNL, 2007). However, that estimate represents a nationwide average, and excess capacity varies throughout the country. For example, while Texas could provide energy for 73 percent of its LDV fleet, the California and Nevada area only could recharge 15 percent of its local fleet with off-peak power. This rate could be problematic given the large number of vehicles present in this region. With larger penetration of PEVs over the coming decade (about 25 percent), it has been suggested that there will be significant strain in regions such as California if the grid does not adapt (Guo et al., 2010).

The local distribution grids of each utility could also be affected by a significant deployment of PEVs (or even by a small number of PEVs if they are concentrated in a small area served by a small number of local transformers). The most likely upgrade required by the addition of PEVs is the replacement of transformers. A study by the Electric Power Research Institute and the Natural Resources Defense Council (EPRI and NRDC, 2007) and discussions by the committee with Pacific Gas and Electric Company (Takemasa, 2011) and previous discussions with Southern California Edison (Cromie and Graham, 2009) indicate that the local grid effects are manageable and within the utilities' normal cost of doing business. See Appendix G, Section G.2, for more discussion and an estimate of the investment cost.

3.3.4 Costs

There are four potential major sources of investment costs beyond the cost of the electricity itself:

- Charging stations to transfer energy from the electric distribution system to the PEVs;
- Necessary upgrades to the transmission and distribution system uniquely associated with charging PEVs;
- Additional generation capacity needed to provide fuel for large numbers of PEVs; and
- Conversion of the electric power system to realize approximately 80 percent lower annual GHG emissions.

These investment costs are estimated in Appendix G.6. The results are summarized in the following sections.

3.3.4.1 Charging Station Costs

Three types of charging stations are available. Level 1 charging stations use normal 110 V circuits and provide AC power to the vehicle. They are relatively low power and require typical charging times of over 20 hours for a 24 kWh battery. Level 2 charging stations provide AC power via a 240 V circuit (typically used today for electric clothes dryers and electric stoves). Because energy flow goes as the square of the voltage, level 2 charging stations will cut the charging time by a factor of about four. So for today's batteries, the charging time will decrease to a few hours. Level 3 charging stations convert AC line voltage and provide high-voltage DC to the vehicle. DC stations are not suitable for home use, and DC will likely be provided at charging stations analogous to gas stations. Level 3 charging stations now can charge a typical battery of an electric vehicle to 80 percent of capacity—the recommended maximum level to avoid damage and hence reduction in battery life—in 15 to 30 minutes. Preliminary data available to date suggests there will be very limited use of DC fast chargers and that the price of charging will be significantly higher than charging at home using a level 1 or level 2 charging station.

The bulk of the charging station investment cost will be borne by the electric-vehicle owner. Longer electric-only driving distances require larger batteries and more powerful charging stations, and so the investment cost is a function of the type of electric vehicle. Appendix G.6 estimates these costs per vehicle for a wide range of electric-vehicle types, assumes appropriate charging station mixes for both home and commercial installations, and includes the reference and low-GHG grid cases to 2050. Current costs for charging stations per vehicle range from about $800 for a PHEV10 to about $4,200 for a BEV. By 2050 the investment costs per vehicle will have dropped from about $450 for a PHEV10 to about $1,950 for a BEV. Appendix G.6 also converts these costs to $/gge per day for comparison with other fuels. These costs do not include a cost for a parking space for access to charging. The parking space for access to charging is a significant additional barrier as the EIA Residential Energy Consumption Survey (2009) reported that 52 percent of households cannot park a car within 20 feet of an electrical outlet.

3.3.4.2 Costs of Additions and Changes to the Transmission and Distribution System

The upgrade costs for high-voltage transmission are included in the next two sections. The investment costs for the distribution system are considered to be relatively small and manageable by the local utilities. They likely will be included in the price of the electricity. Therefore, no additional capital costs are included.

3.3.4.3 Cost of Additional Generation Needed for Large Numbers of PEVs

The additional energy demand from 100 million PEVs in 2050 is estimated to be about 286 billion kWh. Meeting that additional demand by new plants will require the addition of the equivalent of about 90 1,000-MWe plants at a cost of about $360 billion for new generating capacity and a total of over $400 billion, including the associated high-voltage transmission system additions.

3.3.4.4 Cost of Conversion of the Power System to 80 Percent Lower Annual GHG Emissions

Beyond the addition of new capacity to provide fuel for PEVs, a large additional investment would be required to reduce the annual GHG emissions from the entire U.S. power system by about 80 percent by 2050. This investment cost is estimated to be about $1 trillion. This cost is required to decarbonize the power sector and is not attributable solely to the LDV sector.

3.3.5 Regional and Local Effects

Regional and local effects for electricity-fueled LDVs influence the method of rolling out the charging infrastructure and changes in distribution system. They also affect the attractiveness of electricity as a fuel because of the pricing and GHG emissions of the local grid and because of dominant local use of vehicles versus electric-vehicle characteristics.

The rollout of a robust charging infrastructure is coupled to robust sales and use of PEVs, especially BEVs as opposed to PHEVs, because PHEVs can make use of liquid fuel if electricity charging is unavailable. Automobile manufacturers offering BEVs and PHEVs reported to the committee that they have found most sales to date occurring in urban areas with high income levels and a high proportion of people who are more environmentally minded (Diamond, 2010). Thus, the logical basis for expansion of the use of PEVs and the associated charging infrastructure is to proceed in urban areas in which vehicle and charging infrastructure builds rapidly and achieves the needed critical density. As time goes on, these centers are likely to expand and connect along major transportation corridors to provide power to the large number of BEVs needed to substantially reduce petroleum use and GHG emissions. Government support should follow this natural growth pattern and concentrate initial resources in limited areas rather than supporting a broad use of BEVs and expanded charging networks at many locations. Once the process is successful in one "center," the support there can be phased out and moved to another fertile area (Electrification Coalition, 2009).

Although the U.S. power grid is interconnected, the flow of electricity from all sources to all loads is not perfect. In effect, the country is divided up into a number of regional networks that, while strongly connected internally, have weaker ties to one another. As a result, there are significant regional and even state-to-state differences in pricing and GHG emissions. Electricity as a fuel costs less than gasoline, but customers in areas with higher electricity prices realize smaller fuel-cost savings. Some regional networks with relatively low electricity prices may emit significantly more GHG emissions than others with higher electricity prices (Anair and Mahmassani, 2012). GHG emissions may also be a function of available margin and peak loading on the local grid. Even if the base-load power generation has low GHG emissions, the older and dirtier power sources will be dispatched as the load rises. Thus, the GHG emission characteristics of the local grid might also affect the attractiveness of PEVs to buyers with strong environmental concerns.

The dominant use of the vehicle interacts with the characteristics of the PEVs, and this is likely to vary regionally. BEVs are used primarily as short-commute passenger vehicles and in fleets as vehicles for light hauling, or for relatively short-distance services. Those uses match the BEV's battery capability and charging time requirements and suggest that BEVs initially, and perhaps permanently, will be concentrated in urban locations. BEVs will not be in wide use in rural areas with longer drives and more widely separated charging locations.

3.3.6 Safety

The electrical safety considerations in providing electricity to the vehicle are generally well in hand. For both residential and business charging, the voltages and power levels are well within the state of practice, and safety provisions are well understood and codified. One of the costs associated with charging station installation is that it must meet the requirements of the national and local electrical codes, which means that it will most likely have to be installed by a licensed electrician and inspected and permitted by the appropriate governmental agency. For DC fast chargers used as public chargers, very high power connections between the charger and the vehicle must be made, and additional care is warranted. There are standards in use now for DC charging stations that fall under the formal jurisdiction and requirements of the national, state, and local electrical codes.

3.3.7 Barriers

There do not appear to be technical barriers in the electrical system upstream of the vehicle. There are, however, several potential financial and societal barriers:

- The investment cost for the charging infrastructure is borne largely by the vehicle owners.
- The capital cost for the full implementation of the needed changes to achieve a low-GHG-emitting electrical power system is large.
- Coordinating the needed investments and infrastructure work will require overcoming the complexity of the power system's unique ownership, management, and regulatory situation. The electric power system is regulated by a large number of local, state, regional, and federal entities. In most cases, the investors and owners of the transmission and distribution infrastructure are not the same as the investors and owners of the generating sources. Further, in some cases no benefits may accrue to some of those that have to make investments, such as states that have neither the loads nor the generation sources, but must support transmission lines between adjacent states that have loads and sources.
- Permitting and construction of new power system assets are very time consuming. Large power plant projects and large transmission and distribution system projects can take several years to over a decade to complete.

Finding: For electricity as a fuel for LDVs to be effective in reducing net GHG emissions, the entire U.S. electric power system has to shift largely to electricity production from sources that emit low GHG emissions (for example, nuclear, renewables, and natural gas with or without CCS).

3.4 HYDROGEN AS A FUEL

3.4.1 The Attraction of Hydrogen

When hydrogen is used as a fuel in fuel cell electric vehicles (FCEVs), the only vehicle emission is water. When hydrogen is used in an internal combustion engine, the emissions are water, some nitrogen oxides, and some trace chemicals mostly as a result of using lubricants. Although CO_2 emissions are absent from vehicle emissions when hydrogen is used as an LDV fuel, varying amounts of GHGs are emitted during hydrogen production. The amount depends on the primary fuel source and the technology used for hydrogen production. Most of the hydrogen on Earth is found in either water or hydrocarbons such as coal, oil, natural gas, and biomass. Because of the diverse primary sources for hydrogen, an amount of hydrogen large enough to fuel the entire LDV fleet could be made with only domestic sources. Different process technologies can be used with different primary sources to make a pathway for delivering hydrogen to consumers at different costs and with varying amounts of GHG emissions. The diversity of supply sources and production technologies is an advantage of hydrogen fuel.

3.4.2 Major Challenges

For more than 10 years, there have been serious efforts in the United States, Europe, and Japan to develop FCEVs and the needed production and delivery technologies to supply hydrogen. As Chapter 2 indicates, there has been considerable success in developing FCEVs, but some challenges remain. There also has been considerable success in developing production, distribution, and dispensing technologies for making and delivering low-cost hydrogen, but major challenges still exist. The two major challenge areas are the following:

- Making low-cost hydrogen with low GHG emissions. At present, the lowest-cost methods for hydrogen production used by industry are based on fossil fuels and have associated GHG emissions of varying amounts. The low-GHG methods are currently more expensive and need further development to become competitive.
- Building the hydrogen infrastructure will be a large, complex, and expensive undertaking. Hydrogen-fueling stations would have to be available before FCEVs can be sold. Until a large number of FCEVs are in use, the cost of hydrogen as a fuel will be high. Because FCEVs are new and hydrogen as a consumer fuel is new, there are many practical concerns such as safety, codes and standards, permitting, and zoning issues that need to be addressed before growth can flourish.

3.4.3 Current Status of the Market

Hydrogen as an industrial commodity is produced in large quantities in the United States and in many other countries. The amount of hydrogen produced is over 50 million tons per year worldwide (Raman, 2004; IEA, 2007) and over 10 million tons per year in the United States (EIA, 2008b). Most of the hydrogen is used in the chemical processing industry and in refining crude oil, and most of it is produced in large facilities closely associated with the end use. Over 95 percent of U.S. hydrogen is made from natural gas, with other sources including refinery off-gases, coal, and water electrolysis. Several hydrogen pipeline systems (Houston, Los Angeles, and Chicago) exist to move large quantities of gaseous hydrogen between nearby industrial users with over 1,200 miles of hydrogen pipelines. Some established industrial gas companies produce, store, and distribute hydrogen as either a gas or a cryogenic liquid to smaller users by truck. The demand for hydrogen for industrial use has increased consistently for several decades.

Even as the infrastructure for producing, delivering, and using large amounts of hydrogen for this industrial market is well developed, the infrastructure for producing, delivering, and dispensing hydrogen for use as a transportation fuel has yet to be developed. For illustrative purposes, if hydrogen were to be used as a transportation fuel, then the current U.S. production level of 10 million tons per year would be enough to fuel about 45 million cars (at 60 mpgge and 12,000 mi/yr). There is, however, little spare capacity in the existing system for this new market. Therefore, a new hydrogen infrastructure is needed before large numbers of FCEVs are produced. This infrastructure will need to be much different from the existing one because it has to focus on wide distribution of small amounts if distributed through retail outlets, similar to what is done for gasoline today.

Academic, industrial, and government efforts over the past 10 years to define this retail-fuel-oriented infrastructure have mapped out the needed technology improvements, established performance criteria for different parts of the infrastructure, estimated the cost of hydrogen and the infrastructure over time, and suggested possible implementation methods. The NRC report *Transitions to Alternative Transportation Technologies—A Focus on Hydrogen* (NRC, 2008) contains an analysis of the technical needs, costs, petroleum savings and GHG emission savings possible by moving towards a hydrogen-fuel infrastructure.

3.4.4 Hydrogen Infrastructure Definition

Rather than being built throughout the entire United States before FCEVs are available, a hydrogen infrastructure likely will first be started in a few markets. Then the infrastructure will be built up in conjunction with increasing local FCEV sales. The concentration of demand will result in a decrease in the high initial cost of hydrogen and the infrastructure as equipment for commercial-scale production is installed and used at commercial rates. This process will then be repeated in additional markets until a critical mass of FCEVs and hydrogen stations is built to a market-sustainable level.

The first hydrogen stations are likely to be supplied by truck delivery from local hydrogen-distribution points. This is a high-cost method that may be largely replaced by hydrogen stations with on-site hydrogen generation capabilities where the hydrogen is made at the retail station rather than supplied from the large plants that now supply the bulk of hydrogen. This approach precludes the need to transport or deliver hydrogen, and the distributed hydrogen generation equipment can be sized for the demand. Several technologies are available for the small hydrogen generators, including natural-gas reforming, water electrolysis, and biofuel reforming.

- *Small natural-gas reforming*—The process is the same as that used in today's large natural-gas reforming facilities. However, the reforming apparatus for fuel is small and packaged such that it looks like a large appliance. These reformers have been demonstrated at a number of hydrogen-fueling stations in the United States, Europe, and Japan. CO_2 produced in the process is released to the atmosphere because capturing it is difficult.
- *Small water electrolysis*—Commercial alkaline water electrolysis units are available and have been demonstrated in small hydrogen stations. GHG releases are associated with the source of electricity and can be high or low depending on how the electricity is produced.
- *Small biofuel reforming*—Ethanol reforming and other biofuel reforming have been demonstrated in laboratories, but research and development (R&D) is still needed to increase hydrogen yields and lower costs to be competitive with small natural-gas reformers and small water-electrolysis methods. GHG releases can be low depending on the source of the biofuel.

As the demand for hydrogen increases in a local market, there will come a point when large centralized facilities similar to today's will produce hydrogen at lower cost than is possible with small distributed generators. These facilities will also offer the opportunity to make low-GHG hydrogen through the use of other primary fuels and CCS technology. Several primary feedstock and technology choices are possible, including natural-gas reforming, coal gasification, biomass gasification, and large-scale wind or solar electrolysis.

- *Natural-gas reforming*—This low-cost process is widely used now for generating large amounts of hydrogen. CCS is possible but has not yet been demonstrated with a hydrogen plant.

- *Coal gasification*—This process has been used commercially for decades, but high CO_2 releases require that CCS be available. CCS has not been demonstrated with coal gasification.
- *Biomass gasification*—This process has been demonstrated in the laboratory, but not yet at large pilot-scale facilities. Further development is needed. If CCS is used, then biomass gasification becomes a CO_2 sink with negative releases.
- *Large centralized electrolysis with wind or solar power*—The process is still being researched to lower costs. This process has low GHG emissions.

Other hydrogen-production methods under research hold long-term promise for making hydrogen at low costs, low GHG emissions, or both, but they are not yet developed enough to understand the availability or the cost implications. Some of these methods include nuclear high-temperature chemical cycles or electrolysis, photoelectrochemical methods, and biological systems.

3.4.5 Hydrogen Dispensing Costs and GHGs

The cost of making, transporting, storing, and dispensing hydrogen at a station has been estimated for all of the primary feedstocks. These estimated costs are highly dependent on many assumptions and can vary considerably depending on future technical advances, feedstock costs, and how quickly the market develops (scale). The estimated costs for some of the different hydrogen pathways based on future technology development are shown in Table 3.11. The estimates are expressed in dollars per gallon of gasoline equivalent ($/gge). A gge of hydrogen contains as much energy (Btu) as a typical gallon of gasoline and is defined as 116,000 Btu/gge in this study. The future price basis and resource requirements used to generate the costs in Table 3.11 are shown in Table 3.12. The hydrogen costs in Table 3.11 are in some cases up to $1.00/gge higher than those determined in prior studies

TABLE 3.11 Hydrogen Costs at the Pump ($/gge), Untaxed

	2010	2020	2035	2050
Distributed natural gas reforming	3.50	3.60	3.90	4.20
Distributed grid electrolysis	5.80	5.40	5.50	5.69
Coal gasification without CCS	3.80	3.80	3.80	3.85
Coal gasification with CCS	4.50	4.50	4.50	4.50
Central natural gas reforming without CCS	3.30	3.40	3.70	4.10
Central natural gas reforming with CCS	3.60	3.60	4.00	4.30
Biomass gasification without CCS	4.10	4.10	4.10	4.10

NOTE: Basis: 2008 H_2A future cases updated to 2009 dollars using CEPCI and Nelson-Ferrer cost indexes and the AEO 2011 price basis. $2.00/gge included for distribution and station costs for central methods and $1.88/gge included for station costs of distributed methods.

TABLE 3.12 Resource Prices and Requirements Used in Table 3.11

	2010	2020	2035	2050
Industrial natural gas, $/million Btu	4.80	5.36	7.21	9.06
Delivered coal, $/ton	45.9	46.1	48.9	50.2
Industrial electricity, $/kWh	0.068	0.061	0.064	0.067
Delivered biomass, $/ton	75.0	75.0	75.0	75.0
Coal needed, kg/gge H_2	9.8	9.8	9.8	9.8
Biomass needed, kg/gge H_2	12.8	12.8	12.8	12.8
Natural gas needed, cubic ft/gge H_2	170	170	170	170
Electricity needed, kWh/gge H_2	45	45	45	45

NOTE: Basis—AEO2011 (EIA, 2011a) resource prices and 2008 H_2A future cases for resource requirements

(NRC, 2008, for example). The increased costs compared to the earlier NRC study result from several factors:

- The costs in the current study are based on the 2008 version of the hydrogen analysis (H_2A) production model developed by DOE, whereas the ones in the previous study (NRC, 2008) were from the 2005 version.
- The distribution costs are estimated to be $2/gge, whereas prior ones were $1.00 to $2.00/gge.
- The capital costs are inflated based on actual construction cost inflation to 2009 dollars.

- The costs for biomass and coal are nearly twice what they were in the 2008 study.

The costs in Table 3.11 represent future costs based on using commercial-scale processes and are possible only after about 10 million FCEVs are on the road. Prior to this, the hydrogen cost will be higher because of underutilized or smaller-scale production facilities. Figure 3.3 shows hydrogen costs versus number of FCEVs.

The GHG emissions associated with producing, delivering, and dispensing hydrogen at a station on a life-cycle basis are shown in Table 3.13. This includes an upstream component related to the emissions associated with production and delivery of the base fuel to the hydrogen production plant and, if used, the energy needed to sequester CO_2 plus a component for conversion, delivery, and dispensing of GHGs.

3.4.6 Hydrogen Infrastructure Needs and Cost

Building the infrastructure for delivering hydrogen over the vast size of the United States is a significant challenge for the use of hydrogen for transportation. It requires developing some new technologies, establishing codes and standards, overcoming the problem of interdependence of establishing a critical mass of hydrogen-refueling stations and FCEV sales, overcoming the high initial cost of hydrogen, and increasing the use of production methods with low GHG emissions.

The total investment costs used to calculate the hydrogen costs in Table 3.11 for future technologies used at commercial-size plants are shown in Table 3.14. These

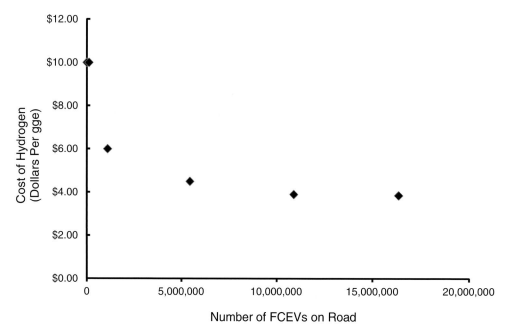

FIGURE 3.3 Hydrogen cost versus number of FCEVs.

TABLE 3.13 Total GHG Emissions (kg CO_2e per gge of hydrogen)

H_2 Production Method	Upstream CO_2e	Plant, Delivery and Dispensing CO_2e	Total CO_2e
Distributed natural gas reforming	2.78	8.66	11.44
Distributed electrolysis, current grid	35.44	0	35.44
Coal gasification without CCS	1.13	24.67	25.81
Coal gasification with CCS	2.77	2.47	5.24
Central natural gas reforming without CCS	2.18	9.28	11.46
Central natural gas reforming with CCS	2.71	0.93	3.64
Biomass gasification without CCS	−24.37	24.57	0.20

NOTE: Basis—H_2A 2008 future cases modified to use GREET 2011 upstream natural gas figures.

costs are normalized to 2009 dollars per gallon of gasoline equivalent per day of produced hydrogen. The station costs appear to be the largest factor for all but coal technology. The station costs include all costs associated with building grassroots new stations that include hydrogen storage, compression, and dispensing and are the same for each technology. The actual hydrogen production investment costs are shown separately. Investment costs for CCS are included for the large coal and natural gas facilities.

The NRC report *Transitions to Alternative Transportation Technologies—A Focus on Hydrogen* (NRC, 2008) outlined one possible hydrogen infrastructure development pathway out to 2050 and estimated the hydrogen cost, GHG emissions, and investment needs over different time periods. The pathway in that report starts with distributed natural gas reforming. As demand increases, coal gasification with

TABLE 3.14 Investment Costs ($/gge per day)

H_2 Production Method	Plant + CCS	Distribution	Stations	Total
Distributed natural gas reforming	700	0	2,345	3,045
Distributed electrolysis, current grid	860	0	2,345	3,205
Coal gasification without CCS	2,250	225	2,345	4,820
Coal gasification with CCS	3,020	225	2,345	5,590
Central natural gas reforming without CCS	400	225	2,345	2,970
Central natural gas reforming with CCS	740	225	2,345	3,310
Biomass gasification without CCS	1,040	225	2,345	3,610

CCS and biomass gasification provide the bulk of increased hydrogen production. This is not the only possible pathway to supply the increasing amount of hydrogen, but it relies on some current, low-cost, and mostly commercially developed processes. With future R&D success, other technologies would likely become part of the transition.

With the increasing amounts of domestically available natural gas and the lower prices for natural gas compared to crude oil projected in the AEO 2011 (EIA, 2011a) study price basis, several other combinations of basic resources and hydrogen-production processes could be viewed as possible in the future with different hydrogen costs and GHG emissions. Some of the many possible pathways for making large amounts of hydrogen are shown in Table 3.15 with the resulting long-term hydrogen cost and GHG emissions.

- *A low-cost case*—The emphasis is on low-cost hydrogen from several resources with little to no emphasis on GHG reductions. Hydrogen is produced from: 25 percent distributed natural-gas reforming, 25 percent coal gasification without CCS, 25 percent central natural-gas reforming without CCS, and 25 percent biomass gasification without CCS.

- *A partial CCS case*—The emphasis is on low-cost hydrogen, but CCS is used for all coal and central natural gas processes. Hydrogen is produced from: 25 percent distributed natural gas reforming, 25 percent coal gasification with CCS, 25 percent central natural gas reforming with CCS, and 25 percent biomass gasification without CCS.

- *A low-GHG case*—The emphasis is on low GHG emissions with less regard to hydrogen cost. Hydrogen is produced from: 10 percent distributed natural-gas reforming, 40 percent central natural gas reforming with CCS, 30 percent biomass gasification without CCS, and 20 percent low GHG grid electricity for electrolysis.

3.4.7 Recent History

More than 200 FCEVs have been demonstrated in the United States over the past 10 years. Several of the auto companies developing FCEVs have gone through multiple iterations to improve performance. Five of these companies—General Motors, Daimler, Toyota, Honda, and Hyundai—have reaffirmed near-term (2015) commercializa-

TABLE 3.15 Alternate Scenario Hydrogen Costs and GHG Emissions

	$/gge H_2	kg CO_2e/gge H_2
Low-cost case	3.85	12.2
Partial CCS case	4.10	5.1
Low-GHG case	4.80	2.6

tion plans for FCEVs. Because these are all multinational companies, the commercialization plans certainly will vary in different markets.

In the United States, there have been about 60 hydrogen fueling stations constructed to service the FCEV demonstration efforts (DOE-EERE, 2012a). Given that the number of vehicles is small, none of these stations is of even small commercial size. They demonstrate, however, the importance of distributed technologies to starting the infrastructure. General Motors has joined 10 companies, government agencies, and universities to build 20 to 25 hydrogen-fueling stations in Hawaii by 2015 (DeMorro, 2010). Several countries have formed much larger infrastructure plans and consortiums than the one in the United States to support early FCEV commercialization. In 2010 Japan announced plans for 1000 hydrogen stations and 2 million FCEVs by 2025 (DOE-EERE, 2011a). To support these goals, a consortium of 13 companies was established to focus on the hydrogen infrastructure. Germany has announced plans to build 150 hydrogen stations by 2013 and up to 1000 by 2017.

3.4.8 Barriers

Although technology is available to provide competitively priced hydrogen from natural gas, technology improvements are needed to provide low-cost hydrogen that is also low in net GHG emissions. Continuous government support for RD&D is required.

The robust performance and the durability of a fueling station with sustained high-volume usage remain to be verified through demonstration.

The high cost of the FCEV is a barrier to wide commercialization for the vehicles and hydrogen. A viable pathway is needed for creating the initial hydrogen infrastructure and for dealing with high initial hydrogen costs. This pathway likely will require government actions.

The lack of an incentive to provide low-GHG fuels in general reduces the benefits for transitioning toward alternative fuels. It also reduces the incentive to make hydrogen from the more costly but lower-GHG methods.

Perceived, real and unknown safety issues with hydrogen production and use especially in a consumer environment could result in delays in acquiring, zoning, and permitting authorizations. There are significant practical challenges of developing sites especially for urban stations within the footprint of existing fueling sites.

Finding: Making hydrogen from fossil fuels, especially natural gas, is a low-cost option to meet future demand from FCEVs; however, these methods result in significant GHG emissions. Making hydrogen with low GHG emissions is more costly (renewable electricity electrolysis) or requires new production methods (e.g., photoelectrochemical, nuclear cycles, biomass gasification, and biological methods) and CCS to man- age emissions. Continued R&D is needed on low-GHG hydrogen production methods and CCS to demonstrate that large amounts of low-cost and low-GHG hydrogen can be produced.

3.5 NATURAL GAS AS AN AUTOMOBILE FUEL

Natural gas can be used for transportation via several pathways, each of which has advantages and challenges (see Appendix G.7). None of them is of much commercial significance in the United States as of 2012.

Less than 3 percent of the natural gas consumed in the United States is for transportation, and most of that is used for powering the transportation pipeline and distribution system for natural gas. Natural gas as an automobile fuel will have to compete with other existing uses of the gas (for electricity generation, and for residential, commercial, and industrial uses). This section addresses the direct use of CNG in internal combustion engines (CNG vehicles, or CNGVs). The other pathways are considered in other sections of this report. Methanol as a transportation fuel is discussed in Appendix G.8.

3.5.1 Current Status

3.5.1.1 *Net GHG Emissions from CNG Use*

Natural gas from production wells is composed mostly of methane (70 to 90 percent), with some ethane, propane, and butane (0 to 20 percent), CO_2 (0 to 8 percent), N_2 (0 to 5 percent), H_2S (0 to 5 percent), traces of O_2, and traces of the noble gases Ar, He, Ne, and Xe (NaturalGas.org, 2011). Natural gas holds promise for providing part of the energy requirements of automobile transportation. Displacing a significant portion of petroleum-based fuels would have large societal and economic benefits by reducing the externalities associated with petroleum importation (e.g., supply and price instabilities, security and defense costs, oil import-related trade and export-import imbalances).

Natural gas vehicles, fueled by CNG or liquid natural gas, are among the most immediately attainable alternative-fueled vehicles. Given methane's molecular structure, natural gas has the highest energy content or hydrogen-to-carbon weight ratio of all fossil fuels. Nevertheless, the use of natural gas, like other forms of primary energy, has associated GHG emissions, including methane emissions, during exploration, well drilling, and the well-to-tank transmission for natural gas. Life-cycle analyses that account for upstream and downstream GHG emissions for natural gas have been published by the DOE's National Energy Technology Laboratory (DiPietro, 2010). In terms of kg CO_2e/million Btu, drilling and extraction generate 19.9 and pipeline transport generates 3.3 (mostly natural gas to power the pumps), for a total upstream (well to tank; WTT) of 23.2. Compression of natural gas into CNG from pipeline pressure to about 3,600 psi adds another 3.5 percent (range 2 to 5 percent), or 0.8 kg CO_2e/million Btu to the GHG emissions.

The Argonne National Laboratory's GREET model uses smaller WTT estimates. For example, the 1.8b version of that model released in September 2008 estimated the upstream emissions to be 9.6 kg CO_2/million Btu (ANL, 2011). The model estimated vehicle tank-to-wheel (TTW) CO_2 emissions of 53.9 kg CO_2/million Btu. Thus, the well-to-wheels CO_2 emissions for CNG as a fuel are 9.6 + 53.9 = 63.5 kg CO_2/million Btu. In 2011 the GREET model estimates were updated to include higher effects of methane leakage and other changes, yielding an upstream estimate of 14.2 kg CO_2/million Btu for shale gas. This estimate is used in this report for all pathways using natural gas as a primary source.[20] Another life-cycle analysis by Burnham et al. (2012) indicated that the life-cycle GHG emissions of natural gas are 23 percent lower than those of petroleum-based gasoline and 43 percent lower than those of coal. Jiang et al. (2011) estimated the life-cycle GHG emissions for producing electricity from shale natural gas to be 20 to 50 percent lower than the life-cycle GHG emissions for producing electricity from coal.

Fugitive natural gas emissions from increasing use of natural gas are the subject of current analyses. In 2010, the EPA reissued its methane emissions guidelines during natural gas extraction, with substantially increased figures versus their previous estimates (EPA, 2010a). Howarth et al. (2011) estimated the leak rate of methane as a percentage of total natural gas produced to be in the range of 3.6 to 7.9 percent. Of the methane leaked, 1.6 percent was attributed to methane escaping from flow-back fluids (1.6 percent) and from drill-out (0.33 percent). The remainder was attributed to venting and equipment leaks, and emissions during liquid unloading, gas processing, and transport, storage, and distribution. The methodologies and data used in the estimates of methane leakage by the EPA and by Howarth et al. were strongly critiqued by an IHS CERA report, *Mismeasuring Methane: Estimating Greenhouse Gas Emissions from Upstream Natural Gas Development* (Barcella et al., 2011). Analysis in that report suggests much lower fugitive methane emissions. Burnham et al. (2012) estimated methane leakage in the range of 0.97 to 5.47 percent for conventional natural gas pathways and 0.71 to 5.23 percent for shale-gas pathways. Methane leakage from the sources mentioned is a concern because of the large global warming potential of methane, but its extent is uncertain (Alvarez et al., 2012). The sources of leakage are amenable to various forms of reduction or control by conventional technologies, representing ongoing considerations in sorting out the environmental aspects of shale gas and conventional natural gas. Several studies are underway to consolidate and define fugitive natural gas

emissions from shale-gas operations as of the writing of this report.

Recognizing that some cost-effective measures exist for reducing methane emitted from producing natural gas, in 2011, the EPA proposed amendments to its air regulations for the oil and gas industry that will reduce GHG and other emissions from exploration, drilling, and production (EPA, 2011c). The final regulation was issued in April 2012. In it, the EPA estimates reductions of 1.0 to 1.7 million tons per year of methane emissions associated with drilling and transportation of natural gas (EPA, 2012b).

3.5.2 Capabilities

3.5.2.1 Natural Gas Supply, Demand, and Prices

The United States used about 98 quads (quadrillion, or 10^{15}, Btu) of energy from the nation's primary energy sources in 2010 (LLNL, 2012). Of the 24 quads of natural gas consumed in the United States in 2010, 98 percent originated from North America and 85 percent was of domestic origin. (In comparison, the United States consumed 37 quads of petroleum, about 50 percent of which was imported.) Transportation used 28 quads of primary energy, 95 percent of which was from petroleum. With a typical 25 percent overall efficiency, a useful energy of about 7 quads is turning the wheels of the U.S. transportation fleet.

Of the 24 quads of natural gas, about 7 quads were used to generate electricity. Natural gas is becoming more attractive for electricity generation than coal, according to recent references quoting numbers from the DOE's Energy Information Administration (Begos, 2012). Electricity generation from natural gas in the United States increased from about 601 billion kWh in 2000 to 981 billion kWh in 2010. During the same period, electricity generation from coal declined from 1,966 billion kWh to 1,850 kWh (EIA, 2011b). Between 2010 and 2035, 80 percent of all newly added electricity generation capacity is expected to come from natural gas-fired plants (EIA, 2011a; NaturalGas.org, 2012). With recently increased concerns about the future of nuclear energy, some of the contemplated future nuclear electric capacity will likely shift to natural gas-fired power plants as well.

According to the June 18, 2009, report of the Potential Gas Committee on the assessment of the year-end 2008 natural gas reserves (Potential Gas Committee, 2009), the United States has 1,836 tcf (trillion, or 10^{12}, standard cubic feet; 1 tcf is equal to approximately 1 quad) of probable natural gas resources, defined as being potentially economically extractable by the use of available technology at the then-current economic conditions. The above number (1,836 tcf) is the sum of 1,673 tcf in traditional reserves and 163 tcf in coal-bed reservoirs. Of the 1,836 tcf of probable reserves, shale gas accounts for 616 tcf (33 percent). In addition to the above probable reserves, the United States also has 238 tcf of proved natural gas resources, defined as deemed to

---

[20]The CNG GHG emissions are estimated as follows: 14.2 kg CO_2/million Btu upstream plus 59.8 kg CO_2/million Btu combustion plus 7 percent of this total for pipelining and compression = 79.2 kg CO_2/million Btu or 9.2 kg CO_2/gge.

be economically extractable (rather than being potentially extractable) or already being extracted economically. The estimated total natural gas reserves of 2,074 tcf (1,836 + 238) represent an increase of 542 tcf (35 percent) over the estimate in the previous biannual assessment. The natural gas consumption of the United States was about 24.1 tcf in 2010 (EIA, 2011b). Dividing the 2008 estimated total of probable and proved natural gas reserves by the 2010 annual consumption gives an estimate of 86 years' worth of natural gas. It has been argued that only a fraction of probable reserves can be recovered economically (Brooks, 2010), so that the "probable technically recoverable resources" would be only 441 tcf, of which 147 tcf is the shale-gas component.

The 2009 report upgraded the probable reserves mainly by reclassifying known shale gas reserves from possible to probable, due to the rapid evolution and deployment of new technology. The new shale gas extraction technology combines two technologies from the oil fields, horizontal drilling and hydraulic fracturing. (See Technology Review, 2009, for a video schematic of these processes.)

The newly reclassified shale gas reserves are located in Louisiana, Texas, the Rocky Mountains, West Virginia, Pennsylvania, and New York. There are large shale gas fields outside the United States as well, and these fields also are likely to be accessible via the new technology. The *BP Energy Outlook 2030* (BP, 2011) stated that in 2009, the world had 6,621 tcf of proved gas reserves, which would be sufficient for 63 years of production at 2010 production levels. Global reserves of unconventional natural gas could potentially add another 30 years to natural gas use.

Most of the natural gas-based transportation fuels are expected to gain new impetus in light of the dramatically upgraded estimates of global natural gas resources. Future natural gas supply and consumption volumes and prices,

broken down to sources and uses, are published yearly by the U.S. Department of Energy's Energy Information Administration, AEO. The AEO 2011 early release projects to the year 2035 (EIA, 2012a). According to AEO (EIA, 2011a), between the years 2010 and 2035, natural gas consumption will grow by 16.8 percent. The share of shale gas will increase from 23 to 49 percent (Figure 3.4). The share of natural gas in transportation will remain at 3 percent, which roughly accounts for the amount of natural gas used for operating the pipelines. In other words, the 2011 AEO is not counting on any significant increase in the use of natural gas for transportation in the United States. This seems to also hold on the global scale. The *BP Energy Outlook 2030* (BP, 2011) projected global use of CNG for transport to be limited to 2 percent of the global demand for transportation fuels.

3.5.2.2 Will There be Enough Natural Gas for LDVs?

In the year 2000, the 110,000 natural gas vehicles in the United States consumed between 8.3 and 12.3 billion standard cubic feet of natural gas, which is between 0.036 and 0.053 percent of the U.S. natural gas consumption (Campbell-Parnell, 2011). According to the 2011 AEO (EIA, 2011a), the U.S. LDV vehicle stock will increase from about 128 million vehicles in 2011 to about 186 million vehicles in 2035. Assuming a 10 percent penetration of CNGVs in 2035 (EIA, 2011a), 45 mpgge, and 14,000 mi/yr, this would translate to a natural gas consumption of 0.73 tcf/yr. Natural gas consumption is forecasted by the AEO 2011 to increase from 24.1 tcf in 2010 to 26.5 tcf in 2035 (with only 6 percent for transportation; mostly natural gas consumed by powering the pipeline system itself). Therefore, a 10 percent CNGV penetration in the 2035 LDV fleet would add only 2.8 percent to the natural gas consumption in that year. Thus, the

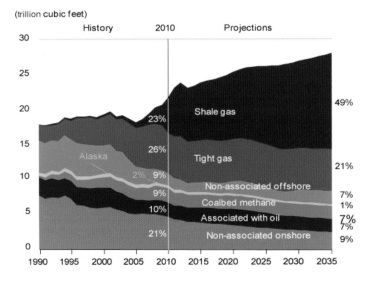

FIGURE 3.4 U.S. natural gas production (trillion standard cubic feet) from 1990 to 2035.
SOURCE: EIA (2012a).

natural gas supply is unlikely to limit the early penetration of CNGVs.

Several studies project that LDVs powered by natural gas will remain a niche for a while in the United States. Those studies include *The Future of Natural Gas* by the Massachusetts Institute of Technology (MIT, 2011), a market analysis of natural gas vehicles by TIAX (Law et al., 2010), an analysis of long-term natural gas demand by Simmons & Company (2011), and an analysis of natural gas demand for transportation by IHS-CERA (IHS, 2010). TIAX (Law et al., 2010) compared the incremental lifetime costs of LDVs using different technologies and found that the direct costs of natural gas vehicles are favorable compared to BEVs, PHEVs, FCEVs, and flex-fuel vehicles. They concluded that CNGVs could become significant with appropriate policy and incentive programs and projected the use of 5.5 billion gge of CNG (still only 0.7 tcf) by 2035.

The reasons for the slow and late development of light-duty CNGVs in the United States transcend the barriers of CNGV and vehicle conversion costs, lack of luggage and tank volume, and the lack of refueling infrastructure. Development of CNGVs also may be significantly hampered by the attractiveness of alternate uses of natural gas, specifically for electricity generation. The AEO 2011 (EIA, 2011a), for example, shows year 2016 levelized costs for electricity generated by 16 different power plant and fuel technologies. Of these, the lowest levelized cost is shown for natural gas-fired combined-cycle power plants (<7 cents/kwh), followed by hydro (8.64), conventional coal (9.48), wind (9.70), biomass (11.25), advanced nuclear (11.39), advanced coal with CCS (carbon capture and storage) (13.62), and photovoltaic solar (21.02). The AEO projections suggest that natural gas will indeed be most attractive for electric power generation because of its low levelized cost.

3.5.3 Costs

3.5.3.1 Natural Gas Fuel Costs and Cost Projections

At filling stations CNG and liquid natural gas are metered and sold on a gallon of gasoline-equivalent basis; the conversion factor of 1 gge = 5.66 lb of natural gas was determined by the National Institute of Standards and Technology (NIST). The prices of natural gas on a gallon of gasoline-equivalent basis are published on the Internet, and they vary by state, region, city, and individual filling station. Natural gas at the time of this writing had a price advantage of about $1 to $2/gge, depending on the particular filling station. For example, overall average U.S. fuel prices reported for the last quarter of 2011 were $3.37/gal for gasoline, and $ 2.13/gge natural gas (DOE-EERE, 2012b).[21]

At a price differential of $1.24/gge in favor of CNG, 30 mpg, 13,000 mi/yr, and 433 gge/yr consumed, the fuel cost savings would be about $540/yr, returning the original investment in a 2012 Honda Civic Natural Gas (versus the LX) in about 13 years (7,500/540 = 14 years). This payback period is not likely to be perceived by the consumer as economically attractive. Various states and the federal government have offered subsidies, which could amount to $4,000 per vehicle. With a $4,000 subsidy, the economic return period would be reduced to 6 years. CNG economics can thus be significantly better in the states that subsidize CNGVs.

Natural gas prices have declined in recent years, whereas oil prices have been rising. With fuel and vehicle subsidies for natural gas, any continued gasoline price increases could eventually make the original equipment manufacturers' natural gas vehicles economically attractive.

The appeal of natural gas as an automotive fuel depends to a large extent on the ratio of oil prices to natural gas prices (Figure 3.5). Long-term future natural gas prices have been forecasted by the 2011 AEO (EIA, 2011a) (Table 3.16).

The price customers would pay at the CNG filling station for filling a vehicle was calculated by taking the average of commercial and industrial prices for natural gas and adding a margin sufficient to generate a 15 percent return on an investment of $1.3 million in a CNG filling station servicing 1,000 cars per week at 10 gge per fill per week. This margin was calculated to be $7.76/million Btu or $0.90/gge NG. (The operating costs and capital expenses of this filing station, excluding fuel costs, were $273,351/yr.) CNG filing station costs and additional natural gas pipeline needs are discussed in Appendix G.9.

The U.S. Department of Energy's Alternative Fuels and Advanced Vehicles Data Center lists 975 public CNG refueling stations as of January 9, 2012 (DOE-EERE, 2012a). Unevenly distributed across the country, they are clustered primarily in California (229 stations), New York (106 stations), Utah (81 stations), Oklahoma (67 stations), Texas (35 stations), and Arizona (30 stations).

The distribution of CNG filling stations corresponds somewhat to the clustering of CNGVs. The EIA (2008a) listed a CNGV count of 113,973 as of 2008, with the largest number in California (35,980 vehicles), followed by Texas (11,032 vehicles), Arizona (10,072 vehicles), and New York

---

[21]In 2011, the quarterly average price ranged from $3.37 to $3.69/gal for gasoline and from $2.06 to $2.13/gge for natural gas (DOE-EERE, 2011b,c,d; 2012b).

TABLE 3.16 Long-term Future Natural Gas Prices ($/million Btu) Forecasted by the 2011 *Annual Energy Outlook*

	2010	2020	2035	2050 (extrapolated)
Commercial natural gas	8.91	8.95	10.98	13.02
Industrial natural gas	4.80	5.36	7.2	9.06
Vehicle natural gas	13.94	14.24	16.81	18.80
in $/gge	1.69	1.73	1.96	2.18

SOURCE: Data from EIA (2011a).

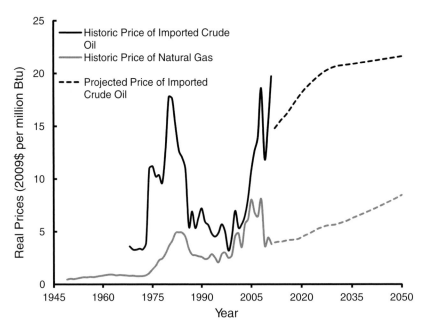

FIGURE 3.5 Historic and projected prices of natural gas and imported crude oil.
NOTE: The prices from 2035 to 2050 were projected by extrapolating the 2030-2035 annual growth rate in EIA (2011a).
SOURCE: Data from EIA (2011a,b).

(10,017 vehicles). The regional clustering of CNG filling stations as a practical model for infrastructure build-up matches the results of models for the clustering of hydrogen filling stations for FCEVs and of public charging stations for BEVs.

CNG prices vary regionally and locally. According to the DOE's Alternative Fuels and Advanced Vehicles Data Center (DOE-EERE, 2011b), average CNG prices per unit gallon of gasoline-equivalent in April 2011 ranged from $1.39 to $2.41 ($2.41 in the Central Atlantic, $2.38 in New England, $2.32 on the West Coast, $1.87 in the Lower Atlantic, $1.84 in the Gulf Coast region, $1.66 in the Midwest, and $ 1.39 in the Rocky Mountain region).

Environmental standards, construction permits, labor costs, natural gas and gasoline costs, vehicle and population density, purchasing power and customer preferences, proximity to natural gas pipelines, the corresponding industrial and commercial natural gas prices, and a host of other factors vary with individual cities, counties, states, and regions, all of which have some effect on the actual and potential extent and rate of penetration of CNGVs. Because of the recent discovery of the U.S. abundance of natural gas, the subject of regional differences needs to be further examined.

3.5.4 Safety of Natural Gas and Compressed Natural Gas Vehicles

Natural gas has a narrow flammability range, which is between 5 and 15 percent by volume in air. Natural gas is lighter than air, and so a gas leak disperses quickly. Unlike gasoline, natural gas will not cause a combustible liquid spill. Its high autoignition temperature means that natural gas does

not easily self-ignite on hot surfaces below 540°C, a property quoted as another safety factor in its favor.

CNGVs meet the same safety standards as gasoline and diesel vehicles, and they also meet the National Fire Protection Association's Vehicle Fuel System Code. CNG tanks meet DOE and other government safety standards and have been certified for that purpose. The Clean Vehicle Education Foundation has published a Technology Committee Bulletin (Clean Vehicle Education Foundation, 2010) that provides a detailed treatise of safety considerations for CNGVs. The Clean Vehicle Foundation actually stated that CNG-powered vehicles are considered to be safer than gasoline-powered vehicles.

The DOE has detailed safety analysis and operating recommendations for natural gas filling stations. Properly designed, maintained, and operated facilities for CNG refueling appear to represent no undue safety problems to the public.

3.5.5 Barriers

Public policies at various government levels have not kept up with the increased abundance of natural gas in the United States and are expected to develop rapidly in the coming years.

The CNG infrastructure (filling stations, gas distribution) is in its early stage of development and requires massive expansion. Regional, clustered development will remain the preferred model.

Finding: With increasing economic natural gas reserves and growing domestic natural gas production mostly

from shale gas, there is enough domestic natural gas to use within the transportation sector without significantly affecting the traditional natural gas markets. The opportunities include producing electricity for PHEVs, producing hydrogen for FCEVs, and using as a fuel in CNGVs.

Finding: CNG used as a transportation fuel is an important near-term transition opportunity that could be exploited because of its ability to economically replace petroleum and to reduce GHG emissions from the LDV fleet.

3.6 LIQUID FUELS FROM NATURAL GAS

3.6.1 Current Status

The production of liquid fuels—diesel, gasoline, or a combination of both—from natural gas has been practiced commercially since the early 1980s. As in the case of coal, the first step in the GTL process is the conversion of natural gas into a mixture of carbon monoxide and hydrogen (synthesis gas). There are two options for using this synthesis gas to produce liquid fuels. One is the production of methanol followed by the conversion of methanol into gasoline (MTG). The other option is the conversion of the synthesis gas via FT chemistry to a broad range of paraffinic hydrocarbons. The hydrocarbon molecules with more than 20 carbons are then hydrocracked into molecules in the diesel (15-20 carbons) and naphtha (6-12 carbons) range. The quality of the diesel fuel is excellent but the naphtha has a low octane value and has to be further processed to be used as gasoline (NAS-NAE-NRC, 2009b).

For nearly 10 years in the 1980s, Mobil Corporation operated a facility in New Zealand that produced gasoline by the MTG process (ExxonMobil, 2009). Today, the facility makes only methanol for chemical use (Tabak, 2006) because converting the methanol to gasoline is not viewed as economical at current gasoline prices. Shell has produced diesel fuel and lubricants since the late 1980s in a facility in Malaysia via FT chemistry and Shell is building a plant in Qatar, based on the same process chemistry. That facility is expected to eventually produce more than 140,000 barrels of diesel fuel per day (Kingston, 2011). Another facility in Qatar that is smaller (about 34,000 bbl/d) and based on the same FT chemistry is coowned by Sasol, Chevron, and the Government of Qatar. Similar facilities have been proposed for gas-rich locations such as Nigeria (Chevron, 2011).

3.6.2 Capabilities

The conversion of natural gas into synthesis gas is significantly simpler when compared to the production of synthesis gas from coal. At present, the preferred pathway uses what is called an auto-thermal reactor (ATR). In an ATR, a portion of the natural gas (methane) is burned with oxygen into CO_2

and water vapor. This reaction is highly exothermic (that is, it releases heat) and results in a mixture of CO_2, unreacted methane, and steam at temperatures close to 2,000°C. This mixture is converted into carbon monoxide and hydrogen in a fixed bed containing a nickel-based catalyst. Although ATRs are very efficient and compact, the design and operation of the feedstock and burner system requires careful attention to the mixing of oxygen, steam, and methane (Haldor Topsoe, 2011).

The processing steps are significantly less complicated than in a coal plant. The natural gas, if needed, is cleaned of sulfur compounds before being fed to the ATR. Because methane has four hydrogen atoms for each carbon atom, the synthesis gas from the ATR has the required ratio of two molecules of hydrogen per molecule of carbon monoxide. Thus, the synthesis gas can be used without further processing to produce either methanol or FT hydrocarbons followed by the conversion of these into gasoline or a diesel/naphtha mixture as discussed above.

3.6.3 Costs

The data presented in Table 3.17 were derived from a report prepared for the Alaska Natural Resources to Liquids LLC and requested by the Alaska legislature (Peterson and Tijm, 2008). The results of that study were in good agreement with data published by various companies (Shell, Sasol, and ExxonMobil) on GTL technology performance and economics.

As in the case of CTL, this committee assumes that the GTL plants built later will benefit from a learning curve. Therefore, the estimated investment required was $5 billion for a 2020 facility, $4 billion for a 2035 facility, and $3 billion for a 2050 facility. These investment costs do not include CCS. Although CCS could be used in a GTL facility, the amount emitted from a GTL facility is significantly less than that for similar-size CTL facilities. Therefore, CCS was not included in GTL facilities for the purpose of this study.

TABLE 3.17 GTL Outlook Process Data

GTL/MTG	2020	2035	2050
Gas, million scf/d	400	400	400
Fuel production, bbl/d	50,000	50,000	50,000
Investment, $billion	5.0	4.0	3.0
Product cost, $/bbl	103.5	106.0	109.0
CO_2e produced by the process, metric tons/d	3,840	3,840	3,840
CO_2 vented, metric tons/d	2,110	2,110	2,110
CO_2 stored, metric tons/d	—	—	—

NOTE: Product cost basis: (1) 20 percent of capital annual charge (financing, return on capital, maintenance), 90 percent capacity utilization; (2) natural gas prices as per AEO 2011 (EIA, 2011a), $5.36/million cubic feet for 2020, $7.21/million cubic feet for 2035 and $9.06/million cubic feet in 2050; (3) CO_2e emissions from gas production are based on GREET estimates for the production and transport of gas.

The cost estimates for GTL are based on the FT process economics (see Table 3.17). There are no published data available for the MTG option. For the purpose of this study, capital cost and overall performance data for the MTG option are expected to be similar to the numbers presented in Table 3.17. The investment required for the GTL processes is lower than the investment estimated for the CTL options. This is expected because CTL requires the greater complexity of coal gasification and the complex cleaning of the synthesis gas, and because of the fact that half of the coal has to be converted to CO_2 (to make hydrogen), which in turn has to be captured and stored (CCS).

The cost for the liquid fuel from a GTL plant is about $106/bbl in 2035, which is less than the price of crude oil in 2035 ($125/bbl) forecasted by EIA (2011a). However, the GTL cost estimate is based on a natural gas price of $7.21/ million cubic feet in 2035, which is lower than natural gas prices in 2008 and earlier. If the price of natural gas were to reach $10.0/million cubic feet, the liquid product cost would increase to $130/bbl. The cost of the liquid product in 2050 is estimated at $109/bbl based on a natural-gas price of $9.06/million cubic feet. If the natural-gas price were $11.0/ million cubic feet, the liquid-product cost escalates to close to $130/bbl.

3.6.4 Implementation

GTL technology has been commercialized in a number of locations where the price of natural gas is low because those locations are far away from markets where the gas can be used directly for power and heat generation. Moreover, all the GTL facilities are based on producing diesel fuel, naphtha, and in some cases high-value lubricants.

When considering the application of GTL technology in the United States, two factors need to be considered. First, the MTG option might be preferred because gasoline is a more widely used transportation fuel than diesel. Second, the price of natural gas will likely be significantly higher in the United States than in other areas of the world where it is readily available (e.g., in the Middle East and in West Africa) because it can be readily used in heating, power generation, petrochemical production, and other industries. The forecasted production of liquid fuels from natural gas (GTL) assuming an optimistic outlook and a more realistic outlook is summarized in Table 3.18.

The estimates for fuel production from GTL are sensitive to natural gas prices. Using the 2011 AEO (EIA, 2011a), the cost of the fuel in 2035 is about $105/bbl, which is lower than the crude-oil price forecasted for that year. However, a 25 percent increase in the price of natural gas would raise the final-product price well above the crude-oil price.

The GHG emissions for the production of GTL fuel are, as in the case of coal, comparable to the emissions from producing petroleum-based fuels. Thus, GTL without CCS for LDVs reduces the consumption of petroleum-based fuels

TABLE 3.18 GTL Outlook Production Estimates

	2020	2035	2050
Optimistic outlook			
GTL/MTG plants	1	4	12
GTL/MTG production, bbl/d	50,000	200,000	600,000
Realistic outlook			
GTL/MTG plants	1	3	6
GTL/MTG production, bbl/d	50,000	150,000	300,000

but does not yield any GHG reduction. Adding CCS to a GTL facility would have a small effect on the life-cycle GHG emissions of the fuel produced because the GHG releases that could be captured at the conversion facility are small compared to the CO_2 release from combusting the liquid fuel.

3.6.5 Infrastructure Needs

Because natural gas is readily available throughout most of the country, there are no major issues with either infrastructure or the location of GTL facilities.

3.6.6 Safety

Although the GTL process includes a complex step for generating synthesis gas, there are no unique safety issues. Natural processing, transmission, and use are widely practiced in the United States. The process of converting natural gas to a liquid fuel for LDVs has many similarities to petroleum-refining processes, and well-known safety practices can be applied.

3.6.7 Barriers

One important barrier to the wide use of natural gas to make liquid fuels is the cost over the life of commercial GTL facilities and the availability of natural gas. Recent technology advances for producing gas from tight shales and other low porosity reservoirs suggest that the natural-gas resources in the United States are significantly greater than previously estimated. The resource availability is a positive factor, but the cost and the environmental impact of producing this tight gas are unclear at present. Moreover, natural gas is used in all sectors of the economy, and the distinct advantage of using natural gas in electricity generation suggests that the demand for gas in this sector could increase dramatically. Use of natural gas directly in LDVs is also being proposed. (See Section 3.5, "Natural Gas as an Automobile Fuel.") The balance between supply and demand for natural gas in the United States depends on the level of consumption in many sectors and the level of production. Therefore, predicting the future price of natural gas is difficult. Because the cost of the gas feedstock is a major factor in the cost of the GTL fuel made, the estimate for total liquid fuels produced from natural gas in 2050 is less than 600,000 bbl/d in the optimistic case. That production level requires an annual consumption of 1.6 tcf of natural gas, or about 8 percent of the present production in the United States.

3.7 LIQUID FUELS FROM COAL

3.7.1 Current Status

Liquid fuels, both gasoline and diesel, have been produced from coal at a significant scale since the 1930s. At present, the CTL facilities with the largest capacity are in South Africa and produce more than 100,000 bbl/d of liquid products. Moreover, a number of proposed facilities are being considered in China.

There are two technology options for the production of liquid fuels from coal: direct and indirect liquefaction. The direct liquefaction of coal involves reacting coal with hydrogen or a hydrogen-donating solvent. This technology option has been the subject of research, development, and pilot-scale demonstration since the late 1970s. The consensus view is that this technology is still in development and that the complexity of the process scheme and the poor quality of the liquid products are major limitations. However, a demonstration facility was built in China, and that facility may provide a definitive assessment of the coal-to-liquid fuels option (NMA, 2005; NPC, 2007; NAS-NAE-NRC, 2009b).

This section focuses on the indirect liquefaction option that involves the gasification of coal to a mixture of carbon monoxide and hydrogen (synthesis gas) followed by the conversion of this gas into liquid products. There are two schemes to make the synthesis gas into liquid-fuel products. One option is to convert the synthesis gas into methanol followed by MTG (Zhao et al., 2008). The second option is to convert the synthesis gas into a broad range of hydrocarbons via FT chemistry followed by the hydrocracking of the molecules with more than 20 carbons into shorter-chain molecules. The FT option results in a mix of liquid products that includes mostly diesel fuel and a significant amount of naphtha that can be upgraded to gasoline.

The commercial-scale facilities in South Africa are producing diesel and gasoline from coal by the FT option. Although the Mobil Corporation operated a facility that used the MTG option, the feedstock was natural gas rather than coal.

In the report *Liquid Transportation Fuels from Coal and Biomass: Technological Status, Costs, and Environmental Impacts* (NAS-NAE-NRC, 2009b), a process scheme labeled coal-and-biomass to liquid fuel (CBTL) is proposed. The process uses a separate gasifier for the coal and the biomass feedstocks. The effluents from these gasifiers undergo a number of separation steps to remove solid and gaseous impurities. The biomass gasifier effluent also includes a thermal cracking step to convert the tar produced from the biomass to lighter products. The clean-up streams are then combined

and undergo the required processing steps to make liquid products from carbon monoxide and hydrogen and remove and compress the CO_2.

A number of cases presented in NAS-NAE-NRC (2009b) include or exclude CCS, and in other cases the proposed facility produces significant amounts of electric power (these are called once-through cases). Although interesting synergies have been identified in these schemes, all process schemes require different gasification reaction systems for the coal and for the biomass. They can be viewed as requiring a separate CTL and BTL gasification plants in a given site. The number of sites in the United States where there are significant amounts of biomass and coal for commercial-scale facilities might be small.

The potential benefits of combining the gas products from the biomass and coal gasification to make liquid fuels and electric power are clear from the studies available. A CBTL facility produces liquid fuels at a higher cost than does a CTL facility but at lower cost than a BTL facility. Moreover, by capturing the CO_2 produced in the biomass portion of the facility, the process drastically reduces the life-cycle GHG emissions of the liquid fuels (the emissions during their combustion are counterbalanced by the CO_2 taken up during plant growth). The potential benefits of CBTL facilities, while significant, will require commercial-scale demonstrations of BTL technology and combining it with CTL technology.

The CBTL process was not included in the case study model runs explained in Chapter 5 because it is a derivative process of two commercially available processes. Coal conversion and biomass conversion to liquids are individually included in all of the model scenarios.

3.7.2 Capabilities

The United States has ample coal resources that can allow the production of significant amounts of liquid fuels such as gasoline and diesel from coal. Most coal produced in the United States (about 1 billion tons per year) is used to generate electricity. In principle, additional coal could be mined to produce liquid fuels because the coal reserves in the United States are estimated to be in the range of 250 billion tons. However, concerns have been raised about the environmental impact of coal mining and of the disposition of mineral ash present in coal. Those concerns apply to all uses of coal (AAAS, 2009; EPA, 2011a,b).

The process to convert coal into a liquid fuel is complex and expensive. The gasification of the coal is the most challenging process step. The coal has to be fed into a reactor that operates at pressures ranging from 20 to 50 atmospheres along with pure oxygen and water. The average reactor temperature is about 800°C. Because coal is a solid and its quality varies, the feed system is complex and sensitive to the coal quality. Moreover, coal contains a number of impurities including mineral ash, sulfur, nitrogen and mercury. A number of process steps are needed to remove the byproducts of the gasification reaction to yield a pure stream of carbon monoxide and hydrogen (KBR, 2011).

The second major challenge in making liquid fuels from coal that applies to both the FT and the MTG options is the fact that chemistry dictates that two molecules of hydrogen react with one molecule of carbon monoxide. Because coal, on average, contains only an atom of hydrogen per atom of carbon, half of the carbon monoxide produced in the gasification step has to be used to make additional hydrogen. This is done using the water gas shift reaction where water and carbon monoxide are converted into carbon dioxide and hydrogen. Thus, this reaction step yields the required 2:1 mole ratio of hydrogen to carbon monoxide needed for the subsequent reaction steps and also produces one molecule of carbon dioxide for each molecule of carbon monoxide. In other words, half of the coal is converted to CO_2 and the other half into the reactants needed for the next process steps. Therefore, CCS is necessary if coal is to be used to make liquid fuels with life-cycle GHG emissions in the range of those from use of petroleum-based fuels. Although there are a few facilities that use CCS, there is consensus that a large-scale demonstration in a variety of geological formations is required before CCS can be deemed commercially acceptable.

The conversion of carbon monoxide and hydrogen via MTG or FT to diesel or gasoline presents less of a technology challenge and, has been done commercially for many years (ExxonMobil, 2009; NAS-NAE-NRC, 2009b). Most of the commercial facilities have used or are using natural gas rather than coal as the feedstock. The use of natural gas to make liquid fuels is discussed in a separate section above in this chapter.

3.7.3 Costs

The data presented in Table 3.19 are derived from *Liquid Transportation Fuels from Coal and Biomass: Technological Status, Costs, and Environmental Impacts* (NAS-NAE-NRC, 2009b), which describes in detail the process schemes briefly reviewed here. It also described the challenges and potential of the various technology options. It includes estimates of the capital and operating costs for CTL facilities.

Here, the cost of the first CTL facility built by 2035 has been estimated to be 20 percent higher than the facilities built later on. The MTG facility is estimated to be lower in capital cost and to require less coal for the same level of production of 50,000 bbl/d of liquid-fuel product than would the FT process. The MTG process is more selective than the FT process as indicated by the higher energy conversion efficiency. Efficiency is the percent of the energy content of the coal that is contained in the liquid produced. The efficiency in the 50 percent range indicates that close to half of the coal has to be converted into CO_2. That amount of CO_2 has to be "stored" via CCS in both cases. The capital cost estimated

TABLE 3.19 CTL Outlook Process Data

CTL/FT	2020	2035	2050
Coal, tons/d	26,700	26,700	26,700
Fuel production, bbl/d	50,000	50,000	50,000
Investment, $billion	6.0	6.0	5.0
Product cost, $/bbl	126.8	122.5	104.7
CO_2 coal production, metric tons/d	2,580	2,580	2,580
CO_2 vented, metric tons/d	5,011	5,011	5,011
CO_2 stored, metric tons/d	29,208	29,208	29,208

CTL/MTG	2020	2035	2050
Coal, tons/d	23,200	23,200	23,200
Fuel production, bbl/d	50,000	50,000	50,000
Investment, $billion	5.0	5.0	4.0
Product cost, $/bbl	105.2	102.5	86.0
CO_2 coal prod, metric tons/d	2,243	2,243	2,243
CO_2 vented, metric tons/d	5,520	5,520	5,520
CO_2 stored, metric tons/d	23,280	23,280	23,280

NOTE: Product cost basis: (1) 20 percent of capital annual charge (financing, return on capital, maintenance), 90 percent capacity utilization (2) $50/metric ton of CO_2 pipelined and stored underground in 2020, $40 in 2035, and $30 in 2050; (3) coal prices as per AEO 2011 (EIA, 2011a), $1.85/million Btu in 2020, $1.98 in 2035, and $2.00 in 2050; (4) CO_2 emissions from the coal production are based on GREET estimates for the production/transport of coal.
SOURCE: Data from NAS-NAE-NRC (2009b).

for a facility with a 50,000 bbl/d capacity is high and thus has a major impact on the cost of the liquid-fuel product.

The cost of the liquid-fuel product made in the CTL facilities is within the range of the cost of a barrel of crude oil forecasted for 2035 in the 2011 AEO (EIA, 2011a) and the cost of a barrel of crude oil extrapolated to 2050. However, the CTL estimate is based on a coal price that remains essentially constant from the 2009 price; a doubling of the coal price will yield product costs of over $150/bbl. Conversely, coal prices could decrease as a result of increasing use of natural gas or other resources for electricity generation. The CTL facilities take a long time to build, and thus their payback requires high product prices for a long period of time.

3.7.4 Infrastructure Needs

The process cost estimate for CTL is based on the facilities using Illinois #6 coal and the CTL plants being built in the Midwest. Therefore, the mining and transport of the coal to the CTL facilities are assumed to be handled within the present infrastructure. The liquid-fuel products from the facilities will be consumed in the Midwest and will be marketed using the present infrastructure. The CO_2 is assumed to be pipelined and stored underground within a 150-mile range because geological studies indicate a significant storage potential in the Illinois Basin (Finley, 2005). Therefore, the main new infrastructure needed will be the pipelines to transport the CO_2, the injection wells to store it in underground formations, and the equipment to monitor CO_2 emissions in the pipelines and from the underground storage formations. All of these costs are included by adding \$50/metric ton of CO_2 stored in 2020, \$40/metric ton of CO_2 stored in 2035, or \$30/metric ton by 2050 to the product cost.

3.7.5 Implementation

As mentioned above, CTL technology is used in South Africa at present. The main reason for its commercialization was the need to provide liquid fuels in a country rich in coal. Another major consideration was the embargo of crude oil and petroleum products imposed on the country because of its Apartheid Policy. Economic considerations were, therefore, secondary. While a number of feasibility studies on CTL have been announced in the last 10 years, none of the facilities have reached commercialization. China has been operating a CTL demonstration project (China Shenhua Coal to Liquid and Chemical Co. Ltd., 2010; Reuters, 2011).

There are major barriers to the widespread commercialization of CTL technology. First, the process is complex and costly. Second, large amounts of CO_2 generated by the facilities need to be captured and stored. The process to capture CO_2 is based on the absorption of the gas in a liquid solvent. A number of solvents have been used, and the process is practiced at a commercial scale. It requires a significant amount of energy, thus reducing the efficiency of the overall process. Third, the transportation and storage of CO_2 add to the cost. The gas would be compressed to a pressure of about 125 atmospheres and then pipelined to a region where there is a porous underground formation for storage. Wells will be used to transfer the gas to the formation zone, where the gas is expected to either dissolve in the formation water or be converted to a carbonate salt. In 2011, DKRW Advanced Fuels LLC announced that its subsidiary, Medicine Bow Fuel and Power LLC, entered into a contract to produce liquid fuels from coal and to sell the carbon captured for enhanced oil recovery (DKRW Advanced Fuels LLC, 2011).

Two estimates for the eventual production of liquid fuels from coal are presented in Table 3.20. One is an optimistic estimate, and the other one is a realistic outlook. Both estimates assume that no CTL facilities would be operational in 2020. The technology requires demonstration that large amounts of CO_2 can be captured, pipelined, and stored safely, and such demonstrations are not expected to be completed until later in this decade. Moreover, the design and construction of CTL facilities are expected to take at least 5-6 years for the first few facilities.

3.7.6 Safety

The actual production of liquid fuels from coal presents the typical safety issues encountered in the handling, gasifi-

TABLE 3.20 CTL Outlook Production Estimates

Optimistic Outlook	2020	2035	2050
CTL/FT plants	—	1	2
CTL FT production, bbl/d	—	50,000	100,000
CTL/MTG plants	—	2	6
CTL/MTG production, bbl/d	—	100,000	300,000
Total production, bbl/d	—	400,000	150,000

Realistic Outlook	2020	2035	2050
CTL/FT plants	—	1	1
CTL/FT production, bbl/d	—	50,000	50,000
CTL/MTG plants	—	2	3
CTL/MTG production, bbl/d	—	100,000	150,000
Total production, bbl/d	—	100,000	200,000

cation, and refining of coal. Thus, CTL safety is expected to benefit from many decades of prior experience. However, there is much less experience with the safety of pipelining and storing large quantities of CO_2 (at least 9 million metric tons per year from one CTL facility). Although 3,900 miles of national CO_2 pipeline infrastructure exist (Dooley et al., 2001) to transport about 65 million metric tons of CO_2 each year for enhanced oil recovery (Melzer, 2012), geologic storage of CO_2 is only in the demonstration phase (NAS-NAE-NRC, 2009b; see Section 3.8, "Carbon Capture and Storage," below in this chapter). The key issue with CCS is to ensure that the CO_2 does not leak from either the pipeline or the formation itself. At concentrations higher than 2 percent in air, CO_2 can asphyxiate humans and animals (Praxair, 2007). Storing CO_2 entails health and ecological risks associated with acute or chronic leaks (NAS-NAE-NRC, 2009b). Clearly, the safety of CCS operations will be a major concern. CCS is being practiced for oil well stimulation in the North Sea, Algeria, and Saskatchewan, Canada, but at a scale much smaller than what is envisioned for a single CTL facility. There are also a number of pilot demonstrations of CTL in the United States (NETL, 2011).

3.7.7 Barriers

An important issue to be considered when estimating the potential supply of CTL liquids is the actual production of coal with its inherent environmental and safety challenges. If only 500,000 bbl/d of liquid-fuel products are to be produced from coal, 85 million tons of coal would have to be mined and transported each year. Locating CTL facilities close to mines would reduce transportation costs. The coal consumption is equivalent to about 10 percent of the U.S. coal production in 2012. There also are environmental and safety issues related to the disposal of coal ash from the coal gasification step. Thus, a major increase in coal consumption to make liquid fuels is not likely.

The most important barrier to the large-scale use of coal to make liquid fuels is the GHG emissions from these facilities. The process eventually yields a liquid fuel for LDVs that has chemical properties substantially similar to those of petroleum-based fuels. Thus, the carbon content of the fuel is the same as the carbon content of petroleum-based fuels. Moreover, the production of CTL fuel with CCS is estimated to emit at least as much CO_2 as the production, transport and refining of the same fuel from petroleum. For CTL fuels to have life-cycle GHG emissions equivalent to those of petroleum-based fuels, an amount of CO_2 equivalent by weight to the weight of the coal used has to be captured and stored. Thus, CTL technology can reduce the amount of petroleum used in LDVs but does not contribute to reducing GHG emissions.

Finding: GTL fuel and CTL fuel with CCS can be used as a direct replacement for petroleum-based fuel. However, the GHG emissions from GTL or CTL fuel are slightly higher than those from petroleum-based fuel. The role of GTL and CTL with CCS in reducing petroleum use will thus be small if the goals of reducing petroleum use and reducing GHG emissions are to be achieved simultaneously.

3.8 CARBON CAPTURE AND STORAGE

3.8.1 Current Status

In carbon capture and storage, CO_2 is captured from various processes, compressed into supercritical conditions to about 125 atmospheres, pipelined, and then injected into a deep (>2,500 ft), porous subsurface geologic formation. Capturing, storing, and transporting CO_2 all have commercial challenges, but, in most cases, the technologies have been demonstrated or are in the demonstration phase. With CCS there are two major options for storage: deep saline formations and enhanced oil recovery.

3.8.1.1 Deep Saline Formations

In the case of a non-hydrocarbon-bearing formation, the CO_2 in supercritical state will be dissolved partially in the subsurface formation's water phase, and the rest will remain in a separate phase. In certain formations, the CO_2 will react over a very long period of time with the solids and form solid carbonates. These are slow reactions, because it takes decades for a significant amount of CO_2 to be converted to a solid carbonate. Experimental work is being conducted to determine the feasibility of extending this concept to storing CO_2 in subsea formations. Currently, demonstrations of deep saline formation CCS of more than 1 million metric tons per year of CO_2 are in progress in a number of locations (Michael et al., 2010). Additional smaller demonstration projects are

planned or underway in the United States and other regions of the world (NETL, 2007, 2011).

3.8.1.2 Enhanced Oil Recovery (EOR)

CO_2 can be injected into already-developed oil fields to recover the oil that is not extracted by initial production techniques. Injected CO_2 mixes with the oil in reservoirs and changes the oil's properties, enabling the oil to flow more freely within the reservoirs and be extracted to the surface. The CO_2 is then separated from the extracted oil and injected again to extract more oil in a closed-loop system. Once economically recoverable oil has been extracted from one area of a given reservoir, an EOR project operator reallocates CO_2 to other productive areas of the same reservoir. Once all economically recoverable oil has been extracted from a given reservoir, the CO_2 remains within the reservoir and the project is plugged and abandoned.

3.8.2 Capabilities

The capture of CO_2 from a gaseous stream has been practiced commercially for many years—for example, CO_2 has been removed from natural gas produced from reservoirs (Statoil, 2010), and the Weyburn project in Saskatchewan, Canada, has used CO_2 captured from a North Dakota coal gasification facility for EOR (Preston et al., 2005, 2009). EOR uses injection of CO_2 into a oil reservoir to assist in oil production. In the United States, typical EOR uses about 5,000 cubic feet of CO_2 per barrel of oil produced (that is, about 160 lb of carbon produce one barrel of oil, which contains about 260 lb of carbon). Oil and gas reservoirs are ideal geological storage sites because they have held hydrocarbons for thousands to millions of years and have conditions that allow for CO_2 storage. Furthermore, their architecture and properties are well known as a result of exploration for and production of these hydrocarbons, and infrastructure exists for CO_2 transportation and storage.

To calculate the largest amount of CO_2 that could be stored by EOR, all the CO_2 used is assumed to remain in the ground. The United States produces about 281,000 bbl/d of crude oil using CO_2 EOR (Kuuskaraa et al., 2011). Based on the best-case scenario for CO_2 use in EOR, this would sequester 0.26 million metric tons per day of CO_2. If all U.S. crude oil was produced by EOR, about 2 million metric tons of CO_2 could be stored per day.

The typical process for capturing CO_2 is by contacting the gaseous stream with a solvent that absorbs the CO_2. A number of solvents have been used. The CO_2 is then desorbed as a concentrated gas and the solvent reused. This process is widely used for processing natural gas streams but much less used with gaseous streams from coal gasifiers or coal combustion units. The key concern is the degradation of the solvent by coal-derived impurities in the process gas. Other processes are being considered and developed to reduce the cost and energy consumption required.

CO_2 compression to about 125 atmospheres for transport and injection is straight forward but consumes a significant amount of energy. High-pressure compression is desirable because it reduces the volume of gas being pipelined, and the supercritical state facilitates injection and retention of the CO_2 (IPCC, 2005).

Pipelining of CO_2 is another conventional and proven step. The key concern is leakage of CO_2 into the atmosphere. An asphyxiant denser than air, CO_2 tends to stay close to the ground and is not easily dispersed. CO_2 is fatal at high concentrations and detrimental to humans at lesser concentrations (Praxair, 2007). Thus, properly designed CCS facilities will include a CO_2 monitoring system and a leak-prevention system.

Specially designed injection wells are required for CCS. Abandoned oil and gas wells will not be used for CO_2 injection into spent oil and gas formations because these wells may not be capable of handling the acidic supercritical CO_2, and they may not be properly cemented to ensure that CO_2 does not leak into aquifers used for drinking water.

3.8.3 Costs

The cost of CO_2 capture is $30-$40/metric ton of CO_2 for a coal gasifier process stream, about $90/metric ton for a natural gas combined-cycle facility (because of a lower concentration of CO_2 compared to coal gasification), and $70-$80/metric ton for coal-fired power facilities (IPCC, 2005). Adding in the cost of compression, pipelining, monitoring and injection into a suitable formation would increase the total cost by $30-40/metric ton (IPCC, 2005). For most CTL facilities, the cost of CO_2 capture is included in the facility design and construction cost. However, additional costs are incurred for compression, pipelining, monitoring, injection, and storage. These costs are estimated at $40/metric ton of CO_2 in the first-mover facilities (2035 timeframe) and $30/metric ton in facilities built later (2050 timeframe). In cases of CTL where the costs of capture are to be included, $80/metric ton of CO_2 for 2035 and $70/metric ton of CO_2 for 2050 are used.

3.8.4 Infrastructure Needs

CCS requires a large infrastructure—primarily the construction of pipelines to transport the CO_2 from where it is captured to injection wells for storage underground. In the United States, potential reservoirs with a capacity for storing more than 100 years' worth of injected CO_2 are available within 100-150 miles of expected sources in most regions of the country (NACAP, 2012).

3.8.5 Barriers

The cost of CCS is significant but probably not the major implementation barrier. The major barrier is the public acceptance of pipelines, injection wells, and storage of large amounts of carbon dioxide in subsurface formations (Court et al., 2012; de Best-Waldhober et al., 2012; Kraeusel and Moest, 2012), especially if these are near population centers. Leakage of stored CO_2 is an issue that is still being investigated through research programs conducted by industry and DOE. Careful design and operation of CCS can likely prevent and mitigate any potential emissions of CO_2, but gaining public acceptance is expected to be difficult given the large quantities of CO_2 to be transported and stored. A single CTL facility producing 50,000 bbl/d of liquid fuels will require CO_2 storage in the range of about 4 million to 9 million metric tons per year.

Finding: CCS is a key technology for meeting the study goals for GHG reductions by 2050. It will be very difficult to make large quantities of low-GHG hydrogen without CCS being widely available. Combining CCS with biofuel production would improve the chances of meeting the study goals.

3.9 RESOURCE NEEDS AND LIMITATIONS

Reducing petroleum consumption and GHG emissions from the LDV fleet will have a significant impact on energy resource use in the United States. Comparing existing resources with the estimated demands on resources for fueling the vehicles in representative scenarios in its analyses, the committee here draws conclusions about whether the projected demands on resources can be met.

Alternative LDV fuels can be produced from natural gas, coal, biomass, or other renewable energy sources, such as wind, solar, and hydro power. The U.S. consumption of natural gas, coal, and biomass in 2010 is shown in Table 3.21. Of the amounts consumed, 976 million tons of coal and 7.378 tcf of natural gas were used for electricity generation (EIA, 2011b). The biomass was used primarily for power in wood-processing plants, with some generated electricity going into the grid.

TABLE 3.21 U.S. Consumption of Natural Gas, Coal, and Biomass in 2010

	Consumption in Quads (higher heating value)	Amount Consumed
Natural gas	24.1	23.4 tcf
Coal	22.1	1,050 million tons
Biomass	4.30	269 million tons

TABLE 3.22 Estimated Amount of Natural Gas Required to Fuel the Entire LDV Fleet via Different Fuel and Vehicle Technologies

Year	Vehicle Miles Traveled (trillion mi/yr)	Natural Gas Required Annually for Different Vehicle-Fuel Combinations (tcf)					
		ICE-CNG	ICE-drop-in	ICE-Methanol	HEV-CNG	Electric	FCEV
2010	2.784	15.6	23.8	22.9	15.1	7.6	11.7
2030	3.727	10.1	15.5	14.9	8.4	7.5	7.3
2050	5.048	10.0	15.4	14.8	7.9	7.8	7.2

Biomass, coal, and natural gas can all be converted into "drop-in" liquid fuels by several routes (e.g., direct liquefaction of biomass or coal, and gasification followed by FT or MTG of all sources). These drop-in fuels will use the existing petroleum products distribution system and existing vehicles. The use of any of these alternative fuels would be transparent to the vehicle owner. The remaining alternative fuel and vehicle combinations include electricity in BEVs and PHEVs, hydrogen in FCEVs, and natural gas as a vehicle fuel, either directly as CNG or through conversion to methanol. All of these fuels can be produced from natural gas via mature technologies, and so a meaningful comparison would be to calculate the amount of natural gas that would be required to fuel the entire LDV fleet via the different fuel and vehicle technologies (Table 3.22). The vehicle efficiencies are assumed to be the mid-range efficiencies outlined in Chapter 2.

Direct use of CNG as a vehicle fuel is more resource efficient and less costly than conversion of natural gas to any liquid fuel. The advantages of conversion to a liquid fuel are the use of the current fuel infrastructure, the ease of onboard storage, and the familiarity of the driving population with liquid fuels. Conversion of natural gas to electricity or hydrogen as an energy carrier is currently more resource efficient than direct use of natural gas, but direct-use efficiency converges with that for PEVs and FCEVs by 2050 because of the differences in efficiency improvements with time. Both electricity and hydrogen carry additional socioeconomic burdens and infrastructure costs as discussed in previous sections. Electricity and hydrogen, as well as GTL and methanol, can be produced from other resources such as coal and biomass. Electricity and hydrogen can also be produced from nuclear, solar, and wind power.

There are two distinct goals for the scenarios evaluated by the committee: one goal targets only petroleum reduction, and the second goal targets reduction of GHG emissions. Both cases use the same vehicle and fuel technologies; however, in the low-GHG cases, the technology and fuels used to generate electricity and hydrogen were modified to reduce GHG emissions. The driving force for the low GHG grid case is discussed above in this chapter. Table 3.23 shows the impact of the low-GHG grid case on the mix of generating sources.

TABLE 3.23 Effect of the Low-Greenhouse Gas Grid on the Mix of Generating Sources

	Total Generation (billion kWh/yr)		
	2009	2050 Reference Grid	2050 Low-GHG Grid
Coal without CCS	1,693	2,368	238
Coal with CCS	0	15	17
Petroleum and natural gas without CCS	871	1,290	1,225
Petroleum and natural gas with CCS	0	0	489
Nuclear	795	855	1,255
Hydroelectric	274	314	323
Biomass	38	159	179
Solar	3	21	56
Wind	71	163	330
Other	34	66	66

The largest changes between the reference grid and the low-GHG grid are an almost 90 percent decline in coal usage, a doubling of natural gas, and a 50 percent increase in nuclear power. Total renewable electricity increases by over a factor of two and rises from 11 percent of total generation to 23 percent.

Table 3.24 shows the fuel usage and resource demands for 10 scenarios: five different vehicle mix scenarios, compounded with the reference grid and the low-GHG grid case and two different resource mixes for producing hydrogen.[22] The implementation of these cases would be driven by various government policies. The reference case scenario is driven by existing and currently proposed policies for LDV CAFE standards and RFS2. The other cases stress increased biofuels, PEVs, FCEVs and CNGVs.

These scenarios have not been optimized to minimize costs, resource use, or GHG emissions. The reference scenario reduces petroleum use by 25 percent, and the others all meet or exceed the goal of an 80 percent reduction in petroleum use. GHG emission reductions are all similar for the reference-grid scenarios. Additional reductions in GHG emissions are possible for the electric and hydrogen cases

---

[22]These scenarios are described in greater detail in Section 5.3.2.

with the use of a low-GHG grid and a change in the mix of resources used to generate hydrogen. Only the FCEV scenario meets the goal of reducing GHG emissions by 80 percent in 2050. The biofuel case can also meet the GHG emissions target if CCS is added to the biorefineries.

The resource demands can be met but involve some challenges. The largest changes are needed to achieve a low-GHG grid. These include an increased use of almost 7 tcf/yr of natural gas (a doubling of the current consumption for electricity), the construction of about fifty 1,000-MW nuclear power plants and about 100,000 new wind turbines and the capture and storage of more than 200 million metric tons/yr of CO_2.

The most challenging related demands concern increased use of biomass and natural gas and public acceptance of the construction of a large number of nuclear power plants. As discussed above in this chapter, the demand for biomass is expected to be achievable and to be less than the biomass availability estimated in other recent analyses. Shipping and handling the mass and volume of biomass involved will be challenging. Natural gas demand doubles over the amount currently used to generate electricity. This increase represents essentially all of the additional natural gas expected to be available for use based on the most recent estimates of future gas availability in the United States.

There are important ancillary impacts from these resource demands on the associated infrastructure:

- Cleaning up the electric grid by 2050, as envisioned in 2011 AEO (EIA, 2011a), the basis for this discussion, will reduce current coal use by 85 percent or about 800 million tons per year, an amount that represents 44 percent of the total annual U.S. railroad freight tonnage. Shipments of biomass could mitigate that impact.
- Most petroleum products are currently shipped long distances by pipeline. Significant increases in hydrogen or electricity as an LDV fuel would idle a large fraction of the petroleum pipeline system.
- The large increase in natural gas consumption would require a significant expansion in natural gas pipelines. Use of hydrogen as an LDV fuel would require construction of an additional hydrogen pipeline system.
- CCS has to be economical and meet stringent performance requirements at large scale. CCS demonstrations at appropriate scale are needed to validate performance, safety, and costs.

Nearly 50 percent of U.S. petroleum refining output is currently used to fuel the LDV fleet. An 80 percent reduction in use of petroleum for LDVs will impact the availability and price of the refining byproducts that are used by other industries.

TABLE 3.24 Fuel Demands for Illustrative Scenarios and Resources Used

Scenario	2005 Actual	Reference	Biofuels	Electric	FCEV	CNG
Petroleum based fuels, billion gge/yr	124.8	93.1	17.2	13.9	3.8	4.1
GTL and CTL, billion gge/yr	0	7.7	7.7	7.7	0.8	0.8
Total biofuels, billion gge/yr	4.9	24.1	55.9	24.1	19.2	19.1
Electricity, billion gge/yr	0	1.3	0	14.4	1.6	1.0
Hydrogen, billion gge/yr	0	0.5	0	1.1	33.5	0.5
CNG, billion gge/yr	0.1	0.1	0.1	0.1	0.1	51.0
Petroleum reduction, %		25.4	86.2	88.9	97.0	96.7
Ethanol, % of liquid fuels	5.6	11.9	17.5	30.9	30.7	33.9
Resources Used to Power Vehicles, Reference Electric Grid						
Corn, million tons/yr	81	165	165	165	84	99
Other biomass, million tons/yr	0	208	703	220	325	208
Natural gas, billion cubic ft/yr	18	1,021	888	1,915	3,038	6,969
Coal, million tons/yr	0	50	39	150	108	14
Net GHG emissions reduction, %	—	11	67	55	60	56
Resources Used to Power Vehicles, Low GHG Electric Grid and Hydrogen Production						
Corn, million tons/yr	81	165	165	165	84	99
Other biomass, million tons/yr	0	209	703	226	358	209
Natural gas, billion cubic ft/yr	18	1,105	890	2,613	4,664	7,039
Coal, million tons/yr	0	41	39	54	15	6
Net GHG emissions reduction, %	—	13	67	72	85	58

3.10 REFERENCES

AAAS (American Association for the Advancement of Science). 2009. Coal-to-liquid technology. AAAS Policy Brief. Available at http://www.aaas.org/spp/cstc/briefs/coaltoliquid/. Accessed December 28, 2011.

Alvarez, R.A., S.W. Pacala, J.J. Winebrake, W.L. Chameides, and S.P. Hamburg. 2012. Greater focus needed on methane leakage from natural gas infrastructure. *Proceedings of the National Academy of Sciences of the United States of America* 109(17):6435-6440.

Anair, D., and A. Mahmassani. 2012. *State of CHARGE. Electric Vehicles' Global Warming Emissions and Fuel-Cost Savings across the United States.* Cambridge, Mass.: Union of Concerned Scientists.

ANL (Argonne National Laboratory). 2011. GREET Model. The Greenhouse Gases, Regulated Emissions, and Energy Use in Transportation Model. Available at http://greet.es.anl.gov/. Accessed July 31, 2011.

Barcella, M.L., S. Gross, and S. Rajan. 2011. *Mismeasuring Methane: Estimating Greenhouse Gas Emissions from Upstream Natural Gas Development.* Englewood, Colo.: IHS CERA.

Begos, K. 2012. Electric Power Plants Shift from Coal to Natural Gas. Available at http://www.huffingtonpost.com/2012/01/16/electric-plants-coal-natural-gas_n_1208875.html. Accessed July 19, 2012.

BP. 2011. *BP Energy Outlook 2030.* London: BP.

Brooks, A. 2010. Separating wheat from the chaff of unconventionals. *Musings from the Oil Patch.* November 23.

Burnham, A., J. Han, C.E. Clark, M. Wang, J.B. Dunn, and I. Palou-Rivera. 2012. Life-cycle greenhouse gas emissions of shale gas, natural gas, coal, and petroleum. *Environmental Science and Technology* 46(2):619-627.

Campbell-Parnell. 2011. Why CNG? April 15, 2011. Available at http://www.usealtfuels.com/education_cng.htm.

Chevron. 2011. Gas-to-Liquids Transforming Natural Gas into Superclean Fuels. Available at http://www.chevron.com/deliveringenergy/gastoliquids/. Accessed December 28, 2011.

China Shenhua Coal to Liquid and Chemical Co. Ltd. 2010. China Shenhua Direct Coal Liquefaction Demonstration Project. Available at http://www.csclc.com.cn/ens/cpyfw/youpin/2010-12-21/320.shtml. Accessed May 2, 2012.

Clean Vehicle Education Foundation. 2010. *How Safe Are Natural Gas Vehicles?* January 19, 2012. Acworth, Ga.: Clean Vehicle Education Foundation.

Court, B., T.R. Elliot, J. Dammel, T.A. Buscheck, J. Rohmer, and M.A. Celia. 2012. Promising synergies to address water, sequestration, legal, and public acceptance issues associated with large-scale implementation of CO_2 sequestration. *Mitigation and Adaptation Strategies for Global Change* 17(6):569-599.

Cromie, R., and R. Graham. 2009. Transition to Electricity as the Fuel of Choice. Presentation to the NRC Committee on Assessment of Resource Needs for Fuel Cell and Hydrogen Technologies, May 18.

de Best-Waldhober, M., D. Daamen, A.R. Ramirez, A. Faaij, C. Hendriks, and E. de Visser. 2012. Informed public opinion in the Netherlands: Evaluation of CO_2 capture and storage technologies in comparison with other CO_2 mitigation options. *International Journal of Greenhouse Gas Control* 10:169-180.

DeMorro, C. 2010. GM Announces Hydrogen Infrastructure Pilot Project in Hawaii. Available at http://gas2.org/2010/05/12/gm-announces-hydrogen-infrastructure-pilot-project-in-hawaii/. Accessed June 27, 2012.

Diamond, R. 2010. Electric Vehicles: Their Potential and How to Achieve It. Presentation to the NRC Committee on the Potential for Light-Duty Vehicle (LDV) Technologies, 2010-2050: Costs, Impacts, Barriers and Timing for Review. December 14.

DiPietro, D. 2010. *Life Cycle Analysis of Coal and Natural Gas-Fired Power Plants.* Available at http://www.netl.doe.gov/energy-analyses/pubs/LCA_coal&NG_plants.pdf.

DKRW Advanced Fuels LLC. 2011. DKRW Advanced Fuels Signs Major Gasoline Contract: Moving Forward on Medicine Bow Project in Wyoming. Available at http://www.dkrwadvancedfuels.com/_filelib/FileCabinet/PDFs/Press_Releases/PR_News_Release_Proof_(2).pdf. Accessed June 26, 2012.

DOE (U.S. Department of Energy). 2011. *U.S. Billion-Ton Update.* Washington, D.C.: U.S. Department of Energy.

DOE-EERE (U.S. Department of Energy-Energy Efficiency and Renewable Energy). 2011a. *2010 Fuel Cell Technologies Market Report.* Washington, D.C.: U.S. Department of Energy-Energy Efficiency and Renewable Energy.

———. 2011b. *Clean Cities. Alternative Fuel Price Report.* April. Washington, D.C.: U.S. Department of Energy.

———. 2011c. *Clean Cities. Alternative Fuel Price Report.* July. Washington, D.C.: U.S. Department of Energy.

———. 2011d. *Clean Cities. Alternative Fuel Price Report.* October. Washington, D.C.: U.S. Department of Energy.

———. 2012a. Alternative Fueling Station Total Counts by State and Fuel Type. Available at http://www.afdc.energy.gov/afdc/fuels/stations_counts.html?print. Accessed January 19, 2012.

———. 2012b. *Clean Cities. Alternative Fuel Price Report.* January. Washington, D.C.: U.S. Department of Energy.

———. 2012c. Flexible Fuel Vehicle Availability. Available at http://www.afdc.energy.gov/afdc/vehicles/flexible_fuel_availability.html. Accessed April 25, 2012.

Dooley, J.J., R.T. Dahowski, and C.L. Davidson. 2001. Comparing existing pipeline networks with the potential scale of future U.S. CO_2 pipeline networks. *Energy Procedia* 1(1):1595-1602.

EIA (Energy Information Administration). 2008a. Estimated Number of Compressed Natural Gas (CNG) Vehicles in Use, by State and User Group. Available at http://www.eia.gov/cneaf/alternate/page/atftables/attf_v9.html. Accessed January 19, 2012.

———. 2008b. *The Impact of Increased Use of Hydrogen on Petroleum Consumption and Carbon Dioxide Emissions.* SR-OIAF-CNEAF/2008-04. Washington, D.C.: U.S. Department of Energy.

———. 2011a. *Annual Energy Outlook 2011 with Projections to 2035.* Washington, D.C.: U.S. Department of Energy.

———. 2011b. *Annual Energy Review 2010.* Washington, D.C.: U.S. Department of Energy.

———. 2012a. *AEO 2012 Early Release Overview.* Washington, D.C.: U.S. Department of Energy.

———. 2012b. Biodiesel Overview. Available at http://www.eia.gov/totalenergy/data/monthly/pdf/sec10_8.pdf. Accessed July 19, 2012.

———. 2012c. Fuel Ethanol Overview. Available at http://www.eia.gov/totalenergy/data/monthly/pdf/sec10_7.pdf. Accessed July 19, 2012.

———. 2009. Residential Energy Consumption Survey. Available at http://www.eia.gov/consumption/residential/data/2009/. Accessed July 19, 2012.

Electrification Coalition. 2009. *Policy Report Fleet Electrification Roadmap.* Available at http://www.electrificationcoalition.org/sites/default/files/EC-Fleet-Roadmap-screen.pdf.

Elgowainy, A., A. Burnham, M. Wang, J. Molburg, and A. Rousseau. 2009. *Well to Wheels Energy Use and Greenhouse Gas Emissions of Plug In Hybrid Electric Vehicles.* ANL/ESD/09-2. Chicago, Ill.: Argonne National Laboratory.

EPA (U.S. Environmental Protection Agency). 2010a. *Greenhouse Gas Emissions Reporting from the Petroleum and Natural Gas Industry.* Washington, D.C.: U.S. Environmental Protection Agency.

———. 2010b. *Renewable Fuel Standard Program (RFS2) Regulatory Impact Analysis.* US EPA 420-R-10-006. Washington, D.C.: U.S. Environmental Protection Agency.

———. 2011a. Coal Ash. Available at http://www.epa.gov/radiation/tenorm/coalandcoalash.html. Accessed December 28, 2011.

———. 2011b. Mid-Atlantic Mountaintop Mining. Available at http://www.epa.gov/region3/mtntop/. Accessed December 28, 2011.

————. 2011c. Oil and natural gas sector: New source performance standards and national emission standards for hazardous air pollutants reviews. Proposed rule. *Federal Register* 76(163):52738-52843.

————. 2012a. Alphabetical List of Registered E15 Ethanols. Available at http://www.epa.gov/otaq/regs/fuels/additive/web-e15.htm. Accessed April 25, 2012.

————. 2012b. Oil and natural gas sector: New source performance standards and national emission standards for hazardous air pollutants reviews. Final rule. *Federal Register* 77(159):49490-49600.

EPRI and NRDC (Electric Power Research Institute and Natural Resources Defense Council). 2007. Environmental Assessment of Plug-In Hybrid Electric Vehicles. Volume 1: Nationwide Greenhouse Gas Emissions. Available at http://my.epri.com/portal/server.pt?open=514&objID=223132&mode=2. Accessed July 31, 2012.

ExxonMobil. 2009. *Methanol to Gasoline (MTG): Production of Clean Gasoline from Coal.* Fairfax, Va.: ExxonMobil Research and Engineering.

Farrell, A.E., R.J. Plevin, B.T. Turner, A.D. Jones, M. O'Hare, and D.M. Kammen. 2006. Ethanol can contribute to energy and environmental goals. *Science* 311(5760):506-508.

Finley, R. 2005. *An Assessment of Geological Carbon Sequestration Options in the Illinois Basin. Final Report.* Champaign, Ill.: Midwest Geological Sequestration Consortium.

Guo, J.Y., G. Venkataramanan, B. Lesieutre, A. Smick, M. Mallette, and C. Getter. 2010. *Consumer Adoption and Grid Impact Models for Plug-in Hybrid Electric Vehicles in Dane County, Wisconsin. Part A: Consumer Adoption Models.* Madison, Wisc.: Public Service Commission of Wisconsin.

Haldor Topsoe. 2011. Autothermal Reforming. Available at http://www.topsoe.com/business_areas/methanol/Processes/AutothermalReforming.aspx. Accessed December 28, 2011.

Hertel, T.W., A.A. Golub, A.D. Jones, M. O'Hare, R.J. Plevin, and D.M. Kammen. 2010. Effects of US maize ethanol on global land use and greenhouse gas emissions: Estimating market-mediated responses. *BioScience* 60(3):223-231.

Hill, J., E. Nelson, D. Tilman, S. Polasky, and D. Tiffany. 2006. Environmental, economic, and energetic costs and benefits of biodiesel and ethanol biofuels. *Proceedings of the National Academy of Sciences of the United States* 103(30):11206-11210.

Howarth, R.W., R. Santoro, and A. Ingraffea. 2011. Methane and greenhouse-gas footprint of natural gas from shale formations. *Climatic Change*, doi:10.1007/s10584-011-0061-5.

IEA (International Energy Agency). 2007. *Hydrogen Production and Distribution.* Paris: International Energy Agency.

IHS. 2010. *Natural Gas for Transportation: Niche Market or More?* Englewood, Colo.: IHS.

IPCC (Intergovernmental Panel on Climate Change). 2005. *Carbon Capture and Storage.* Geneva: Intergovernmental Panel on Climate Change.

Jiang, M., W.M. Griffin, C. Hendrickson, P. Jaramillo, J. VanBriesen, and A. Venkatesh. 2011. Life cycle greenhouse gas emissions of Marcellus shale gas. *Environmental Research Letters* 6(3).

KBR. 2011. Coal Gasification. Available at http://www.kbr.com/Technologies/Coal-Gasification/. Accessed December 28, 2011.

Kingston, J. 2011. Pearl's Gas-to-Liquids Project Rolls Out, Taking ITS Cues from Many World Markets. Available at http://www.platts.com/weblog/oilblog/2011/06/17/pearls_gas-to-l.html. Accessed December 28, 2011.

Kraeusel, J., and D. Moest. 2012. Carbon capture and storage on its way to large-scale deployment: Social acceptance and willingness to pay in Germany. *Energy Policy* 49:642-651.

Kuuskaraa, V.A., T. Van Leeuwen, and M. Wallace. 2011. *Improving Domestic Energy Security and Lowering CO_2 Emissions with Next Generation CO_2-Enhanced Oil Recovery (CO_2-EOR).* Pittsburgh, Pa.: National Energy Technology Laboratory.

Law, K., M. Chan, W. Bockholt, and M.D. Jackson. 2010. *US and Canadian Natural Gas Vehicle Market Analysis.* Cupertino, Calif.: TIAX, LLC.

LLNL (Lawrence Livermore National Laboratory). 2012. Energy Flow. Available at https://flowcharts.llnl.gov/. Accessed January 19, 2012.

McAllister, E. 2012. AAA Calls for Suspension of E15 Gasoline Sales. Available at http://www.reuters.com/article/2012/11/30/us-usa-gasoline-idUSBRE8AT0JF20121130. Accessed December 3, 2012.

Melzer, L.S. 2012. *Carbon Dioxide Enhanced Oil Recovery (CO_2 EOR): Factors Involved in Adding Carbon Capture, Utilization and Storage (CCUS) to Enhanced Oil Recovery.* Midland, Texas: Melzer Consulting.

Michael, K., A. Golab, V. Shulakova, J. Ennis-King, G. Allinson, S. Sharma, and T. Aiken. 2010. Geological storage of CO_2 in saline aquifers—A review of the experience from existing storage operations. *International Journal of Greenhouse Gas Control* 4(4):659-667.

Milhench, C., and I. Kurahone. 2011. Energy E&P Spending to Reach Record High. Available at http://www.reuters.com/article/2011/12/05/us-barclays-capex-idUSTRE7B428A20111205. Accessed April 25, 2012.

MIT (Massachusetts Institute of Technology). 2011. *The Future of Natural Gas.* Cambridge, Mass.: Massachusetts Institute of Technology.

Mullins, K.A., W.M. Griffin, and H.S. Matthews. 2010. Policy implications of uncertainty in modeled life-cycle greenhouse gas emissions of biofuels. *Environmental Science and Technology* 45(1):132-138.

NACAP (North American Carbon Atlas Partnership). 2012. *The North American Carbon Storage Atlas.* Available at http://www.netl.doe.gov/technologies/carbon_seq/refshelf/NACSA2012.pdf. Accessed July 31, 2012.

NAS-NAE-NRC (National Academy of Sciences, National Academy of Engineering, National Research Council). 2009a. *America's Energy Future: Technology and Transformation.* Washington, D.C.: The National Academies Press.

————. 2009b. *Liquid Transportation Fuels from Coal and Biomass: Technological Status, Costs, and Environmental Impacts.* Washington, D.C.: The National Academies Press.

NaturalGas.org. 2011. Background. Available at http://www.naturalgas.org/overview/background.asp. Accessed July 31, 2011.

————. 2012. Electric Generation Using Natural Gas. Available at http://naturalgas.org/overview/uses_eletrical.asp. Accessed July 19, 2012.

NBB (National Biodiesel Board). 2010. NBB Member Plant Locations. Available at http://www.biodiesel.org/buyingbiodiesel/plants/default.aspx?AspxAutoDetectCookieSupport=1. Accessed November 17, 2010.

NETL (National Energy Technology Laboratory). 2007. *Carbon Sequestration Technology and Program Plan.* Pittsburgh, Pa.: National Energy Technology Laboratory.

————. 2011. *2010 Carbon Sequestration Atlas of the United States and Canada.* Pittsburgh, Pa.: National Energy Technology Laboratory.

NMA (National Mining Association). 2005. *Liquid Fuels from U.S. Coal.* Washington, D.C.: National Mining Association.

NPC (National Petroleum Council). 2007. *Coal to Liquids and Gas.* Washington, D.C.: National Petroleum Council.

NRC (National Research Council). 2008. *Transitions to Alternative Transportation Technologies: A Focus on Hydrogen.* Washington, D.C.: The National Academies Press.

————. 2011. *Renewable Fuel Standard: Potential Economic and Environmental Effects of U.S. Biofuel Policy.* Washington, D.C.: The National Academies Press.

Oil and Gas Journal. 2011. Deloitte: MENA NOCs to Invest $140 Billion in 2011. Available at http://www.ogj.com/content/ogj/en/articles/print/volume-109/issue-45/general-interest/deloittemena-nocs-to-invest.html. Accessed June 16, 2012.

Peterson, R., and P. Tijm. 2008. *An Alaska North Slope GTL Option.* Anchorage, Alaska: Alaska Natural Resources to Liquids, LLC.

PNNL (Pacific Northwest National Laboratory). 2007. Impacts Assessment of Plug-In Hybrid Vehicles on Electric Utilities and Regional U.S. Power Grids. Part 1: Technical Analysis. Available at http://www.ferc.gov/about/com-mem/wellinghoff/5-24-07-technical-analy-wellinghoff.pdf. Accessed July 31, 2012.

Potential Gas Committee. 2009. *Potential Supply of Natural Gas in the United States (December 31, 2008).* Golden, Colo.: Colorado School of Mines.

Praxair. 2007. Product: Carbon Dioxide. Praxair Material Safety Data Sheet. Available at http://www.praxair.com/praxair.nsf/AllContent/D3CBF4EF4186073485256A860080D221/$File/p4574j.pdf. Accessed May 2, 2012.

Preston, C., M. Monea, W. Jazrawi, K. Brown, S. Whittaker, D. White, D. Law, R. Chalaturnyk, and B. Rostron. 2005. IEA GHG Weyburn CO_2 monitoring and storage project. *Fuel Processing Technology* 86(14-15):1547-1568.

Preston, C., S. Whittaker, B. Rostron, R. Chalaturnyk, D. White, C. Hawkes, J.W. Johnson, A. Wilkinson, and N. Sacuta. 2009. IEA GHG Weyburn-Midale CO_2 monitoring and storage project—moving forward with the final phase. *Energy Procedia* 1(1):1743-1750.

Raman, V. 2004. Hydrogen production and supply infrastructure for transportation—Discussion paper. In *The 10-50 Solution: Technologies and Policies for a Low-Carbon Future.* Washington, D.C.: Pew Center on Global Climate Change and the National Commission on Energy Policy.

Reuters. 2011. China Shenhua Coal-to-Liquids Project Profitable-Exec. Available at http://www.reuters.com/article/2011/09/08/shenhua-oil-coal-idUSL3E7K732020110908. Accessed May 2, 2012.

Simmons & Company. 2011. *Contemplating Long-term Natural Gas Demand Opportunities.* Houston, Texas: Simmons & Company.

Statoil. 2010. Leading the World in Carbon Capture and Storage. Available at http://www.statoil.com/en/TechnologyInnovation/NewEnergy/CO$_2$MANAGEMENT/pages/carboncapture.aspx. Accessed December 28, 2011.

Tabak, S. 2006. ExxonMobil methanol to gasoline. Commercially proven route for production of synthetic gasoline. In *The U.S.-China Oil and Gas Industry Forum.* Hangzhou City, China.

Takemasa, R. 2011. Electric Utility Preparations for Electric Vehicles. Presentation to the NRC Committee on the Potential for Light-Duty Vehicle (LDV) Technologies, 2010-2050: Costs, Impacts, Barriers and Timing, June 28.

Technology Review. 2009. Drilling for Shale Gas. Available at http://www.technologyreview.com/video/?vid=439. Accessed January 19, 2012.

Wald, M.L. 2012. In Kansas, Stronger Mix of Ethanol. *The New York Times.* July 11. Available at http://www.nytimes.com/2012/07/12/business/energy-environment/at-kansas-station-e15-fuel-reaches-the-masses.html?_r=2. Accessed July 31, 2012.

Wright, M.M., J.A. Satrio, R.C. Brown, D.E. Saugaard, and D.D. Hsu. 2010. *Techno-Economic Analysis of Biomass Fast Pyrolysis to Transportation Fuels.* Golden, Colo.: National Renewable Energy Laboratory.

Zhao, X., R.D. McGihon, and S.A. Tabak. 2008. Coal to clean gasoline. *Hydrocarbon Engineering.* September.

4

Consumer Attitudes and Barriers

The preceding chapters demonstrate that there is great potential for new generations of advanced-technology vehicles, fuels, and fueling infrastructure to advance the nation toward the twin goals of significantly reducing greenhouse gas (GHG) emissions and petroleum use from the light-duty vehicle (LDV) fleet by 2050. But technological advances alone are insufficient to promote success. Consumers must embrace the new designs and new fueling systems discussed in Chapters 2 and 3, or LDVs and fuels will never achieve the market penetration rates necessary for successful achievement of the petroleum and GHG reduction goals of this study. While highly efficient internal combustion engine vehicles (ICEVs) and "drop-in" biofuels would differ little in most characteristics that consumers consider (other than cost), alternative-fuel vehicles (AFVs) operating on electricity or hydrogen will appear very different to consumers. Given that most of these vehicles will come with a so-called technology premium that, initially at least, will make them more expensive than the vehicles they will seek to replace, winning consumer acceptance will be challenging, likely requiring substantial policy intervention.

Consumer purchasing patterns have been studied for decades. Although many vehicle attributes influence car-purchasing decisions (Box 4.1), the common conclusion is that buyers' economic concerns are one of the primary drivers of almost all transactions (Caulfield et al., 2010; Egbue and Long, 2012): money talks; most of the rest is window dressing. Thus, when dealing with the task of selling vehicles whose primary purpose is to help reduce petroleum consumption and the related environmental impacts, appeals to consumers' environmental and social sensibilities are not likely to move much metal after the thirst of the relatively small groups of innovators and early adopters is satiated.

Attracting members of these two groups, part of a hierarchy established by Everett Rogers in his seminal *Diffusions of Innovations* (Rogers, 1962), is critical, however. Rogers (2003) estimated that they collectively make up just 16 percent of the consumer base, but their acceptance or rejection

of innovations guides the remaining consumer groups. They set the stage by removing uncertainty about new products, policies, or technologies and by establishing a level of peer acceptability that makes more risk-adverse consumers comfortable with accepting them as well.

The initial group, the innovators, is the smallest, estimated by Rogers at just 2.5 percent of the consumer base. Their role is to launch new ideas, products, and technologies. They typically are younger and more financially sound than the general population and are characterized by a desire to be first to possess or use something new and different in the market. They are willing to take risks and can use their financial well-being to soften the impact of the occasional failed venture. Early adopters are the next group to adopt an innovation. They constitute approximately 13.5 percent of the consumer base. The group includes a high percentage of opinion leaders, but

---

BOX 4.1
Attributes that Could Affect Car-Purchasing Decisions

CO_2 emissions
Comfort
Ease of fueling
Fuel consumption
Initial and operating costs
Performance or power
Reliability
Safety
Size of car or internal and cargo space
Style or appearance or image
Travel range

NOTE: The attributes are listed in alphabetical order.

its members are less risk-averse than the general population and more selective than innovators in their enthusiasm for innovations to adopt. Like innovators, they tend to be younger and have higher income levels and social status than other consumers. Early adopters tend to be opinion leaders in their communities and are in the group most looked-to by other consumers for validation of or information about new things. In the automotive arena, Deloitte Development LLC (2010) characterized early adopters for one combination of alternative vehicle and fuel technologies—the battery electric vehicle—as young individuals with annual household incomes of $200,000 or more who consider themselves to be environmentally sensitive and politically involved.

Not all innovators and early adopters will embrace the same products, ideas, or technologies, so technology and policy developers cannot count on the groups as a monolithic 16 percent of the market. Still policy makers and the private auto and fuel industry companies must work together in pursuit of the nation's GHG and petroleum-use reduction goals. They must be able to attract the interest of a significant portion of these two groups to make inroads with the general consumer base, which Rogers further divided into the early and late majority adopters, each constituting an estimated 34 percent of the consumer base, and the laggards, or last to adopt, constituting the remaining 16 percent of consumers. Rogers determined that innovations achieve peak market penetration with the early majority adopters.

Each of the various groups can be further subdivided into smaller market categories defined by factors such as age, gender, geography, income, social status, and political leanings. Thus, the automotive innovator group might include dedicated environmentalists, older empty-nesters, and "first on the block" ego gratification seekers. The environmentalists would be willing to pay a premium and accept reduced travel range, cargo and passenger capacity, and limited refueling opportunities to acquire vehicles and/or use fuels that they believe would help reduce GHG emissions; the empty nesters might simply wish to free themselves of the expense of purchasing petroleum-based gasoline (and recognize that they no longer need a vehicle that can travel long distances); and the first-on-the-block innovators may simply be those whose egos are gratified by being seen as out in front of the pack in their vehicle choices and whose incomes can support their desires. The success of a new automotive and/or fuel technology or idea will require that the needs of such disparate subgroups be met.

Meeting the needs of all subgroups or selling these new automotive ideas to the early majority will not be easy. Increased utility and convenience cannot be counted on as selling points. The automobile became a successful new technology in the early 20th century because it demonstrated superiority to the horse- and ox-drawn vehicles it would replace. It offered greater speed, greater range, and greater utility than animal-drawn vehicles and promised the individual a new level of freedom of movement (Morris, 2007).

With an engine that demanded combustible fuel, the auto also gave the oil industry a whole new market for its product.

If policy makers determine that AFVs are essential to meeting the nation's oil and GHG reduction goals, then consumers will have to be asked to consider adopting another significant change in personal transportation, but it is one that—at least in the formative stages—means sacrifice, not improvement. The contemplated change is not replacing the horse-drawn buggy with a motorized carriage that can carry its own fuel for hundreds of miles and be refueled in minutes. Rather, it is the swapping of a sizeable portion of conventional, internal-combustion LDVs that run on liquid hydrocarbon fuels and the accompanying nationwide system of fueling stations for a variety of new vehicles and fuels that will require development of massive new production, distribution and retailing systems. In addition, many of these new AFVs use powertrains—such as plug-in hybrid electric (PHEV) systems—that typically cost more and offer no improvements other than increased fuel efficiency, reduced emissions, and, in the case of plug-in vehicles, cheaper fuel costs for the electricity used to charge the batteries. Battery electric vehicles (BEVs) offer less range, and along with PHEVs would require large GHG emissions reductions in the electricity production system to deliver meaningful net GHG reductions for the LDV sector. Some options, however, such as the drop-in biofuels described in Chapter 3, entail few if any customer acceptance challenges for the vehicles, which can still use internal combustion engines. In this case, the technology challenges are upstream in the fuel supply sector, with implications for the fuel costs experienced by LDV consumers.

This chapter examines demonstrated results and stated preference surveys, with stated preference surveys in the forefront because, as many of the vehicle and fuel types under consideration are not yet in the market, there has been little opportunity for researchers to conduct studies of demonstrated preferences. The preference surveys, particularly in environmental matters, have a certain level of bias engendered by respondents' wish to appear environmentally responsible even if economic conditions rather than environmental beliefs ultimately determine their actions (Kotchen and Reiling, 2000), but the impact of such biases—which remains unquantifiable (Hensher, 2010)—does not materially affect their value in illustrating general trends over time.

4.1 LDV PURCHASE DRIVERS

There is no big mystery at work in the LDV-buying decision process. Consumers typically acquire things for a range of reasons. In the case of LDVs, research has shown that the bulk of purchases revolve around perceived need—to replace an aging vehicle, for instance. "Desires," whether for a different color or body style, improved "infotainment" content, a more prestigious nameplate, or simply a newer model, still account for a significant minority of purchase decisions,

TABLE 4.1 Car-Buying Motivations, 2005 and 2011

Motivation to Consider Buying New Car	April 2005 (%)	August 2011 (%)
Old car had high mileage	34.3	25.7
Old car needed frequent repairs	17.3	14.3
Needed additional vehicle for family	18.0	11.9
Needed vehicle with more room	12.0	12.3
Lease expired	9.7	9.5
Wanted new vehicle	6.5	8.0
Wanted better fuel economy	21.9	16.7
Not sure/other	18.3	22.4
Liked styling of new models	16.3	12.3
Wanted vehicle with better safety features	14.6	11.5
Financing deals/incentives too good to pass up	13.8	11.7
Significant other wanted new car	17.6	16.1
Wanted car with new infotainment equipment (navigation, DVD player, etc.)	11.8	7.9

NOTE: Sum of totals exceeds 100 percent because respondents could provide multiple responses.
SOURCE: BIG Research, Consumer Intentions and Reactions, April 2005, August 2011, proprietary information prepared for the committee by request.

however. Table 4.1 shows surveys of retail consumers taken in two periods—2005 and 2011—representing different economic conditions.

A large number of LDVs are purchased each year for commercial and government fleets, and those purchases are not reflected in Table 4.1 or in Figure 4.1, both of which examine trends among retail consumers. Yet the fleet segment is one in which a substantial number of AFVs will be sold in the future per private and governmental policies encouraging greater use of highly fuel-efficient vehicles. It is too early to tell how those sales might affect the overall success of any particular AFV or alternative fuel.

As these surveys show, replacing a vehicle for reasons including high mileage (age), the frequency of repairs, expired leases, and/or the perceived need for a vehicle of a different size account for more than half the stated reasons for buying a new vehicle. Reasons stated as "wants" or desires rather than needs ran a close second. The need to acquire a new vehicle because the old one was wearing out remains a strong motivation but has diminished in importance among those who purchase their vehicles as vehicle reliability and quality have improved—providing for longer-lived cars and trucks in our garages. Lessees, of course, replace their vehicles more frequently, and typically for reasons other than age-related wear. But leasing accounts for just 20 percent of the new-vehicle market (Automotive News, 2012). The need or desire for a vehicle with better fuel economy, however, has concurrently increased in importance over the past few decades as primary motivation for new-car purchase. (Note: The decline in stated importance of fuel economy between 2005 and 2011 as shown in Table 4.1, is a result of the unusually high level of importance attached to fuel economy that was shown in the April 2005 BIG Research survey and was spurred by gasoline price increases at the time.) The trend of fuel efficiency rising in importance along with fuel prices

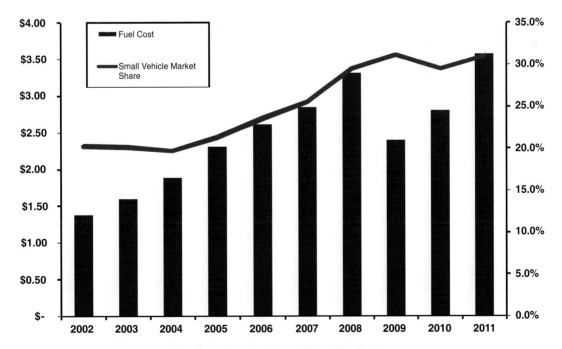

FIGURE 4.1 Small vehicle market share (retail sales only) and fuel cost (in 2011 dollars).
NOTE: Recession-driven sales of less-expensive models helped keep small-vehicle market share high despite fuel price declines in 2009 and 2010.
SOURCE: Data provided by Edmunds.com's AutoObserver.com Data Center; chart prepared for committee by Edmunds.com.

continues: *Consumer Reports* magazine reported recently that in an April 2012 telephone survey of 1,702 adult consumers who were asked to state what they believed would be the most important factors in their next new-car purchase, 37 percent cited fuel economy as their top consideration (Consumer Reports, 2012). While altruistic reasons for purchasing a new vehicle—to help improve air quality, reduce oil use, cut GHG emissions, improve the environment—score highly in some special-interest group surveys (Consumer Reports, 2011b), in broader whole-market surveys that allow respondents to list their own reasons for purchase, they appear, at worst, to be not considered at all or, under the best of interpretations, to be secondary, hidden constituents of the more selfish, economics-driven stated reasons such as "wanted better fuel economy" or "wanted new vehicle." A motivator not mentioned in the surveys cited but known to be a purchase driver of certain AFVs in the state of California is single-occupant access to high-occupancy vehicle (HOV) lanes, also called carpool lanes. Although most states that provide such lanes limit access to vehicles carrying at least two people, California currently permits drivers of most battery electric vehicles, some plug-in hybrids, and all fuel-cell electric vehicles to use the lanes even if there are no passengers in the vehicles when an authorized, state-issued access sticker is displayed. Access to HOV lanes in a state noted for its crowded rush-hour freeway traffic is believed to be an important selling point for those vehicles. Indeed, General Motors has released a television advertisement for its 2013 Volt PHEV that highlights the fact that a specially tuned version of the vehicle qualifies for HOV lane access in California.

Achieving a considerable reduction in LDV fleet GHG emissions and petroleum use through adoption of alternative fuels and powertrains is not likely to be accomplished by appealing to altruism. Once early adopters have made their choices, the remaining 84 percent of consumers are going to have to be persuaded either that the alternative fuels and vehicles offer them an improvement over their present preferences, or that there is a pretty immediate economic benefit to be had in making the switch. Environmental benefits simply do not appear to be a determinant for consumers in large purchases, such as motor vehicles. "Economic concerns are consumers' priority," researchers at the Mineta Transportation Institute have found (Nixon and Saphores, 2011, pp. 10-11).

4.2 WHAT DO CONSUMERS WANT?

Conventional wisdom holds that American consumers want big cars and trucks with large and powerful engines and that fuel economy just is not that important because gasoline and diesel prices in the United States are so much lower than in much of the rest of the world. Those attitudes certainly have shaped U.S. automakers' marketing and product planning agendas for most of the time since World War

II. As recently as April 2011, in an editorial in the influential trade journal *Automotive News*, publisher Keith Crain bluntly stated that while the auto industry has responded to rising gasoline prices and increased regulatory demand for better fuel economy with a number of cars that achieve an EPA highway-cycle rating of 40 mpg, "the trouble is, no one wants to buy them" (Crain, 2011). Gloria Bergquist, the Alliance for Automobile Manufacturers' vice president for communications, repeatedly has pointed out that in 2010 a single pickup model—the Ford F150—outsold all 30 gas-electric hybrid cars and sport utility vehicles (SUV)s offered for sale in the United States by mainstream auto manufacturers (Harder, 2011). Those statements reflect consumer choices influenced at least in part by continued low pricing of gasoline. In the past year, however, sales of smaller cars with high fuel efficiency have increased as a percentage of the market, as have sales of larger cars, crossovers, and light-duty trucks that use smaller, more efficient engines to replace "gas guzzler" V6s and V8s (Drury, 2011). History, however, has shown that the march toward efficiency stops when fuel prices have stabilized or dropped after a run-up (see Figure 4.1).

Still, such attitudes may be generational. Most Americans under 40 have now been exposed to smaller vehicles, mainly from the import brands, and, as sales trends show, acceptance of compact cars in the U.S. market is growing. The recession of 2008-2009 and the continued economic slump that has followed certainly have influenced that growth, as have increasing fuel prices in recent years. However, there is evidence indicating that potential savings from fuel efficiency improvements is not a significant factor in consumers' vehicle purchase choices, indicating that consumers are becoming inured to gasoline price increases because they inevitably have been followed by price decreases. (See details in Section 4.6 below.)

4.3 FACTORS IN CONSUMERS' CHOICES

Numerous studies have attempted to quantify the needs and desires that drive LDV-purchase decision making. Their findings are fairly consistent and are exemplified by a recent stated-preference study by Capgemini (2010) that ranked the most important factors gathered from 2,600 online respondents in the United States, Europe, and Asia and found reliability, safety, vehicle price, fuel economy, and the variety and cost of options all in the top 10. Consumers who identified themselves as planning to purchase a new vehicle within the next 15 months were asked to rank the most important factors they would apply to their car-purchase decision making (see Table 4.2).

In addition,, respondents were asked about their interest in so-called green vehicles, and 72 percent of U.S. respondents (versus 57 percent overall) cited fuel economy as the number-one reason they would consider a fuel-efficient petroleum or alternative fuel car or truck. Only 13 percent

TABLE 4.2 Importance of Factors in Consumers' Choice of Vehicle

Factors in Consumer Choice	Percent Respondents Saying Important/ Very Important	
	Mature Markets	Developing Markets
Brand reliability	89	90
Safety	89	91
Price	86	85
Fuel economy	82	85
Quality of exterior styling	77	84
Quality of interior styling	77	85
After-sales service	71	83
Vehicle availability (take it home versus wait for special order)	71	82
Extra options at no cost	70	74
Features and options	66	79
Low emissions	64	75
Financing at 0% or low %	62	73
Brand name	55	80
Cash-back incentive	46	69
Hybrid or other alternative fuel system	36	66

SOURCE: Capgemini (2010).

of U.S. respondents (versus 23 percent overall) cited making a positive impact on the environment as a significant reason for acquiring an AFV, while just 1 percent (versus. 3 percent overall) said tax credits would be an important factor in their purchase decision (Capgemini, 2010). That contrasts rather sharply with the 46 percent (69 percent overall) who said they would prefer a cash-back incentive. The preference for cash-in-hand at time of purchase versus an end-of-year tax credit has important implications when considering incentive policies.

4.4 SUBSIDIES

Capgemini is not the only one finding that income tax credits, although currently the preferred federal policy for incentivizing AFV purchases via subsidies, may not be the best route to take. A number of studies prepared since hybrid-electric vehicles achieved sufficient market penetration to figure as a potentially valuable tool in the effort to reduce the nation's GHG emissions and petroleum use have found that while subsidies work, those that directly place cash in the hands of the consumer are more effective than those—like income tax rebates—that require the consumer to pay the full price up front and wait until tax time for the subsidy payment (Gallagher et al., 2008; Diamond, 2008; Beresteanu et al., 2011).

In addition to providing immediate gratification, direct rebates, sales-tax credits, or other types of cash subsidies, including subsidies enabling the manufacturer to lower the retail price of the vehicle, would enable consumers to rationalize that the cost of the vehicle is less than its so-called

sticker price. When applied to the amount being financed, such direct subsidies lower the monthly payment and can help a greater number of consumers qualify for loans to purchase new AFVs. Tax credits, in contrast, do not affect the qualifying terms or monthly payments for purchasers (although they may be used to lower monthly lease costs, as has been the case with the Chevrolet Volt PHEV and Nissan Leaf BEVs). One argument against tax credits such as the present "up to $7,500" federal credit on BEVS and some PHEVS (depending on battery size) is that they tend to reward higher-income consumers—who arguably are least needful of subsidies—and do not provide the full potential reward for consumers with lower incomes and thus lower tax liabilities.

4.5 ICEVs STILL TOPS

Even in the aftermath of publicity about the possibility of future oil shortages and the need for increased national energy security, gasoline as a fuel is not seen by most car-buying consumers as a negative. Indeed, there is a consensus in consumer preference surveys that unless there is intervention through government policy, internal combustion engines powered by petroleum or a competitively priced drop-in biofuel (if such a fuel is commercialized) are likely to remain the predominant powertrain in LDVs in the United States for decades to come. A sampling of recent studies bears this out.

In its June 2011 report on AFV preferences, the Mineta Institute found that "in general, gasoline-fueled vehicles are still preferred over AFVs," with 36 percent of the study's respondents ranking conventional ICEVs as their first choice (Nixon and Saphores, 2011, p. 1). Hybrid-electric vehicles (HEVs) were second in popularity, with 26 percent of respondents identifying them as their first choice, followed by compressed natural gas vehicles (CNGs) at 13 percent, hydrogen fuel cell electric vehicles (FCEVs) at 18 percent, and BEVs at just 9 percent. The responses exceed 100 percent because each is the average of respondents' choices in a variety of scenarios. A stated-preference survey of 3,000 consumers in the United States, Germany, and China, conducted in the first quarter of 2011 by Gartner, Inc. (Koslowski, 2011), presents similar findings, with 78 percent of respondents ranking gasoline-fueled vehicles as the type they "definitely" would consider for their next new-vehicle purchase, followed by HEVs, 40 percent; CNGVs, 22 percent; and BEVs, 21 percent. Respondents in the Gartner study were permitted to make more than one selection and FCEVs were not included in the choices. Although such surveys have value in indicating trends, they do not reflect present realities. Hybrid vehicles, for instance, still account for less than 3 percent of annual U.S. new-car sales more than 12 years after their introduction in the market. J.D. Power and Associates found in its most recent "green" vehicles study that its research into consumer attitudes over the years shows that "while most consumers say they want to create a smaller personal carbon footprint . . . this consideration carries relatively low weight in the vehicle-purchase decision" (Humphries et al., 2010, p. 10).

Reasons for the strong preference for continued use of gasoline-powered vehicles appear to be based strongly on up-front cost—they are demonstrably less costly to purchase than alternatively fueled vehicles. The cost efficiencies realized by the tens of millions of internal combustion engines produced each year make petroleum-fueled cars and trucks far less expensive to purchase than any of the new crop of alternatively fueled/powered LDVs.

Convenience, especially the ready availability of fuel, is the second most-stated reason for preferring petroleum. The United States has a widespread gasoline service station network that serves even the smallest communities, and gasoline prices in the United States remain among the lowest in the world. Both factors make it incredibly convenient for consumers to continue purchasing and using gasoline-fueled vehicles. Perceived reliability of ICEVs versus alternative vehicles is another key factor, with some researchers finding that consumers believe conventional ICEVs are far more reliable than alternative vehicles (Synovate, 2011).

4.6 HOW CONSUMERS VALUE FUEL ECONOMY

Many consumers responding to attitudinal surveys say that they place fuel economy at or near the top of the list of factors they will consider when buying their next vehicle. But when it comes to applying potential fuel economy savings to the purchase decision, most research has shown that consumers just do not do it. So even though a case can be made for long-term fuel and maintenance savings making some AFVs less costly to own than gasoline vehicles over a period of years, a tendency by consumers to ignore such savings potential would make it more difficult for manufacturers and policy makers to persuade consumers to consider alternative fuels and vehicles with higher prices than conventional ICEVs. Researchers at the University of California, Davis, Institute for Transportation Studies, for instance, have found that consideration of a payback period for higher-priced AFVs "is not part of the vehicle purchase decision-making even in the most financially skilled households" (Turrentine and Kurani, 2007, p. 1220).

This tendency of consumers to fail to modify the up-front acquisition cost of AFVs by the long-term value of reduced fuel and other ownership costs (maintenance, repairs, and insurance chief among them) can be explained by applying behavioral economics' principle of loss or risk aversion. In general, increasing a vehicle's fuel economy through improved technology requires paying a higher initial cost. Future fuel savings, however, are uncertain due to the unpredictability of future fuel prices, the fact that the fuel economy consumers will achieve in actual use will differ from the government's ratings, and potential variations in

vehicle use, lifetime, and other factors. Given the uncertainty in future fuel savings it is reasonable for a consumer to be reluctant to pay more for higher fuel economy. One of the most well established findings of behavioral economics is that when faced with a risky bet, typical consumers count potential losses approximately twice as much as potential gains and exaggerate the probability of loss. This approach can result in an undervaluing of future fuel savings by half or more relative to what would otherwise be expected (Greene, 2010a). Other possible explanations have been proposed, including shortsightedness and the lack of information or the necessary skills to estimate future energy savings. There is not an established consensus on this subject, however, and the published literature contains evidence to support both views—that consumers accurately value and that they undervalue future fuel savings (Greene, 2010b). Anderson et al. (2011) found that consumers typically take no position and merely consider future fuel prices to be the same as today's because they cannot accurately predict. Because the evidence for undervaluing appears to be stronger, the analyses and modeling in Chapter 5 assume that consumers behave as though they required a simple 3-year payback for an expenditure on higher fuel economy.

Overall, there is little doubt that a significant portion of consumers are interested in fuel efficiency. A variety of recent studies and surveys have shown that fuel economy is a top concern of 60 to 80 percent of prospective auto buyers (Consumer Reports, 2011a). Just how important, however, seems to depend on what it will cost the consumer to achieve a higher degree of efficiency. J.D. Power and Associates consumer research over the years has shown that "many may consider it, but when the time comes to put their money where their mouth is, very few follow up," the research firm's senior manager of global powertrain forecasting, Michael Omatoso, said in an interview (personal communication, M. Omatoso, Troy, Michigan, September30 2011). There have been a number of studies that include attempts to discern the premium consumers are willing to pay for AFVs, and they find it most typically is in the range of $1,600 to $2,000 (Boston Consulting Group, 2011; Deloitte Touche Tohmatsu Ltd. Global Manufacturing Industry Group, 2011). But as more AFVs come into the marketplace, the issue seems to remain a fertile field for future research.

4.7 INTEREST IN AFVs LIMITED

There is interest in AFVs, but it is limited by a number of factors including a general unwillingness to abandon a fuel and powertrain combination that has shown itself to be quite effective in providing for consumers' transportation needs over the decades, even if that effectiveness is not accompanied by the levels of environmental cleanliness necessary to achieve the nation's present goals. In a 2010 survey of consumer adoption literature, researchers at the University of Wisconsin found broad agreement that there is considerable interest in AFVs if performance characteristics remain comparable to those of ICEVs (Guo et al., 2010). Now that there are some of these vehicles in the marketplace (most notably conventional hybrids, although at this writing there is one compressed natural gas passenger car, two BEVs, and one PHEV in the market, pricing for several more BEVs and PHEVs has been announced, and there are several test programs utilizing fuel-cell electric vehicles), it has become clear that initially these vehicles will cost more and in most cases provide a reduced user experience—based on range and fueling convenience issues—than conventional ICEVs. As a result, more recent studies have predicted relatively slow and low adoption rates for AFVs, typically—in the aggregate—below 20 percent of the U.S. market by 2025 (Humphries et al., 2010).

4.8 BARRIERS

Although cost and convenience are the most-often cited reasons for anticipated low adoption rates, they are but are two of several significant barriers to AFV adoption cited when consumers are asked to list, or to pick from a prepared list, those things that most concern them about alternatively fueled vehicles (Table 4.3). All of these concerns must be addressed via public policy and/or manufacturers' marketing efforts if the best fleet mixes necessary to meet the goals set out in the committee's statement of task are determined—as indicated by the modeling results in Chapter 5—to be those requiring large numbers of AFVs. Such efforts will be needed to help overcome objections to vehicles that at least initially could offer less performance, range, utility, and fueling

TABLE 4.3 Principal Barriers to Adoption of AFVs

Reason That Could Influence Purchase Decision of an Alternative-fuel Vehicle	Percent Respondents in Each Study Citing Reason as a Concern			
	Auto Techcast	Gauging Interest[a]	Green Auto[b]	Mineta[c]
Cost vs. comparable conventional vehicle	NA	74	35	53
Fuel availability	NA	75	32	55[c]
Fuel cost	30	NA	17	46
Payback period	46	49	18	NA
Performance		49	16	
Range (BEVs)	43	75	12	49
Refueling/Recharging time/convenience	38	NA	NA	55[c]
Reliability	26	57	17	NA[d]
Size/Seating capacity	17	33	NA	NA

[a]Gauging Interest responses are from U.S. participants only.
[b]Green Auto responses are only from consumers who said they would not purchase an AFV.
[c]Mineta survey, by Nixon and Saphores, combines fuel availability and refueling time.
[d]NA = Not asked.
SOURCE: Data from Harris (2011); Ernst & Young (2010); J.D. Power (2011); Nixon and Saphores (2011).

convenience and will cost consumers more to purchase than conventional ICEVs with advanced-technology gasoline powertrains that will not have the higher initial costs.

In its recent "Drive Green" study (Humphries et al., 2010), J.D. Power and Associates set out to determine the perceived drawbacks to specific types of AFVs. Researchers found that while there are differences in degree and in rankings, the top reasons in all cases (HEVs, clean diesel, PHEVs, and BEVs [fuel-cell electric vehicles were not asked about]) were the so-called initial cost premium consumers attached to most AFVs and the perceived long-term cost of ownership (exclusive of the purchase price premium), which some respondents believed to be higher for an AFV than for a conventional ICEV.

In the case of BEVs and PHEVs, concerns about driving range on a single battery charge also ranked high. This should not be an issue with PHEVs because they can be driven using their gasoline engines or engine-generators and are not solely dependent on batteries, showing continuing consumer confusion about the differences among the advanced powertrain technologies.

Range also could be an issue with AFVs using compressed natural gas. The only factory-built model currently in the market is the Honda Civic Natural Gas. Its design retrofits the CNG fuel storage and delivery system into a vehicle designed for petroleum-based gasoline. The pressurized tanks needed for the CNG occupy much of the vehicle's trunk area and even then hold only the usable equivalent of 7.5 gasoline gallons. While the CNG Civic attains almost the same EPA combined city-highway fuel economy rating as the gasoline model (32 mpg vs. 33 mpg), its smaller-capacity fuel tank limits its range to about 240 miles versus the gasoline Civic's estimated range of 430 miles. However future CNGV are likely to be designed from the ground up and could better house larger fuel tanks, thus enabling them to deliver improved range.

The move to more efficient, lower-emission LDVs almost certainly means that cars and trucks, regardless of the fuel source or powertrains, will have to be lighter than they are today. Present and proposed federal Corporate Average Fuel Economy (CAFE) policy is devised to enable larger vehicles to continue to meet the standards and does not necessarily lead to downsizing of the fleet to go along with the lightweighting. But downsizing has occurred, principally for economic reasons stemming from the recession of 2008-2010 and subsequent slow economic recovery and prolonged period of high unemployment. While that raises concern among those who find that consumers today do not want to give up size for efficiency, it might not be as big an issue in the future. Sales of larger vehicles could begin climbing as the economy improves in the future. But as younger consumers who today are in the used-car market or still are too young to be car purchasers begin replacing Baby Boomers and Gen-Xers in the new-car market, there may be a generational shift toward a preference for smaller cars.

In the past decade, according to sales data from online automotive information provider Edmunds.com (see Figure 4.1), the U.S. market share for small cars—a category including compact and subcompact cars, vans, SUVs, and compact pickup trucks—has increased by 53 percent from 20.3 percent in 2002 to 31.1 percent in 2011 (Edmunds.com, 2011); at the same time, the average price of a gallon of regular-grade unleaded gasoline has increased by 88 percent when adjusted for inflation.

For decades, sales activity for small cars and trucks seemed to correspond closely to fluctuations in retail gasoline prices. But as Hughes et al. (2008) found in their study of gasoline price inelasticity, driver behavior triggered by increases in gasoline prices has changed considerably in the past decade. Price run-ups may no longer lead as rapidly as in the past to the behavior changes once commonly associated with periods of unusually high gas prices—driving less and buying smaller and more efficient vehicles are two examples. In addition, fleet fuel efficiency has increased in recent years, dampening the impact of rising gasoline prices. Small vehicles' share of the LDV market keeps gradually increasing, but this could be a sign of increased general market acceptance as well as a reaction to several years of a weak national economy. It also could be related to the downsizing of aging Baby Boomers' households and transportation needs.

Both Edmunds.com and auto industry consulting firm AutoPacific, Inc. track consumer consideration of compact and subcompact vehicles. (Edmunds derives shopper consideration rates from details gleaned from consumer searches on its website—repeated, lengthy, and detailed research into a specific model equates to "consideration" of that specific vehicle type versus casual browsing; AutoPacific uses a bimonthly internal online consumer intent survey that asks approximately 1,000 respondents what types of LDVs they are considering for their next purchase.) Each recently compared small-car consideration rates to fluctuations in gasoline prices. Both indicate that while consideration rose sharply and in lockstep with price run-ups in the first half of 2007 and the last half of 2008, consumers may not be increasing their consideration of small cars at the same pace in the most recent series of gasoline price hikes, which began in September 2010 (Figure 4.2). That data and the previously mentioned small-vehicle sales versus fuel price data (see Figure 4.1) appear to further validate the results of Hughes et al. (2008), but also could mean that while fuel price still matters, price increases have to be very large in order to elicit significant movement toward smaller, more efficient vehicles. This would mean that policies based on only modest increases in fuel taxes or other fuel-efficiency related fees would be less likely to succeed than policies such as CAFE standards, or

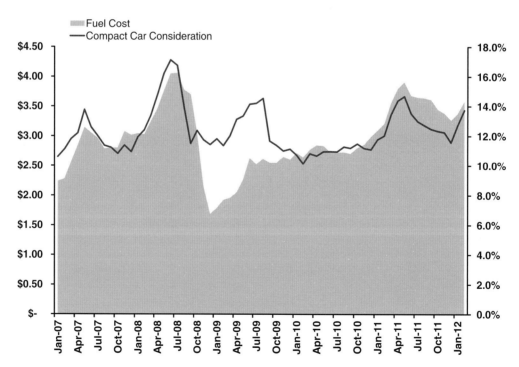

FIGURE 4.2 Compact car consideration and fuel cost.
NOTE: "Normal" consideration level is in the range of 10 to 12 percent. Consideration spike in the period of February to September in 2009 corresponds to the U.S. "Cash for Clunkers" economic stimulus program in which consumers received funds to apply to the purchase of new, more efficient vehicles in return for junking older, less-efficient models.
SOURCE: Data provied by Edmunds.com Data Center; chart prepared for this report by Edmunds.com.

fee systems aimed at making the use of inefficient and/or high-emissions vehicles prohibitively expensive.

"People remember the gas (price) spike in 2008 and how a lot of people panicked and downsized their vehicles, only to see (gas) prices drop. So now they are taking a wait-and-see approach," said market researcher George Peterson, president of AutoPacific (personal communication with G. Peterson, Troy, Michigan, August 25, 2011). He said the so-called tipping point at which consumers say they would change their new-vehicle buying goals and shop for more efficient vehicles has steadily increased and now is about $5.50 a gallon, up from $3 a gallon just a decade ago. Undoubtedly, the tipping point will continue to increase with economic recovery and improving fuel efficiency for ICEVs.

4.9 PEER INFLUENCE CRITICAL

Advanced alternative fuels and powertrains are still rare and consumers have had very little real-world experience with them. Thus there's little solid information available to help determine what consumers will accept in the way of alternatives to gasoline- and diesel-fueled vehicles.

In fact, there is some concern that this lack of knowledge has led to confusion in the marketplace about the characteristics, values, and drawbacks of the various types of AFVs and has caused some degree of consumer paralysis (Synovate, 2011). Researchers on both sides of the country, however, have found that word of mouth can be a powerfully influential tool, pointing to the potential value both of public demonstration and deployment programs and of public information campaigns. Axsen and Kurani argue that the mere presence of greater numbers of AFVs on the nation's roads will increase both public awareness and public acceptance as the real-world experiences of many drivers are communicated to friends, neighbors, family members, and co-workers (Axsen, 2010; Axsen and Kurani, 2011). Zhang et al. (2011) found that positive word of mouth increases the perceived value of AFVs and leads to a higher willingness by consumers to pay a premium for them. Such studies show that getting AFVs into the market, even in small numbers, would generate word-of-mouth reports that could help put to rest (although there is also the possibility that some will reinforce) the negative concerns about barriers that appear to be limiting AFV acceptance at this point. Price disparity, however, still can be a strong disincentive, as has been shown by the slow market penetration of conventional hybrid vehicles, which still account for less than 3 percent of the U.S. LDV market more than a decade after introduction. Consumers do not have many negative attitudes about hybrids any longer. But because most HEVs still have a price premium when compared to comparably sized and equipped ICEVs, sales have risen and fallen with gasoline prices in recent years but overall have leveled off in the range of 2.5 to 3 percent.

It should be pointed out again that these early positive reports are coming from a unique and generally accept-

ing group of AFV purchasers, the so-called early adopters whose interest in and desire to possess advanced technologies invariably make them prone to acceptance. Engineers at Nissan Motor Company, for example, told the committee that early Nissan Leaf owners were adapting to the Leaf's characteristics in ways that mainstream buyers might not. For example, the heating system on a BEV is a significant drain on the battery charge, reducing range when in use. As a result, many early Leaf owners have developed the technique of using the car's seat heaters—which draw much less charge from the battery—rather than the cabin heater. It is uncertain whether a potential mainstream buyer would see that as a plus or a minus.

4.10 INFRASTRUCTURE AVAILABILITY

The availability of fuel, including battery-charging facilities for BEVs, is also a major issue affecting consumer willingness to acquire AFVs. There are so few of the vehicles and so little infrastructure available at present that it is not possible to determine the necessary balance. One exception is E85 fuel (which is a blend of 85 percent ethanol and 15 percent gasoline) and the "flex-fuel" vehicles built to use either gasoline or E85. There often is no financial incentive for the owner of a flex-fuel vehicle to purchase E85. While a gallon of E85 may cost less than a gallon of gasoline, it delivers significantly fewer miles.

Earlier studies of consumer adoption in Canada and New Zealand of flex-fuel, or dual-fuel, vehicles using CNG as the alternate fuel found that the presence of refueling infrastructure was a significant factor in consumers' decisions to acquire such vehicles. Greene (1990) concluded after reviewing a Canadian government survey of consumers in the provinces of Quebec, Ontario, and British Columbia that a "substantial refueling network is a pre-condition for the markets accepting alternative fuel vehicles and . . . essential if dual- or flexible-fuel vehicles are to use the new fuel a significant fraction of the time." In their study of buyers of CNG vehicle conversions in New Zealand in the 1980s, Kurani and Sperling (1993) found that successful achievement of the government's goal of pushing 150,000 converted vehicles into the market between 1979 and 1986 (that goal was not met; the total number of conversions by 1986 was 110,000) depended in large part on two types of government subsidies: those that helped consumers defray or earn back the cost of acquiring the converted vehicles, and those that helped underwrite new CNG fueling stations so that consumers would perceive that a fueling infrastructure was being installed and that they would have access to the fuel. The CNG conversion program ended—dropping from 2,400 a month in 1984 to 150 a month in 1987—following a 1985 change of administrations that saw significant curtailment of government subsidies for the program.

From these studies and from consumers' stated concerns in the more recent studies cited earlier in this chapter, it is clear that policies aimed at promoting increased use of AFVs will have to address adequate provision of infrastructure.

4.11 IMPLICATIONS

To painlessly achieve any necessary transition to alternative light-duty cars and trucks, the new-generation vehicles intended to replace petroleum-burning LDVs will have to provide utility, value, creature comforts, style, performance, and levels of convenience in fueling and repair and maintenance service that closely replicate those of the liquid-fueled vehicles being phased out. They are going to have to fulfill consumers' needs and desires, or consumers will have to be presented with disincentives to continued purchase of conventional ICEVs or offered various incentives to make up for the things they perceive they would lose in a switch to an alternative vehicle or fuel. Most people do not want to pay more for a green vehicle, and of those who are willing, most would expect fuel and other savings to recoup the additional purchase expense over their period of ownership. Boston Consulting Group recently found in a survey of 6,593 consumers in the United States, Europe, and China that 40 percent of U.S. and European car buyers say they would be willing to pay up to $4,000 more for an AFV but would expect full "payback" over the first 3 years of ownership (Boston Consulting Group, 2011). Only 6 percent of U.S. respondents said they would be willing to pay a premium—the average was $4,600—without expecting to earn back the money during their full ownership period (Boston Consulting Group, 2011).

So although consumers overwhelmingly say that they want fuel efficiency and energy security, they have not demonstrated a willingness to pay much extra for it or to accept inconvenience in order to attain it. Vehicle purchase price, the long-term cost of ownership, the time it takes to refuel, the availability and cost of fuels, and the perceived need to downsize and to surrender performance attributes such as speedy acceleration and cargo and towing capacity all are cited in various studies as reasons people are not interested in AFVs. Some of this is due to lack of information, and studies such as those conducted by Axsen and Kurani (2011) and Zhang et al. (2011) have shown that word of mouth and demonstrated use by neighbors, friends, and relatives all have a positive impact on consumers' willingness to consider AFVs. That, of course, requires getting the vehicles into people's garages and onto the roads.

Some of these barriers, of course, are likely to change over time. As additional advanced-technology vehicles are placed into service, public familiarity with and knowledge of their advantages, and will improve, perhaps mitigating perceived disadvantages. AFVs also will develop a track record for resale value—a key component in determining overall cost of ownership and one that is missing now because few of the vehicles have been in the market long enough to develop a resale value history. Early estimates published by the manufacturers and a few ratings companies and analysts show that BEVs and PHEVs are thought to have lower lease residual values, an indicator of marketplace resale value. Pike Research analyst David Hurst estimated in 2011 that both the Nissan Leaf and the Chevrolet Volt would have residuals of around of 42 percent at 3 years—lower than either the popular Toyota Prius, which has a 60 percent residual value at 3 years, or corresponding conventional ICEVs such as the Nissan Versa (a Leaf counterpart) or the Chevrolet Cruze (a Chevrolet Volt counterpart), both at 52 percent (Hurst, 2011).

The relatively rapid rate of performance improvement and cost reduction that is characteristic of some new technologies can both help and harm rapid adoption of AFVs, fostering a larger market by lessening both cost and convenience barriers. Rising production volumes for biofuels could bring down their costs and make them more widely available, similarly addressing two barriers in ways that can accelerate expanding demand. Improved batteries and battery-charging rates could help reduce or even eliminate BEV range anxiety, fostering a larger market by lessening both cost and convenience barriers. Rising production volumes for biofuels could bring down their costs and make them more widely available, similarly addressing two barriers in ways that can accelerate expanding demand. Improvements in materials and engineering could make it possible to produce AFVs that are competitive with gasoline vehicles with respect to cargo capacity, towing ability, and other performance characteristics, and without the cost premiums that would inhibit widespread adoption. Conversely, rapid rates of technology advancements could inhibit diffusion beyond an early-adopter segment. Such progress would hasten the obsolescence of earlier generations of an advanced AFV technology and also suppress residual values. For example, if ongoing improvements in battery technology, such as steadily decreasing costs and rising performance, reduce the purchase price of a newer BEV relative to older BEVs still operating within their battery life expectancies (see Chapter 2), then early AFV models could depreciate more rapidly than is typical in the car market. This could lead to expectation among consumers of additional advances in the future, and a corresponding uncertainty about how well new generations of BEVs would hold their value if additional advances do indeed occur. This uncertainty could inhibit purchases by consumers concerned about resale value or could result in unfavorable lease terms.

However, because of the time it takes for automakers to bring new technologies into their fleets and for the national LDV fleet to turn over, these barrier modifications would have to be in place by or before 2030 to have a great impact on the fleet in 2050.

Absent a national emergency that requires consumers to abandon the gasoline or diesel ICEV, achieving the volumes needed to realize sufficient consumer acceptance in the early years of a planned transition to AFVs is unlikely without significant government policy intervention.

The simulations described in Chapter 5 suggest that the types of AFVs that might be needed to achieve the desired levels of petroleum and GHG reduction are those that initially will carry a large price premium because of their technology content. Once advanced vehicle technologies have become widely diffused, the vehicles in which they are incorporated will become much closer in cost to the advanced "conventional" vehicles that then would be available. In fact, the committee's midrange case shows that both BEVs and FCEVs could cost less than advanced ICEVs by 2050. (See Figure 2.8 in Chapter 2.) In addition, the superior energy efficiency of those alternative vehicles would return more than enough benefit to consumers, in terms of reduced fuel consumption, to offset any cost premium that did exist. The trick will be to persuasively convey this information to consumers.

Accomplishing this is likely to require increased understanding of consumers' attitudes about issues of sustainability, climate change, and environment and of how to motivate consumers in these arenas. The President's Council of Advisors on Science and Technology has recently recommended that the Department of Energy incorporate societal research in its programs to gain an understanding of how energy programs succeed in the market (PCAST, 2010).

Broadening such research to include a focus on understanding consumer attitudes, expectations, and past behaviors relative to alternative automotive and fuel choices as well as to other technologies introduced to increase fuel efficiency and reduce emissions would seem essential to successful achievement of the petroleum use and GHG reduction goals set out for the 2030 and 2050 time periods in the committee's statement of task.

4.12 REFERENCES

Anderson, Soren, R. Kellogg, and J.M. Sallee. 2011. What Do Consumers Believe About Future Gasoline Prices? NBER Working Paper 16974. Cambridge, Mass.: National Bureau of Economic Research.

Automotive News. 2012. Leasing Boom? Not So Fast. March 7, 2012. Available at http://www.autonews.com/article/20120307/finance_and_insur ance/12030990. Accessed April 3, 2012.

Axsen, J. 2010. Interpersonal Influence within Car Buyers' Social Networks: Observing Consumer Assessment of Plug-in Hybrid Electric Vehicles (PHEVs) and the Spread of Pro-Societal Values. Davis, Calif.: University of California, Davis.

Axsen, J., and K.S. Kurani. 2011. Interpersonal Influence Within Car Buyers' Social Networks: Developing Pro-Societal Values Through Sustainable Mobility Policy. Available at http://www.internationaltrans portforum.org/2011/pdf/YRAAxsen.pdf.

Beresteanu, A., and S. Li. 2011. Gasoline prices, government support, and the demand for hybrid vehicles in the U.S. International Economics Review 52(1):161-182.

Boston Consulting Group. 2011. Powering Autos to 2020: The Era of the Electric Car? Boston, Mass.: Boston Consulting Group.

Capgemini. 2010. Cars Online 10/11. Listening to the Voice of the Consumer. Capgemini. Available at http://www.capgemini.com/m/en/tl/Cars_Online_2010_2011__Listening_to_the_Consumer_Voice.pdf.

Caulfield, B., S. Farrell, and B. McMahon. 2010. Examining individuals' preferences for hybrid electric and alternatively fuelled vehicles. Transport Policy 17(6):381-387.

Consumer Reports. 2011a. Survey: Cost Savings Are Driving Shoppers to Better Fuel Economy. Available at http://news.consumerreports.org/cars/2011/05/survey-cost-savings-are-driving-shoppers-to-better-fuel-economy.html. Accessed on May 30, 2011.

———. 2011b. Survey: Car Buyers Want Better Fuel Economy and Are Willing to Pay for It. Available at http://news.consumerreports.org/cars/2011/05/survey-car-buyers-want-better-fuel-economy-and-are-willing-to-pay-for-it.html. Accessed on August 12, 2011.

———. 2012. High Gas Price Motivate Drivers to Change Direction. May. Consumer Reports.org. Available at http://www.consumerreports.org/cro/2012/05/high-gas-prices-motivate-drivers-to-change-direction/index.htm.

Crain, K., 2011. If you build it, they will come. Automotive News 85(April 4):12.

Deloitte Development LLC. 2010. Gaining Traction. A Customer View of Electric Vehicle Mass Adoption in the U.S. Automotive Market. New York: Deloitte Global Services, Ltd. Available at http://www.deloitte.com/assets/Dcom-UnitedStates/Local%20Assets/Documents/us_automotive_Gaining%20Traction%20FINAL_061710.pdf.

Deloitte Touche Tohmatsu Ltd. Global Manufacturing Industry Group. 2011. Unplugged: Electric Vehicle Realities Versus Consumer Expectations. New York: Deloitte Global Services Ltd. Available at http://www.deloitte.com/assets/Dcom-UnitedStates/Local%20Assets/Documents/us_auto_DTTGlobalAutoSurvey_ElectricVehicles_100411.pdf.

Diamond, D. 2008. Public Policies for Hybrid-Electric Vehicles: The Impact of Government Incentives on Consumer Adoption. Ph.D. Dissertation. Fairfax, Va.: George Mason University. Available at http://mars.gmu.edu:8080/dspace/bitstream/1920/2994/1/Diamond_David.pdf.

Drury, I. 2011. Four-Cylinder Engines, Mpg on the Rise. Available at http://www.autoobserver.com/2011/10/four-cylinder-engines-mpg-on-the-rise.html. Accessed on March 23, 2012.

Egbue, O., and S. Long. 2012. Barriers to widespread adoption of electric vehicles: An analysis of consumer attitudes and perceptions. Energy Policy 48:717-729.

Ernst & Young. 2010. Gauging Interest for PHEVs and EVs in Select Markets.

Gallagher, K.S., and Erich Muehlegger. 2008. Giving Green to Get Green: Incentives and Consumer Adoption of Hybrid Vehicle Technology. Harvard Kennedy School Faculty Research Working Paper RWP08-009.

Greene, D.L. 1990. Fuel choice for multifuel vehicles. Contemporary Policy Issues VIII(4):118-137.

Greene, David L. 2010a. Why the Market for New Passenger Cars Generally Undervalues Fuel Economy. Discussion Paper 2010-6. International Transportation Forum, OECD, Paris. Available at http://cta.ornl.gov/cta/Publications/Reports/Why_the_Market_for_New0001.pdf. Accessed May 23, 2012.

———. 2010b. How Consumers Value Fuel Economy: A Literature Review. EPA 40-R-10-008. Available at http://www.epa.gov/oms/climate/regulations/420r10008.pdf. Accessed May 23, 2012.

Guo, J.Y., G. Venkataramanan, B. Lesieutre, A. Smick, M. Mallette, and C. Getter. 2010. Consumer Adoption and Grid Impact Models for Plug-in Hybrid Electric Vehicles in Dane County, Wisconsin. Part A: Consumer Adoption Models. Madison, Wisc.: Public Service Commission of Wisconsin.

Harder, A. 2011. Obama Claims Connection Between Fuel Standards, Jobs, but Reality Is Complicated. Available at http://www.nationaljournal.com/energy/obama-claims-connection-between-fuel-standards-jobs-but-reality-is-complicated-20110811. Accessed on August 12, 2011.

Harris Interactive. 2011. Auto TECHCAST Study. June.

Hensher, D.A.H.B. 2010. Hypothetical bias, stated choice experiments and willingness to pay. Transportation Research Part B 44(6):735-752.

Hughes, J.E., C.R. Knittel, and D. Sperling. 2008. Evidence of a shift in short-run price elasticity of gasoline demand. *The Energy Journal* 29(1):93-114.

Humphries, J., D. Sargent, J. Schuster, M. Marshall, M. Omotoso, and T. Dunne. 2010. *Drive Green 2020: More Hope Than Reality?* Westlake Village, Calif.: J.D. Power and Associates.

Hurst, D. 2011. The Plug-In Vehicle Residual Value Conundrum. Available at http://www.pikeresearch.com/blog/articles/the-plug-in-vehicle-residual-value-conundrum. Accessed on May 4, 2012.

Koslowski, T. 2011. *Strategic Market Considerations for Electric Vehicle Adoption in the U.S.* Stamford, Conn.: Gartner, Inc.

Kotchen, M.J., and S.D. Reiling. 2000. Environmental attitudes, motivations and contingent valuation of nonuse values: A case study involving endangered species. *Ecological Economics* 32:93-107.

Kurani, K.S., and D. Sperling. 1993. *Fuel Availability and Diesel Fuel Vehicles in New Zealand.* TRB Reprint Paper No. 930992. Davis, Calif.: University of California, Davis.

Morris, E. 2007. From horse power to horsepower. *Access* 30(Spring):2-9.

Nixon, H., and J.-D. Saphores. 2011. *Understanding Household Preferences for Alternative-Fuel Vehicle Technologies.* Cambridge, Mass.: Massachusetts Institute of Technology. Available at http://transweb.sjsu.edu/PDFs/research/2809-Alternative-Fuel-Vehicle-Technologies.pdf.

PCAST (President's Council of Advisors on Science and Technology). 2010. *Report to the President on Accelerating the Pace of Change in Energy Technologies Through an Integrated Federal Energy Policy.* Washington, D.C.: Executive Office of the President.

J.D. Power and Associates. 2011. 2011 U.S. Green Automotive Study. April.

Rogers, E.M. 1962. *Diffusion of Innovations.* New York: Free Press.

———. 2003. *Diffusion of Innovations.* 5th Edition. New York: Free Press.

Synovate. 2011. Synovate Survey Reveals Whether Consumers Will Stay Away from Electric Powertrain Vehicles Because They Don't Understand How They Work. Available at http://www.ipsos-na.com/news-polls/pressrelease.aspx?id=5482. Accessed January 29, 2013.

Turrentine, T.S., and K.S. Kurani. 2007. Car buyers and fuel economy? *Energy Policy* 35:1213-1233.

Zhang, T., S. Gensler, and R. Garcia. 2011. A study of diffusion of alternative fuel vehicles: An agent-based modeling approach. *Journal of Product Innovation Management* 28:152-156.

5

Modeling the Transition to Alternative Vehicles and Fuels

5.1 INTRODUCTION

Achieving the goals of reducing light-duty vehicle (LDV) petroleum use and greenhouse gas (GHG) emissions by 80 percent by 2050 and petroleum use by 50 percent by 2030 is likely to require a transition from internal combustion engines powered by fossil petroleum to alternative fuels or vehicles or both. There is also potential for significant technological advancement both in the LDV fleet and in the fuel and fueling infrastructure that will power vehicles over the next 40 years. Which of these technologies will actually enter the market depends on a range of factors, including the extent of progress in the different vehicle and fuel technologies, market conditions in gasoline and other fuels markets that will affect cost and competiveness, consumer preferences over vehicle and fuel characteristics, and government policies toward this sector. Government policies are likely to be particularly important because the benefits of both petroleum and greenhouse gas reductions accrue to the public as a whole, and so market forces alone cannot be relied on to provide sufficient reductions.[1]

Two different models were used by the committee to assess the potential and opportunities for achieving the goals of this study. The first was the VISION model developed by Argonne National Laboratory (Singh et al., 2003). This spreadsheet model was an ideal starting point for the committee's analysis because it has been widely used in the past for light-duty vehicle (LDV) sector forecasts of energy use and GHG emissions. All inputs must be specified, including future rates of penetration of vehicle and fuel types and

the costs of each. VISION does not, however, attempt to estimate how markets will react to alternative vehicles and fuels or to the policies that may be needed to successfully introduce them.

The second model, the Light-duty Alternative Vehicle Energy Transitions (LAVE-Trans) model, incorporates market decision making and reflects the most significant economic barriers to the adoption of new vehicles and fuels. It therefore allows for assessment of policies and possible transition paths to attain the goals. Penetration rates of different vehicle and fuel types are determined in this model in response to price, costs, and vehicle fueling characteristics; they are not simply assumed as they are in VISION. Moreover, LAVE-Trans includes a consistent and comprehensive assessment of the benefits and costs of different policy and technology pathways over time.

It is important to emphasize the nature and extent of the uncertainties that lie behind all of the analyses in this chapter. First, the analysis uses estimated improvements to fuel efficiency and fuel carbon content, and the associated costs, for vehicles up to the 2050 model year as provided by expert members of this committee, evidence from the literature, and consultation with experts outside the committee. (Detailed descriptions can be found in Chapters 2 and 3.) Both models use the same GHG emissions, fuel economy, and vehicle cost estimates. These estimates by necessity reflect numerous assumptions, most of which are highly uncertain, particularly when such forecasts are made far into the future. One way the committee represents this uncertainty is to include both "midrange" and "optimistic" estimates for important variables such as vehicle fuel efficiency and fuel carbon intensities. However, it is difficult to reflect the full range of uncertainty. Thus, a "pessimistic" case is not included here for vehicles in which either technology does not progress very rapidly or costs do not come down over time and with volume as expected.

There is, in addition, uncertainty in the assumptions about consumer preferences for different vehicle characteristics,

---

[1] Both petroleum use reduction and GHG emissions reduction are types of public goods in that once they are reduced, all members of society benefit through greater security and reduced risk of global climate change. No one is excluded from these benefits. The private sector will tend to underprovide such goods because private individuals must pay the costs of reductions but do not get all of the benefits—the benefits are shared by all. When there are public goods, then, government action may be essential for attaining amounts of the public goods that are economically efficient for society (Boardman et al., 2011).

including range and limited fuel availability for alternatives such as hydrogen fuel cell vehicles.[2] A sensitivity analysis illustrating uncertainties about the market's response to alternative vehicles and fuels is described in Section 5.7. There is also controversy about the magnitude of the social cost of GHG emissions and the social cost of the United States' reliance on oil and petroleum-based gasoline. The estimates used in this report are drawn from the most recent literature but do not reflect the full range of uncertainty. Finally, it is extremely difficult to model all of the feedback effects that will inevitably result over time as technology development and markets interact.

Despite the inherent uncertainties in attempting to forecast four decades into the future, the committee's modeling effort here uses the best available evidence and information and makes plausible assumptions where sound data are missing. Analysis of the results from the two models then provides useful insights about what various vehicles and fuel combinations can achieve, the nature of the processes by which changes will occur, and the general magnitude of potential costs and benefits of different policy options.

5.2 MODELING APPROACH AND TOOLS

5.2.1 VISION Model

VISION is designed to extend the transportation sector-specific component of the National Energy Modeling System (NEMS) used by the Energy Information Administration (EIA). It provides longer-term forecasts of energy use and GHG emissions than does NEMS. While not as detailed or comprehensive as the NEMS model, VISION provides greater flexibility to analyze a series of projected usage scenarios over a much longer timeframe. It has been used extensively in the literature.

For the purposes of this study, VISION has been modified in a number of ways. The most up-to-date assumptions from the committee about vehicle efficiencies, fuel availability, and the GHG emissions impacts of using those fuels have been included. It is assumed that new-technology vehicle sales ramp up slowly and that new sales for a particular vehicle type never increase by more than about 5 percent of total new LDV sales in a given year. In addition, only one plug-in hybrid electric vehicle (PHEV), a PHEV-30 with a real-world all-electric driving range of 25 miles, is included. It is assumed that because of their limited range, battery electric vehicles are to be driven 1/3 fewer miles per year than other vehicles (Vyas et al., 2009) and that any decrease in miles driven by electric vehicles will be offset by increased mileage from other vehicles. Total new car sales and annual vehicle miles traveled (VMT) are assumed to be the same

as in the projections from the *Annual Energy Outlook 2011* (AEO; EIA, 2011a), and there is no assumption of a "rebound effect"[3] if the cost of driving a mile declines. Adjustments to VMT can be included separately in any VISION run assessment.[4] Finally, GHG estimates from biofuels include both emissions from production and from indirect land-use changes (see Chapter 3).

The committee uses the VISION model to explore how a focus on specific technologies or alternative vehicle and fuel types has the potential to reduce oil use and GHG emissions to achieve the study goals. The committee then turns to the LAVE-Trans model to shed light on how policies might be used to achieve the needed transitions.

5.2.2 LAVE-Trans Model

The Light-duty Alternative Vehicle Energy Transitions (LAVE-Trans) model uses a nested, multinomial logit model[5] of consumer demand to predict changes in the efficiency of vehicles and fuels over time, including a possible transition to alternatively fueled vehicles. Any transition to these advanced vehicles faces a number of barriers, including high costs due to the lack of scale economies and lack of learning, consumer uncertainty about safety or performance, and the lack of an energy supply infrastructure. Each of these barriers has been incorporated into the LAVE-Trans model so that the costs of overcoming them and, alternatively, the benefits of policies needed to do so can be measured (subject to the limits of current knowledge).

The model incorporates an array of factors that affect and are derived from consumer behavior, including the rebound effect; "range anxiety" and perceived loss of utility, particularly as it pertains to the availability of a fueling infrastructure; aversion to new technology and its reciprocal effect, early adoption; and the significant discounting of future fuel benefits over the lifetime of the vehicle. Nine variables influence the market shares of the alternative advanced technologies:

---

[3]Improvements in the efficiency of energy consumption will result in an effective reduction in the price of energy services, leading to an increase of consumption that partially offsets the impact of the efficiency gain in fuel use. This is known as the "rebound effect."

[4]If a 5 percent reduction in vehicle miles traveled is plausible under certain policies, then the estimates of GHG emissions and oil use can be reduced by 5 percent.

[5]A multinomial logit model is a standard model often used to represent consumer choice where there is a finite set of discrete options. The probability of choosing among the set of available options is governed by representative parameters for a particular class of consumer. A nested model refers to multiple layers of choice (see Daly and Zachary, 1979; McFadden, 1978; Williams, 1977). For example, the first level of choice in the LAVE-Trans model is between choosing whether or not to buy an LDV. If a consumer chooses to buy an LDV, the next level of choice is between purchasing a passenger car or a light truck. Then, within a particular class of vehicle there are multiple options, such as whether to purchase an ICEV, FCEV, or BEV. Further description of the LAVE-Trans nested multinominal logit model can be found in Section H.2 in Appendix H.

---

[2]Thanks to recent research, such issues are better understood than they were a decade ago (e.g., UCD, 2011; Bastani et al., 2012), yet much remains to be learned.

1. Retail price equivalent (RPE),
2. Energy cost per kilometer,
3. Range (kilometers between refuel/recharge events),
4. Maintenance cost (annual),
5. Fuel availability,
6. Range limitation for battery electric vehicles (BEVs),
7. Public recharging availability,
8. Risk aversion (innovator versus majority), and
9. Diversity of make and model options available.

It also includes policy options that affect consumer choices, including new-vehicle rebates, incentivized infrastructure development, and fuel-specific taxation. Although both the LAVE-Trans and VISION models use the same committee-developed technology and cost assumption for different vehicles and fuels over time, the LAVE-Trans model represents a significant improvement over the VISION model in several ways. First, because it includes consumer behavior in the vehicle market, it is able to predict the shares of different vehicles that enter the market in response to policy and market changes, whereas VISION must assume these shares over time. Thus, LAVE-Trans is much better able to assess the types of policies that may be necessary to achieve the goals addressed in the present study. Second, LAVE-Trans can be used to assess the full range of benefits and costs of different policies. The committee's approach to measuring benefits and costs is discussed more fully below.

5.3 RESULTS FROM RUNS OF VISION MODEL

Forecasts of the penetration rates of different types of vehicles using the VISION model must be compared to some alternative outcome in which there are no further policy actions and limited technological advances. In this analysis, two such cases are presented. One is the business as usual (BAU) case. It closely follows the AEO 2011 reference case

projection to 2035 and from there is extrapolated to 2050. In this case, NHTSA CAFE and EPA GHG emission joint standards for LDVs are set out to 2016, with fuel economy continuing to increase to 2020 per the Energy Independence and Security Act of 2007. Renewable fuel production increases in response to RFS2 (the amended Renewable Fuel Standard), but it is assumed that financial and technological hurdles facing advanced biofuel projects will delay compliance. The other case is the Committee Reference Case. It adds to the BAU case the CAFE rules that have been set through the 2025 model year, and the levels of advanced biofuels production required under RFS2 are assumed to be fully met by 2030 through the production of thermochemical cellulosic biofuel.

5.3.1 Baseline Cases

5.3.1.1 Business as Usual (BAU)

In the BAU case, new-vehicle sales increase to 22.2 million in 2050 from 10.8 million units in 2010 (a year in which sales were severely depressed due to the recession). Diesel, hybrid, and plug-in hybrid vehicles make modest gains in market share (Figure 5.1). The total stock of LDVs increases from about 220 million in 2010 to 365 million in 2050.

Fleet average on-road fuel economy improves from 20.9 miles per gallon gasoline equivalent (mpgge; equivalent to a consumption of 4.8 gge/100 mi) in 2005 to 34.7 mpgge (or 2.9 gge/100 mi) in 2050. This is consistent with the Energy Independence and Security Act of 2007, which requires a fleetwide fuel economy test value of at least 35.5 mpg in 2020 and includes modest improvements in vehicle efficiency thereafter. This is enough to offset most of the forecasted increase in vehicle travel from 2.7 trillion to 5.0 trillion miles. Energy use increases to 159 billion gallons gasoline equivalent (billion gge) from 130 billion gge. Com-

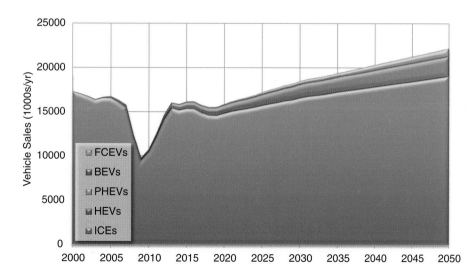

FIGURE 5.1 Vehicle sales by vehicle technology for the business as usual scenario.

pared to 2005 levels, petroleum use remains unchanged, the result of increased use of corn-based ethanol (to 12.0 billion gge/yr in 2050) and the addition of 8.9 billion gge/yr of cellulosic ethanol and 8.1 billion gallons of gasoline produced from coal. The net effect of increased overall energy use and the shift to a somewhat less carbon-intensive fuel mix is a 12 percent increase in 2050 GHG emissions.

Oil prices in this scenario are expected to gradually increase to $123/bbl by 2035 (in 2009$) according to AEO 2011, resulting in a pre-tax gasoline price of $3.16 in 2035. Gasoline prices are then extrapolated out to 2050 assuming the compound rate of growth modeled in AEO 2011 from 2030 to 2035, yielding a pre-tax price of $3.37. The current gasoline tax of $0.42/gal is assumed to remain the same (in constant dollars) out to 2050. Gasoline prices in this scenario are shown in Figure 5.2. The pre-tax fraction of these gasoline prices is assumed in all modeling scenarios.

5.3.1.2 Committee Reference Case

The committee further defined its own reference case to include all of the midrange assumptions it developed about vehicle efficiencies, fuel availability, and GHG emission rates up to 2025 (summarized in Chapters 2 and 3). This Committee Reference Case assumes that the 2025 fuel efficiency and emissions standards for LDVs will be met. The committee interprets the standards to require that new vehicles in 2025 must have on-road fuel economy averaging around 40 mpg (given a fleetwide CAFE rating of 49.6 mpg for new vehicles, the difference between on-road and test

values, and the likely application of various credits under the CAFE program). See Box 5.1 for an explanation of on-road fuel economy compared to tested fuel economy ratings.

This case also assumes that the RFS2 goals will be met by 2030. As a result, corn ethanol sales rise to almost 10 billion gge/yr by 2015 and then remain at that level. Based on the analysis in Chapter 3, it is also assumed that all cellulosic biofuels will be thermochemically derived gasoline. The RFS2 requirements result in annual production of 13.2 billion gallons of such biofuels by 2030 and roughly constant levels thereafter.

Under the assumptions of the Committee Reference Case, the fuel economy (fuel consumption) of the stock of LDVs in use improves to 35.5 mpgge (2.8 gge/100 mi) in 2030 and to 41.6 mpgge (2.4 gge/100 mi) in 2050, up from 20.8 mpgge (4.8 gge/100 mi) in 2005 (Figure 5.3). This improvement is largely due to efficiency improvements in internal combustion engine vehicles (ICEVs) as well as increasing sales of hybrid electric vehicles (HEVs). Hybrids are more successful in this scenario compared to the BAU case, increasing their share of new-vehicle sales to 33 percent (7.3 million units) by 2050.

Greenhouse gas emissions are 30 percent below 2005 levels in 2030, at 1,057 million metric tons CO_2 equivalent (MMTCO$_2$e) per year, but rise again and are just 22 percent below in 2050 (1,121 MMTCO$_2$e/yr) as VMT continues to rise while the efficiency of the on-road fleet remains approximately constant (Figure 5.4). Petroleum use is 36 percent below the 2005 level in 2030 (1.91 billion bbl/yr) and 30 percent below in 2050 (2.09 billion bbl/yr), also rising with

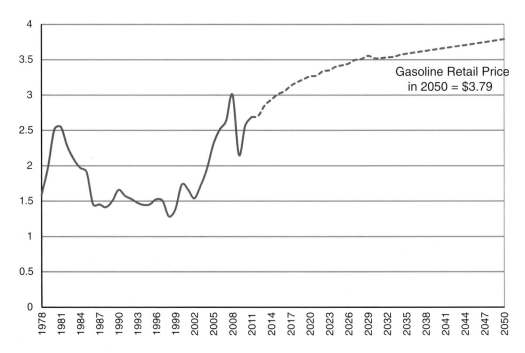

FIGURE 5.2 Retail gasoline fuel prices (1978-2050), including federal and state taxes. Projected values shown as dotted line. SOURCE: Data from *Annual Energy Review 2010* [1978-2010] (EIA, 2011b), *Annual Energy Outlook 2011* [2010-2035] (EIA, 2011a), and extrapolation by the committee using the compound annual growth rate for 2030-2035 (0.42%) [2035-2050].

BOX 5.1
The Distinction Between "As Tested" and "Actual In-Use" Fuel Consumption

A large difference exists between the fuel economy (miles per gallon, or mpg) figures used to certify compliance with fuel economy standards and those experienced by consumers who drive the vehicles on the road and purchase fuel for their vehicles. The numbers used to certify compliance with the Corporate Average Fuel Economy (CAFE) standards are based on two dynamometer tests. These test values are also the numbers discussed and presented in the tables and figures of this report. A different 5-cycle test procedure is used to compute the Environmental Protection Agency (EPA) "window-sticker" (label) fuel economy ratings that are used in automotive advertising, most car-buying guides, and car-shopping Websites. Neither procedure accurately reflects what any given individual will achieve in real-world driving. Motorists have different driving styles, experience different traffic conditions, and take trips of different lengths and frequencies. Realized fuel economy also varies with factors including climate, road surface conditions, hills, temperature, tire pressure, and wind resistance. The impacts of air conditioning, lighting, and other accessories on fuel consumption are not included in the two-cycle tests.

Both CAFE mpg and "window-sticker" mpg were based on the values determined via standardized city- and highway-cycle procedures that were codified by law in 1975. The divergence between test-cycle values and real-world experience was recognized and in 1985 the EPA revised calculation procedures for the window-sticker ratings in order to bring them more in line with the average performance motorists were reporting in real-world driving. From 1985 through 2007, the window sticker values averaged about 15 percent lower than the unadjusted values used for CAFE regulation. The label values were updated starting in model year 2008, and the update further increased the difference between CAFE and "window sticker" values by factoring in additional adjustments, so that the current window sticker values average about 20 percent lower than those used for regulation.

The results can be confusing. For example, the 2017-2025 CAFE rules envision a 49.6-mpg "fleet average new LDV fleet fuel economy" for the 2025 model year, but acknowledge that real-world fuel economy will be significantly lower—probably somewhere below 40 mpg. A further complication is that the "National Plan" (the joint rulemaking by NHTSA and EPA) regulates greenhouse gas (GHG) emissions in addition to fuel economy. Because some technologies for reducing LDV GHG emissions do not involve fuel economy, EPA now also reports a "mpg-equivalent" value representing the CAFE fuel economy that would be needed to achieve a similar degree of GHG emissions reduction. That type of number is the one given as the 54.5 mpg "equivalent" stated in many discussion of the 2025 target; it reflects special credits for various technologies that can help in achieving fleet average GHG emissions of 163 grams per mile by 2025.

The CAFE numbers represent a higher fuel economy than most consumers are likely to experience on the road. The estimates of actual fuel consumption and associated GHG emissions presented in this report, however, reflect a downward fuel economy adjustment for approximating real-world impacts. Although there is no universally agreed-upon method for converting test values to on-road values, the committee has determined that an appropriate estimate for analytic purposes can be obtained by adjusting the CAFE values downward by about 17 percent (i.e., multiplying by 0.833). That factor is used whenever the report discusses "average" on-road values.

VMT. Thus, the Committee Reference Case, which assumes current policies included in the AEO BAU case augmented by the proposed 2025 fuel economy and emissions standards and RFS2 compliance, does not come close to meeting the 2030 or 2050 goals.

5.3.2 VISION Cases

To explore possible paths to attain the goals addressed in this study, VISION was run for a range of cases. The predominant characteristic of these runs is to focus on a market dominated by a particular vehicle type and alternative fuel (e.g., electric vehicles and grid with reduced GHG emissions, or fuel cell vehicles and hydrogen generated with CCS). To assess the range of possibilities, the committee looked at runs that used the midrange vehicle efficiencies as well as at runs that used the optimistic efficiencies representing technological progress proceeding more rapidly than expected, as described in Chapter 2. From the fuels side, the committee considered both present methods of producing a fuel as

well as fuel supply technologies with reduced GHG impacts as described in Chapter 3. Each of the possible fuel types is shown in Table 5.1. A brief description below of each of the scenarios modeled with VISION identifies the important assumptions and variation in those assumptions. Section H.1 in Appendix H provides further detail.

- *Emphasis on ICEV efficiency.* These runs continue the reference case's focus on LDV fuel efficiency improvements through the period to 2050. Shares of advanced ICEVs and HEVs increase to about 90 percent of new-vehicle sales by 2050. Two runs are included that differ only in their assumptions about the fuel efficiency improvements of vehicles over time. The first assumes the midrange assumptions for fuel efficiency for all technologies (Chapter 2, Table 2.12), and the second assumes optimistic fuel efficiency for ICEs and HEVs while maintaining midrange values for the small numbers of other types of vehicles in the fleet. It is assumed that the RFS2

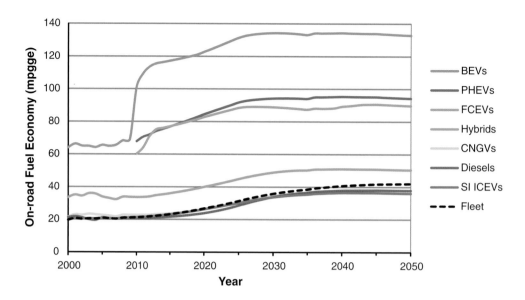

FIGURE 5.3 Average on-road fuel economy for the Reference case light-duty vehicle stock. In most cases, the average efficiency plateaus as the fleet gradually turns over to vehicles that meet the 2025 model year CAFE standard. There are small reductions over time with rising use of advanced technologies in trucks.

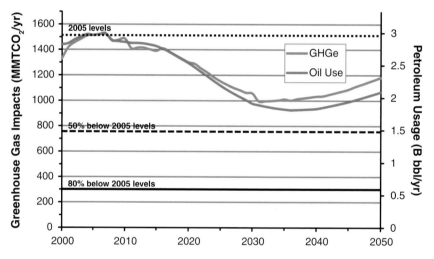

FIGURE 5.4 Petroleum use and greenhouse gas emissions for the Committee Reference Case.

requirements described in the Committee Reference Case, above, are still in place. These increased vehicle efficiency cases require much less liquid fuel over time and assume that gasoline is the fuel reduced.

- *Emphasis on ICEV efficiency and biofuels.* These two runs are similar to the case described above. The difference is that more biofuels are brought into the market after 2030, as described in Table 5.1. The modeling runs assume this additional biofuel, largely in the form of drop-in gasoline that displaces petroleum, and the only difference in the two runs is the assumption of vehicle fuel efficiency. The first run assumes all vehicles are at the midrange efficiency, and in it the share of petroleum-based gasoline as a

liquid fuel falls to about 25 percent by 2050. The second run assumes optimistic fuel efficiency for ICEVs and HEVs. In this case, bio-based ethanol, bio-based gasoline, and a small amount of coal-to-liquid (CTL) and gas-to-liquid (GTL) fuels make up all liquid fuel, with almost no petroleum-based gasoline.

- *Emphasis on fuel cell vehicles.* This case comprises 4 different runs of VISION, to capture variation in both vehicle efficiency and fuel carbon content. In all of these runs, the share of fuel cell electric vehicles (FCEVs) increases to about 25 percent of new car sales by 2030 and then to 80 percent by 2050, modeled on the maximum practical deployment scenario from *Transition to Alternative Transportation*

TABLE 5.1 Description of Fuel Availabilities Considered in Modeling Light-Duty Vehicle Technology-Specific Scenarios

Fuel Type	Description (values reflect annual production in 2050)
AEO 2011	AEO 2011 projection extrapolated to 2050; 12.0 billion gge corn ethanol; 8.9 billion gge cellulosic ethanol; 8.1 billion gal CTL gasoline
Reference	RFS2 met by 2030: 10 billion gge corn ethanol; up to 13.2 billion gge cellulosic thermochemical gasoline; up to 3.1 billion gge CTL; up to 4.6 billion gge GTL
Biofuels	Includes Reference biofuel availability plus additional drop-in biofuels: Up to 45 billion gge cellulosic thermochemical gasoline; 10 billion gge corn ethanol
AEO 2011 Electricity Grid	AEO 2011 Electricity Grid: 541 g CO_2e/kWh; 46% coal, 22% natural gas, 17% nuclear, and 12% renewable
Low-C Electricity Grid	AEO 2011 Carbon Price Grid: 111 g CO_2e/kWh; 6% coal, 25% natural gas, 12% natural gas w/CCS, 30% nuclear, and 23% renewable
Low-Cost H_2 Production	Lowest Cost: $3.85/gge H_2; 12.2 kg CO_2e/gge H_2; 25% distributed natural-gas reforming, 25% coal gasification, 25% central natural-gas reforming, and 25% biomass gasification
CCS H_2 Production	Added CSS: $4.10/gge H_2; 5.1 kg CO_2e/gge H_2; 25% distributed natural-gas reforming, 25% coal gasification w/CCS, 25% central natural-gas reforming with CCS, and 25% biomass gasification
Low-C H_2 Production	Low CO_2 emissions: $4.50/gge H_2; 2.6 kg CO_2e/gge H_2; 10% distributed natural-gas reforming, 40% central natural-gas reforming w/CCS, 30% biomass gasification, and 20% electrolysis from clean electricity

NOTE: CCS H_2 case analyzed by LAVE model, not VISION.

Technologies: A Focus on Hydrogen (NRC, 2008). There are two runs with the midrange vehicle fuel efficiencies, the first with low-cost hydrogen production (Low-Cost H_2 Production) and the second with low-GHG hydrogen production (Low-C H_2 Production), described in Table 5.1. Finally, there are two additional runs with optimistic assumptions about the fuel efficiency of FCEVs, each with the different assumptions for the GHG emissions from hydrogen production.

- *Emphasis on plug-in electric vehicles.* There are 4 VISION runs emphasizing plug-in electric vehicles (PEVs) to account for differences in assumptions about vehicle efficiency as well as GHG emissions impacts of the fuel. In all runs, the share of BEVs and PHEVs increases to about 35 percent of new LDV sales by 2030 and 80 percent by 2050, in line with the rates put forth in *Transitions to Alternative Transportation Technologies: Plug-In Hybrid Electric Vehicles* (NRC, 2010a). Relatively greater sales of PHEVs than BEVs in all years are assumed (see Table H.3 in Appendix H for details). Each of the two runs in

each pair of runs—midrange and optimistic—uses a different assumption about GHG emissions from the electricity grid (AEO 2011 Grid and Low-C Electric Grid, Table 5.1). The low-emissions grid is assumed to emit 25 percent of GHGs per unit of generation compared to the BAU grid by 2050.

- *Emphasis on natural gas vehicles.* These runs assume that sales of compressed natural gas vehicles (CNGVs) are 25 percent of the market by 2030 and 80 percent by 2050. In both midrange and optimistic cases, CNG fuel use rises over time, and so little liquid fuel is needed by 2050 that it is assumed that no CTL and GTL plants are ever built. It is further assumed that RFS2 must be met by 2030, and so the liquid fuel that is used is primarily biofuels in both of these runs.

5.3.3 Results of Initial VISION Runs

Figures 5.5 to 5.7 indicate the results of the VISION model runs described above. The total amount of each type of fuel used in each scenario is shown in terms of energy use (billions of gallons of gasoline-equivalent). For the hydrogen and electricity cases, the fuels are not broken down by carbon content. Figure 5.5 shows results of the assumptions about fuel use that were made for the different VISION runs. For example, the total amount of liquid fuels used is the same for the Efficiency and Efficiency + Biofuels scenarios—it is assumed that it is the fraction of that fuel generated from biomass that is different. Higher prices for biofuels are likely to drive liquid fuel prices up over time and could result in less total liquid fuel used, but that type of market feedback cannot be accounted for in the VISION model runs.

Some ethanol and cellulosic biofuels are used in all of the scenarios because of the assumptions that they will be required under regulations such as RFS2. Over all of the scenarios, fuel energy use is lowest for the Plug-in Electric Vehicle, Hydrogen Fuel Cell Vehicle, and Optimistic Efficiency for ICEV and HEV cases.

Figure 5.6 shows that the long-term petroleum reduction goal of 80 percent by 2050 could occur if there is either (1) a major increase in biofuel availability with high-efficiency ICEVs (including HEVs) or (2) a large increase in alternatively fueled vehicles. All of the cases involving a transition to alternatively fueled vehicles meet or nearly meet a midterm petroleum reduction goal of 50 percent by 2030; in addition, optimistic ICEV efficiencies and widespread availability and use of biofuels could meet this interim goal as well. It is important to note that all of these scenarios assume very aggressive deployment of the specific vehicles and fuels being emphasized. The VISION model cannot address how these vehicle shares would be achieved. The model tells us nothing about how market conditions or policies would produce such results in vehicle and fuel shares.

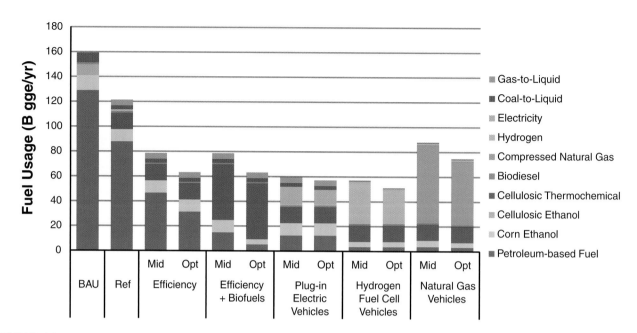

FIGURE 5.5 Fuel usage in 2050 for technology-specific scenarios outlined in Section 5.3.2. Midrange values are the committee's best estimate of the progress of the vehicle technology if it is pursued vigorously. Optimistic values are still feasible but would require faster progress than seems likely. No GTL or CTL fuel is used in the fuel cell and natural gas scenarios.

Figure 5.7 shows GHG emissions results for each scenario. It is noteworthy that all of the scenarios show substantial emissions reductions from the Committee Reference Case. However, meeting the 80 percent reduction goal is extremely difficult. Even given the aggressive deployment of advanced vehicle technologies and fuel supply technolo-

gies assumed in these runs, only two scenarios meet the 80 percent goal, the FCEV-dominated fleet powered by very low GHG-emitting hydrogen fuel and the optimistic case for vehicle efficiency plus biofuels. Several other scenarios come close to meeting the goal, and small reductions in VMT that could be expected with strict policies to reduce

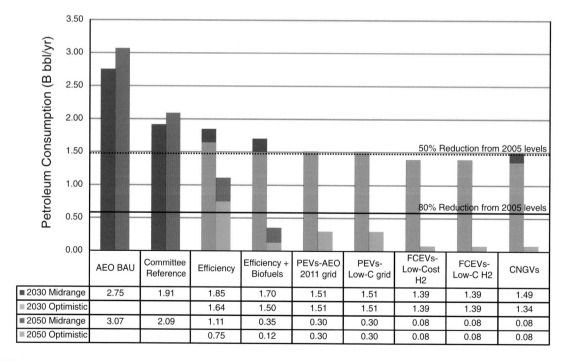

FIGURE 5.6 U.S. light-duty vehicle petroleum consumption in 2030 and 2050 for technology-specific scenarios outlined in Section 5.3.2. Midrange values are the committee's best estimate of the progress of the vehicle technology if it is pursued vigorously. Optimistic values are still feasible but would require faster progress than seems likely.

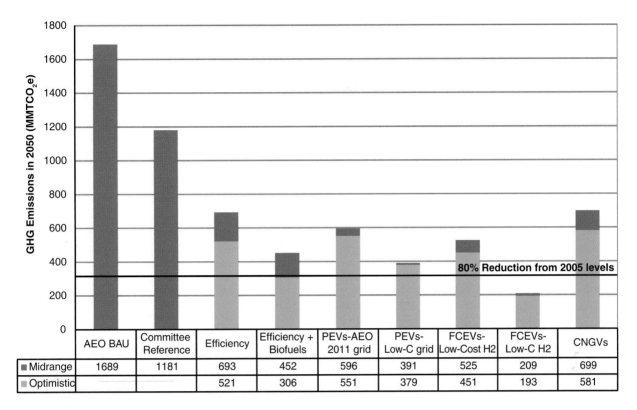

	AEO BAU	Committee Reference	Efficiency	Efficiency + Biofuels	PEVs-AEO 2011 grid	PEVs-Low-C grid	FCEVs-Low-Cost H2	FCEVs-Low-C H2	CNGVs
■ Midrange	1689	1181	693	452	596	391	525	209	699
■ Optimistic			521	306	551	379	451	193	581

FIGURE 5.7 U.S. light-duty vehicle sector greenhouse gas emissions in 2050 for technology-specific scenarios outlined in Section 5.3.2. Midrange values are the committee's best estimate of the progress of the vehicle technology if it is pursued vigorously. Optimistic values are still feasible but would require faster progress than seems likely.

GHGs might be sufficient to push them to the 80 percent goal as well.

Although these model results illustrate penetration levels of certain vehicles and fuels that may achieve the petroleum usage and/or greenhouse gas emissions reductions desired, the VISION model does not estimate the cost or the policy actions that would be necessary. For this, an alternative approach is needed.

5.4 LAVE-TRANS MODEL

The LAVE modeling builds on the VISION analyses, illustrating how market responses may influence the task of achieving the petroleum and greenhouse gas reduction goals as well as providing a sense of the intensity of policies that may be required and measuring, very approximately, the costs and benefits. The committee recognizes that such estimates will be neither certain nor precise. Both market and technological uncertainty are very substantial, as is illustrated in Section 5.7, a fact that requires an adaptable policy process. However, ignoring market responses and the costs of necessary policies would be a mistake. The policy options included in the LAVE model are briefly summarized in Box 5.2 and described in greater detail in Section 5.4.2.

The analyses using the LAVE-Trans model proceed as follows. First, the LAVE-Trans and VISION model projec-

tions of the BAU case are compared to establish the general consistency of the two models. The LAVE-Trans model is then used to approximately replicate the VISION model scenarios, which again shows broad consistency but also some differences between the two models. The strategy and approach to policy analysis using the LAVE-Trans model are described next, including how costs and benefits have been measured. All of the policy scenarios described below include strict CAFE standards that are tightened over time, and also some policy approach to bring alternative fuels into the market, such as RFS2. In addition all policy scenarios below also include the Indexed Highway User Fee (IHUF).

- The first set of policy analyses explore what might be achievable by means of continued improvement of energy efficiency beyond 2025 and introduction of large quantities of "drop-in" biofuels with reduced greenhouse gas impacts produced by thermochemical processes. To provide incentives for greater efficiency from ICEs and HEVs, the first feebate policy in Box 5.2, the Feebate Based on Social Cost (FBSC) is introduced.
- A second set explores the potential impacts of policies that change the prices of vehicles and fuels to reflect the goals of reducing GHG emissions and petroleum use. In these model runs, stronger feebates

---

BOX 5.2
Policies Considered in the LAVE-Trans Model

Feebates Based on Social Cost (FBSC)—An approximately revenue neutral feebate system that precisely reflects the assumed societal willingness to pay to reduce oil use and GHG emissions (see Boxes 5.3, 5.4, and 5.5 on feebates and the values of GHG and oil reduction).

Indexed Highway User Fee (IHUF)—A replacement for motor fuel taxes, the IHUF is a fee on energy indexed to the average energy efficiency of all vehicles on the road and is designed to preserve the current level of revenue for the Highway Trust Fund (see Chapter 6 for details).

Carbon/Oil Tax—A gradually rising tax levied on fuels to reflect the societal values of their carbon emissions and petroleum content (see Boxes 5.4 and 5.5).

Feebates Based on Fuel Savings—A feebate system that compensates for consumers' undervaluing of future fuel savings. This feebate reflects the discounted present value of fuel costs (excluding the social cost fuel tax) from years 4 to 15, discounted to present value at 7 percent per year.

Transition Policies (Trans)—Polices that consist of subsidies to vehicles and fuel infrastructure designed to allow alternative technologies to break through the market barriers that "lock in" the incumbent petroleum-based internal combustion engine vehicle-fuel system. These could be either direct government subsidies or subsidies induced by governmental regulations, such as California's Zero-Emissions Vehicle standards.

---

assumptions of technological progress and market behavior.

5.4.1 Comparing LAVE-Trans and VISION Estimates

As shown in Table 5.2, the BAU cases from the LAVE-Trans and VISION models confirm the general consistency of the two models. Each was calibrated to match in all years with respect to total vehicle miles of travel and total vehicle sales. There are differences in new-vehicle and vehicle stock fuel economies, the distributions of stock by age, and in the starting year GHG emissions rates due to the use of two different starting base years.[6] These lead to differences between the models of about 5 percent in energy and GHG emissions estimates in 2010, with the differences declining in subsequent years. This decline reflects the fact that the differences are chiefly due to the starting-year data for vehicle stocks and LDV energy efficiency and usage.

LAVE-Trans models vehicle purchase decisions and vehicle use in ways that VISION does not, enabling it to include market responses to improvements in vehicle technologies. If vehicles have fuel economy gains that are more than paid for by their fuel savings, for example, consumers will purchase more vehicles and the size of the vehicle stock will increase. If vehicle efficiency improves but fuel prices do not increase proportionately, vehicle use will increase. Market shares of vehicle technologies are not assumed in LAVE-Trans as they are in VISION but are based on a model of consumer choice that accounts for the prices, energy costs, and other attributes of the different technologies. All of these factors change a great deal over time in all cases.

The purchase prices and energy efficiencies of future vehicles strongly affect their market acceptance. In the LAVE-Trans model, novel technologies start out at a significant disadvantage relative to ICEV and HEVs because millions of these latter vehicles have already been produced and can access a ubiquitous infrastructure of refueling stations. Novel technologies must progress down learning curves by accumulating production experience and acquire scale economies through high sales volumes. As a result, the initial costs of BEVs, PHEVs, CNGVs, and FCEVs are much higher than the long-run costs projected in the midrange and optimistic scenarios. The long-run costs for passenger cars in Figure 5.8 show what is estimated to be technologically achievable in a given year at fully learned, full-scale production. In the midrange assessment, these potential costs converge between 2030 and 2040, with FCEVs and BEVs becoming slightly less expensive than ICEVs but with PHEVs remaining several thousand dollars more expensive. The optimistic assessment trends are similar but the convergence occurs more rapidly and the advantages of FCEVs and

on vehicles, those based on fuel savings, are included, and carbon and petroleum taxes are added that reflect estimates of the full social cost of using those fuels.

- The third, fourth, and fifth sets explore transitions from ICEVs fueled by petroleum to plug-in electric vehicles, fuel cell vehicles powered by hydrogen, and compressed natural gas vehicles, respectively. These all include transition policies tailored to the particular vehicle and fuel type being considered.

- Two final groups of cases consider combinations of PEVs and FCEVs and the implications of more optimistic technological progress. These also include the appropriate transition policies.

- Finally, the implications of uncertainty about technological progress and the market's response to advanced technologies and transition policies are considered. These cases include the IHUF, FBSC feebate, and transition policies while examining varied

---

[6]The LAVE-Trans model has a starting year of 2010, while VISION uses a base year of 2005. Instead of reprogramming or recalibrating the models, it was checked simply that their estimates were consistent.

TABLE 5.2 Comparison of Business as Usual Projections of the VISION and LAVE-Trans Models

		2010		2030		2050	
		LAVE	VISION	LAVE	VISION	LAVE	VISION
Energy use	billion gge	132	126	137	129	158	159
Petroleum use	billion gge	124	120	118	115	129	129
Greenhouse gas emissions	MMTCO$_2$e	1,431	1,498	1,467	1,487	1,645	1,689
Vehicle sales	thousands	10,797	10,797	18,502	18,502	22,219	22,219
Vehicle stock	thousands	222,300	236,310	255,603	281,976	314,538	365,199
Vehicle miles traveled	trillion miles	2.73	2.73	3.75	3.75	5.05	5.05
New light-duty vehicle fuel economy	mpg	22.5	22.6	29.8	30.3	33.8	34.8
Stock light-duty vehicle fuel economy	mpg	20.6	21.2	27.4	27.8	32.0	31.7

BEVs are greater (see Figures 2.10 and 2.11 as compared with Figures 2.8 and 2.9).

The energy efficiencies of new vehicles are shown in the midrange case to continue to improve at a rapid rate beyond 2025 (see Table 2.12 for details). The new-vehicle fuel economy numbers are inputs to the LAVE-Trans model and are taken from the estimates presented in Chapter 2 after accounting for the difference between on-road and test-cycle values. Internal combustion engine cars (both gasoline and CNG) increase to over 90 mpg by 2050, while HEVs exceed 120 mpg. PHEV fuel economy is the same as HEV mpg when operating in charge-sustaining mode and the same as BEVs when operating in charge-depleting mode. Such large increases in energy efficiency mitigate the effects of fuel prices over time.

The prices of energy are also important and vary substantially among the cases examined below. Figure 5.9 shows the different assumptions about what influences the price of gasoline. The price depends not only on the level at which gasoline is taxed but also on the quantities of biofuel blended into it. Some included cases reflect the use of an

IHUF on energy which increases very gradually over time as the average energy efficiency of all vehicles on the road increases. The greatest effect on pump prices, however, is with the introduction of a tax on the social value of carbon emissions and petroleum use, as described in Box 5.3, Box 5.4, and Box 5.5, assumed to be phased in over a period of 5 years. It is important to note that policies that greatly reduce the amount of oil used in the transportation sector, such as a number of those considered here, are likely to reduce both the demand for petroleum and its price. Less domestic use will mean fewer imports from insecure sources, which will likely reduce the magnitude of the social costs of using petroleum.

Figure 5.10 and Figure 5.11 show prices of other fuels under different assumptions. The price of electricity to consumers is affected by the de-carbonization of the grid, the IHUF, and the social value tax. Hydrogen prices start at more than $10/kg at low volumes and decrease as production approaches 6,000 tons/d. When and how quickly the decline occurs varies by scenario according to the level of hydrogen demand.

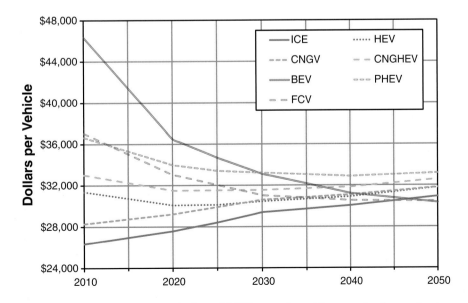

FIGURE 5.8 Fully learned, high-volume retail price equivalents (2009$) assuming midrange technology estimates.

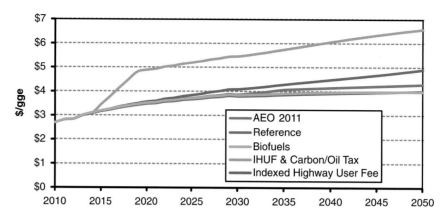

FIGURE 5.9 Retail prices of gasoline (in 2009$) under various policy assumptions.

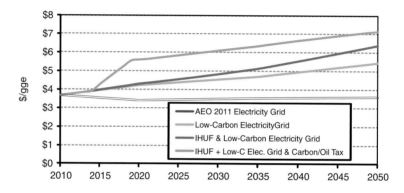

FIGURE 5.10 Retail prices of electricity (in 2009$) under various policy assumptions.

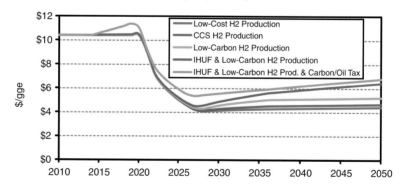

FIGURE 5.11 Retail prices of hydrogen (in 2009$) under various policy assumptions.

BOX 5.3
Feebates

Feebates are a fiscal policy aimed at influencing manufacturers to produce and consumers to purchase vehicles that are more energy efficient or produce fewer GHG emissions or both. A feebate system consists of a metric (e.g., g CO_2/mi, gge/mi), a benchmark, and a rate. Each vehicle is compared to the benchmark and is assigned a fee or a rebate according to the difference between its performance on the metric and the benchmark, multiplied by the rate. For example, if the metric is g CO_2/mi, the benchmark is 250 and the rate is $20/(g CO_2/mi), a vehicle emitting 300 g CO_2/mi would pay a fee of $1,000, whereas a vehicle emitting only 150 g CO_2/mi would receive a rebate of $2,000. By carefully choosing the benchmark, the feebate system can be made approximately revenue neutral. Benchmarks can be defined in various ways, including as a function of a vehicle attribute, such as the footprint measure (wheelbase × track width) used in the current CAFE standards.

BOX 5.4
The Social Cost of Carbon Emissions

Twelve government agencies conducted a joint study of the social cost of carbon (SCC) to allow agencies to incorporate the social benefits of reducing carbon dioxide emissions into cost-benefit analyses (Interagency Working Group, 2010). The agencies used three well-known economic integrated assessment models (IAMs) to produce the estimates and considered a broad range of factors that affect the damage estimates. Their estimates for the years 2010 to 2050 (Table 5.4.1) represent the present value, in the year in question, of the discounted future damage resulting from a 1 metric ton increase in CO_2 emissions. Estimates are given for three different discount rates (5%, 3%, and 2.5%), and for a 95th percentile (5% probability) estimate from the models at a 3 percent discount rate.

The group provided the higher 95th percentile estimate because of the following important limitations of the current state of knowledge concerning future damage due to climate change:

1. Incomplete treatment of non-catastrophic damage
2. Incomplete treatment of potential catastrophic damage
3. Uncertainty in extrapolation of damage to high temperatures
4. Incomplete treatment of adaptation and technological changes, and
5. Assumption that society is risk neutral with respect to climate damage.

The interagency study strongly recommends using the full range of estimates in assessing the potential damage from climate change (p. 33). The range is an order of magnitude: from $4.70 to $64.90 per metric ton in 2010, rising to $15.70 to $136.20 per metric ton in 2050. In the committee's judgment, the 80 percent greenhouse gas mitigation goal reflects a societal willingness to pay that is most consistent with the highest, 95th percentile estimates. This is the value the committee refers to as the social value of reducing greenhouse gas emissions.

TABLE 5.4.1 Social Cost of CO_2, 2010-2050, in 2007 Dollars

Discount Rate Year	5% Avg	3% Avg	2.5% Avg	3% 95th
2010	4.7	21.4	35.1	64.9
2015	5.7	23.8	38.4	72.8
2020	6.8	26.3	41.7	80.7
2025	8.2	29.6	45.9	90.4
2030	9.7	32.8	50.0	100.0
2035	11.2	36.0	54.2	109.7
2040	12.7	39.2	58.4	119.3
2045	14.2	42.1	61.7	127.8
2050	15.7	44.9	65.0	136.2

SOURCE: Interagency Working Group (2010).

The market responses included in the LAVE model should make it somewhat more difficult to meet the GHG and oil reduction goals. To illustrate this, the LAVE model was used to approximately replicate the VISION model cases shown in Figures 5.6 and 5.7. The approach was to solve for the subsidies to alternative technologies that cause the LAVE model to predict the same market shares assumed in the corresponding VISION model run.[7] This solution method results in a net subsidy to vehicle sales which over time will increase the size of the vehicle stock and thereby increase vehicle travel and energy use. In reality, the same market shares could be achieved by cross-subsidizing vehicles, which would reduce the impact on vehicle sales (e.g., via feebates; see Box 5.3). In that respect, the method will tend to exaggerate the greater difficulty of meeting the GHG and petroleum goals as a consequence of market responses.

In most cases the models produced very similar reductions in petroleum use and GHG emissions (Figure 5.12) with the LAVE-Trans model predicting somewhat smaller reductions, as expected. In most cases the differences are on the order of 5 percentage points. The VISION and LAVE-Trans CNGV

---

[7]Only the key market shares were carefully matched. For example, in the PEV cases the market shares for battery electric and plug-in hybrid electric vehicles were matched; the remaining technologies' market shares were as predicted by the LAVE model. In the FCEV cases only the market shares of fuel cell vehicles were closely matched.

BOX 5.5
Social Costs of Oil Dependence

The costs of oil dependence to the United States are caused by a combination of:

1. The exercise of monopoly power by certain oil-producing states,
2. The importance of petroleum to the U.S. economy, and
3. The lack of ready, economical substitutes for petroleum.

Costs exceed those that would prevail in a competitive market due to the use of market power chiefly by nationalized oil exporters. The direct economic costs of oil dependence can be partitioned into the following three, mutually exclusive components (Greene and Leiby, 1993):

1. Disruption costs, reductions in gross domestic (GDP) due to price shocks,
2. Long-run GDP losses due to higher than competitive market oil prices,
3. Transfer of wealth from U.S. oil consumers to non-US oil producers via monopoly rents.

When the U.S. takes actions to reduce its oil demand the world demand curve contracts resulting, other things equal, in lower world oil prices.[1] Such use of monopsony power counteracts the use of monopoly power, increasing U.S. GDP and reducing the transfer of U.S. wealth to non-U.S. oil producers. Individuals will generally not consider the fact that reducing one's own oil consumption produces benefits to others via lower oil prices. As a consequence the social benefits of reducing oil consumption exceed the private benefits. Although this appears to be similar to an externality, it is not an externality. The National Research Council (2009a) report *Hidden Costs of Energy: Unpriced Consequences of Energy Production and Use* considers only external costs and thus provides no relevant guidance on the value of reducing oil consumption.

Sudden, large movements in oil prices can temporarily reduce U.S. GDP by creating disequilibrium in the economy, leading to less than full employment of capital and labor (Jones et al., 2004). A substantial econometric literature on this subject has identified an important impact of price shocks on U.S. economic output (e.g., Huntington, 2007; Brown and Huntington, 2010). Reducing oil consumption reduces vulnerability to price shocks.

The Environmental Protection Agency and the National Highway Traffic Safety Administration (EPA and NHTSA, 2011) have published estimates of disruption costs as well as the monopsony effect. The estimates, based on Leiby (2008), recognize uncertainty about future oil market conditions and other parameters and are therefore specified as ranges that vary by year (Table 5.5.1). The range of total social costs per barrel is approximately $10 to $30, with the midpoint estimates lying close to $19 per barrel. If U.S. petroleum use decreases over time in accord with the reduction goals set for this study, the value of the monopsony benefit will also decrease. It is assumed that it will be halved by 2050.

---

cases differ a good deal, chiefly because the VISION model included both ICE and HEV CNGVs while the LAVE model was capable of including only ICE CNGVs. In both models BEVs are assumed to be used only 2/3 as much as other vehicles. The "missing miles" are allocated 60 percent to other existing vehicles, 30 percent to trips not taken (reduced VMT), and 10 percent to increased vehicle sales.

The vehicle and infrastructure subsidies estimated to be necessary to achieve the market shares assumed for the VISION model are very large (subsidies are shown in Figure 5.13 as negative values). The LAVE-Trans model was used to estimate the per-vehicle subsidies required to achieve the market shares for alternative technologies assumed in the VISION runs. No assumption was made about who would pay for the subsidies. For the CNGV and FCEV cases, it was assumed that 300 subsidized refueling stations would be deployed to support initial vehicle sales. Inferred subsidies for five runs using midrange technology assumptions are in the range of $35 billion to $45 billion annually by 2050

(values discounted to present value at 2.3 percent per year[8]). Cumulative subsidies run to hundreds of billions of dollars. Although per-vehicle subsidies are larger in the earlier years, fewer vehicles are being sold so that total subsidies are smaller. The VISION CNGV sales through 2030 are somewhat lower than the LAVE-Trans model would predict in the absence of subsidies, and so small taxes on CNGVs (positive values in Figure 5.13) are needed to match the VISION assumptions. For the most part, the very large subsidies are a consequence of assuming market shares for the 2030 to 2050 period that are substantially higher than the LAVE-Trans model estimates the market would sustain without continuing subsidies. The next section explores what might be possible with temporary subsidies that are sufficient to break down the transition barriers but can be quickly phased out once those barriers have been breached.

---

[8]OMB Circular No. A-94 specifies discount rate for projects up to 30 years, whereas the time-frame for this analysis is 40 years. The recommended rate for 20-year projects is 1.7 percent and for 30-year projects is 2.0 percent (OMB, 2012).

TABLE 5.5.1 Oil Security Premiums, Midpoint, and (Range) by Year (2009 $/barrel)

Year	Monopsony	Disruption Costs	Total
2020	$11.12 ($3.78–$21.21)	$7.10 ($3.40–$10.96)	$18.22 ($9.53–$29.06)
2025	$11.26 ($3.78–$21.48)	$7.77 $3.84–$12.32)	$19.03 ($9.93–$29.75)
2030	$10.91 ($3.74–$20.47)	$8.32 ($4.09–$13.34)	$19.23 ($10.51–$29.02)
2035	$10.11 ($3.51–$18.85)	$8.60 ($4.41–$13.62)	$18.71 ($10.30–$28.20

SOURCE: EPA and NHTSA (2011), Table 4-11.

The estimates in Table 5.5.1 do not include military costs (EPA, NHTSA, 2011, p. 4-32), yet access to stable and affordably priced energy has traditionally been considered a critical element of national security (e.g., McConnell, 2008, p. 41; Military Advisory Board, 2011, p. xi). Estimates of the national defense costs of oil dependence range from less than $5 billion per year (GAO, 2006; Parry and Darmstadter, 2004) to $50 billion per year or more (Moreland, 1985; Ravenal, 1991; Kaufmann and Steinbruner, 1991; Copoulos, 2003; Delucchi and Murphy, 2008). Assuming a range of $10 billion to $50 billion per year, and dividing by a projected consumption rate of approximately 6.4 billion barrels per year (EIA, 2012, Table 11) gives a range of average national defense cost per barrel of $1.50 to $8.00 per barrel (rounded to the nearest $0.50).

Adopting the EPA-NHTSA estimates indicates a range of about $9 to $30 per barrel, with a midpoint of $19. A reasonable range of national defense and foreign policy costs appears to be $1.50 to $8 per barrel, with a midpoint of about $5 per barrel. Adding these numbers produces a range of $10.50 to $38 per barrel with a midpoint of $24 in 2009$, or about $25 per barrel in current dollars. This is the value adopted by the committee to reflect the social value of reductions in petroleum usage.

---

[1]Since OPEC is not a competitive supplier, there is no world oil supply function in the usual sense. The response of world oil prices to a reduction in U.S. demand will therefore depend on how OPEC reacts. OPEC's options, however, are not unlimited. If OPEC does not reduce output, oil prices will fall. If OPEC does reduce output, it loses market share which diminishes its market power. Greene (2009) has shown that in terms of economic benefits to the U.S. there is very little difference between the two strategies.

5.4.2 Analysis of Transition Policy Cases with the LAVE-Trans Model

Given the committee's fuel and vehicle technology scenarios, the LAVE model was used to estimate what might be accomplished by policies that reflect the social value of reducing GHG emissions and petroleum use combined with additional but temporary policies to induce transitions to alternative vehicles or fuels or both. Policies that reflect the value of reducing GHG emissions and petroleum use are initiated in 2015 or 2017 and remain in effect through 2050.[9] The current subsidies for electric and fuel cell vehicles are assumed to end by 2020 and be replaced by the new policies. Policies to induce transitions to alternative vehicle and fuel combinations begin at various dates and are phased out once the alternative technologies achieve a sustainable market share. Their intended function is to overcome the barriers to a transition from the incumbent energy technology to an alternative. Transition policies consist of explicit or implicit vehicle and infrastructure subsidies. Implicit subsidies would result from policies such as California's Zero Emission Vehicle (ZEV) mandates that require manufacturers to sell ZEVs regardless of market demand and therefore to cross-subsidize ZEVs. Or, requiring fuel providers to provide refueling outlets for alternative fuels would induce cross-subsidies from petroleum fuels to low-carbon alternatives. Similarly, policies such as RFS2 to require certain amounts of biofuels are an example of an implicit subsidy for alternative fuels.

At present, there is both uncertainty and disagreement about the value of reducing petroleum consumption and the value of reducing greenhouse gas emissions. The committee's approach is to value these reductions according to society's willingness to pay, as reflected in the stringency of the reduction goals. For example, carbon emissions should be valued at a cost consistent with the cost of de-carbonizing the electric utility sector as discussed in Chapter 3 and described in greater detail in Box 5.4. For GHG mitigation, the commit-

---

[9]The feebate system reflecting the social value of reductions in CO_2 emissions and petroleum use begins in 2017 while all other fiscal policies, if used, begin in 2015.

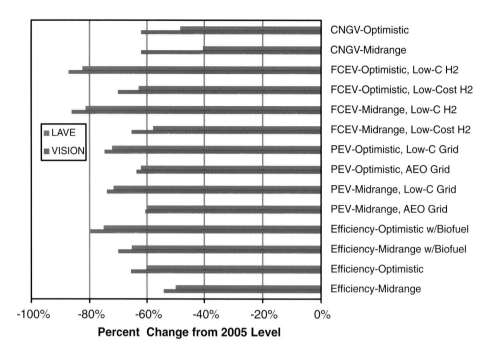

FIGURE 5.12 Comparison of LAVE-Trans and VISION model-estimated GHG reductions in 2050 given matching deployment.

tee elected to adopt the highest estimates of the Interagency Working Group on the Social Cost of Carbon (2010), and for petroleum reduction the committee derived its own estimates based on research by Leiby (2008) and others (see Box 5.5). These assumed values are shown in Figure 5.14.

Policies consistent with a strong commitment to reduce oil use and GHG emissions are included in all the policy cases. Specifically, a steady tightening of CAFE/GHG emissions standards combined with associated policies is assumed to ensure that they are met and enforced, which would yield efficiency improvements of both the midrange and optimistic vehicle technology scenarios, as explained in Chapter 2. Because the fuel economy and emissions standards will almost certainly be a binding constraint on manufacturers' technology and design decisions, they will induce manufacturers to price the different drive train technologies so as to reflect their contributions to meeting the standards. This is represented by an approximately revenue-neutral feebate system that precisely reflects the social value of reductions in petroleum use and GHG emissions (see Boxes 5.3, 5.4, and 5.5 on feebates and the values of GHG and oil reduction).

Policies such as the RFS2, Low Carbon Fuels standards, or equivalent will be needed to bring drop-in biofuels to market, and additional policies will be required to ensure that electricity or hydrogen is produced via methods with

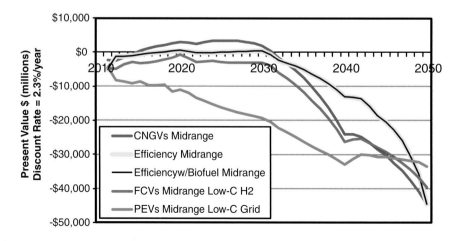

FIGURE 5.13 Annual subsidies to alternative fuels vehicles required to match five VISION cases. Negative values represent a net cost. The two efficiency curves are overlapping but not identical because the vehicle costs are the same and fuel costs nearly the same.

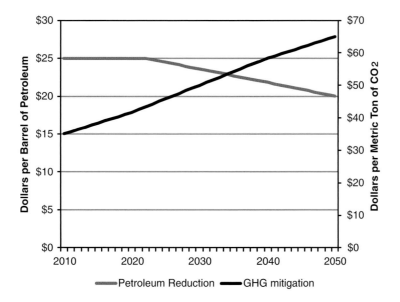

FIGURE 5.14 Assumed social values of reductions in GHG emissions and petroleum usage (in 2009$).

reduced greenhouse gas impacts, as explained in Chapter 3. These policies are implicit in all model runs except the BAU and Reference Cases. Their costs are reflected in the prices of the fuels for those cases assuming fuels produced with reduced GHG impacts (e.g., "+ Low-C Grid"). In addition, the very large improvements in energy efficiency included in all the policy runs will severely reduce Highway Trust Fund revenues unless measures are taken to prevent it. All policy cases assume that motor fuel taxes will be replaced by a user fee on energy (IHUF), indexed to the average energy efficiency of all vehicles on the road (see Chapter 6 for details).

Two additional fiscal policies were considered. A tax can be levied on fuels reflecting the social costs of their carbon emissions and petroleum content. When this tax is used, the feebates reflecting the social value of carbon emissions reductions are reduced. Since the vehicle choice model includes the first 3 years of fuel costs, the fuel taxes paid in those years will be taken into account by consumers in their vehicle purchase decisions. Thus, the feebate rates are adjusted to include only the social values of reductions in carbon emissions and oil use in the remaining years of the vehicle's life. The impact of the fuel tax is therefore on vehicle use rather than vehicle choice. The remaining fiscal policy is an additional feebate system that compensates for consumers' undervaluing of future fuel savings. This feebate reflects the discounted present value of fuel costs (excluding the social-cost fuel tax) from years 4 to 15, discounted to present value at 7 percent per year.[10]

_____
[10]OMB Circular No. A-94 recommends a discount rate for private return on capital of 7 percent (OMB, 2012).

5.4.2.1 Transition Policies

A transition to an alternative vehicle and fuel combination such as fuel cells and hydrogen or plug-in electric vehicles may be necessary to meet the reduction goals. This section focuses on such a transition away from the incumbent petroleum-based, ICEV-fuel system. As seen in the VISION results and again below, it may also be possible that the goals can be met without a transition to hydrogen- or electricity-powered vehicles. A shift away from petroleum fuel toward drop-in biofuels, combined with much more efficient ICEV and HEV engines, also offers an opportunity for significant greenhouse gas and petroleum reductions by 2050, although the 2030 petroleum reduction target remains difficult to achieve in all cases. With the data and model available, the committee is not able to fully explore the transition to large-scale low-carbon biofuels production here but does examine this case with the available information below.

In the LAVE model, transition polices consist of subsidies to vehicles and fuel infrastructure. These could be either direct government subsidies or subsidies induced by governmental regulations, such as California's ZEV standards. The function of these subsidies is to allow alternative technologies to break through the market barriers that "lock-in" the incumbent petroleum-ICEV vehicle-fuel system. The transition policies used in the policy cases have been constructed by following these rules:

1. Annual sales in the first 3 to 5 years of a transition should number in the thousands to tens of thousands of units.
2. The increase in sales in any year should not be more than 6 percent of total light-duty vehicle sales.

3. The growth of sales should avoid abrupt increases or decreases.
4. Subsidies should be phased out as sales approach the level the market will support without subsidies.

In reality, a transition policy would need to be more comprehensive. Transition policies could potentially offer a greater variety of incentives, such as access to high occupancy vehicle lanes, free parking in congested areas, and so on. In the LAVE model the vehicle and fuel subsidies are intended to measure the cost of inducing transitions rather than to describe the specific polices by which they should be accomplished.

5.4.2.2 Transition Costs and Benefits

The costs and benefits of each of the policy cases presented below are measured relative to a Base Case that includes identical assumptions about technological progress and all other factors but does not include new policies to induce a transition to alternative vehicles or fuels or both. This was done to better measure the incremental costs and benefits of accomplishing transitions to alternative vehicles and fuels, as distinguished from the obvious benefits of having better technology. In general, this means that if the midrange technology assumptions are used in a transition case, the transition case will be compared to a Base Case that also uses the midrange technology assumptions. If a transition case uses optimistic assumptions for some technologies and midrange assumptions for others, its Base Case will make identical assumptions about technological progress. The transition cases differ from their respective Base Case only in terms of the transition policies. Except for the BAU and Reference Cases, all Base Cases assume that fuel economy and emissions standards are continuously tightened through 2050.

Costs and benefits are measured[11] as changes from the respective Base Case in the following five quantities:

1. Costs of subsidies,
2. Additional fuel costs or savings,
3. Changes in consumers' surplus,
4. The social value of GHG reductions, and
5. The social value of reduced oil use.

- *Subsidy costs* include the implicit or explicit vehicle subsidies due to the higher costs of more efficient vehicles with lower greenhouse gas emissions, and they include the cost of subsidized infrastructure for public recharging of plug-in electric vehicles or refueling hydrogen or CNG vehicles.

- *Additional fuel costs or savings.* Since consumers are assumed to consider only the first 3 years of fuel savings in making their vehicles choices, it is necessary to account for the additional costs or savings over the remainder of each vehicle's lifetime. Additional fuel costs or savings are private costs or benefits that accrue to the vehicle user that (by assumption) are not capitalized by the vehicle purchaser at the time of purchase.

- *Consumers' surplus* is an economic concept that measures consumers' welfare in dollars. Two changes in consumers' surplus are measured: (1) satisfaction with vehicle purchases and (2) satisfaction with fuel purchases. The LAVE model includes a widely used method of modeling consumer choice that recognizes that not all consumers have the same tastes or preferences. Some may prefer the attributes of electric drive while others prefer internal combustion engines. If electric-drive vehicles become available at competitive prices as a result of successful transition policies, the satisfaction of those with a preference for electric drive will increase. Those who prefer ICEVs will still have that option and so will be no better or worse off than before the plug-in vehicles became available. Consumers' surplus measures that increased value in dollars. Vehicle subsidies increase consumers' surplus but by less than the gross amount of the subsidies. This results in a net economic cost, at least in the early years of a transition. Taxing the energy consumers must purchase to operate their vehicles creates a loss of consumers' surplus, in addition to a transfer of wealth from consumers to the taxing entity. The surplus loss over and above what is counted in the vehicle purchase decision is also measured when changes in tax policies are considered.

- *The social value of reducing greenhouse gas emissions and oil use.* These values are measured by multiplying the changes in estimated annual quantities times the social cost of emissions per unit assumed by the committee consistent with the goals of the study (see Boxes 5.4 and 5.5). Hydrogen and fuel cell vehicles will also have zero tailpipe emissions of other pollutants, and may have lower full fuel cycle emissions, as well. The committee has not attempted to estimate those potential benefits, and they are not included in the cost and benefit estimates.

The net present value (NPV) of a policy case is the sum of all costs and benefits from 2010 to 2050, plus the fuel, GHG, and petroleum costs and benefits of vehicles sold through 2050 that will still be in use beyond that date. From an economic perspective, an optimal policy strategy would be one that maximized NPV. NPV depends strongly on the discount rate assumed, and there may be widely differing opinions about the appropriate discount rate. A 2.3 percent rate for

---

[11]All costs and benefits are measured in constant dollars, discounted to present value using an annual discount rate of 2.3 percent.

all years is used, which is consistent with the most recent guidance of the U.S. government (OMB, 2012); however, the appropriate discount rate is yet another source of uncertainty.

Sections 5.4.3 to 5.4.9 present results from transition policy cases and compare them to their respective Base Cases. In general, all cases (except BAU and Reference) assume fuel economy/emissions standards to 2050. All cases (except BAU and Reference) include feebates and the IHUF. All transition cases assume vehicle subsidies or mandates and infrastructure subsidies or mandates. A few of the cases add special policies as noted in the text.

Rather than enabling us to reach definitive conclusions, the committee's modeling suggests the extent of technological progress and the kinds and stringency of policy measures that are likely to be needed to bring about transitions. It provides useful insights about the interactions between policy, the market, and technological changes. It also provides a general indication of the costs and benefits of achieving the GHG and petroleum reduction goals conditional on the many assumptions that must be made. Uncertainty will be an inherent property of the process of energy transition: uncertainty about technological change, uncertainty about the market's response to technologies and policies, and uncertainty about the future state of the world. The extent of uncertainty about both future technologies and the market's response to them is illustrated by means of sensitivity analysis in Sections 5.6 and 5.7 below.

5.4.3 Energy Efficiency Improvement and Advanced Biofuels

The cases described in this section explore what may be possible given the midrange and optimistic technology pro-

jections, continued tightening of fuel economy and emissions standards, and large-scale production of thermo-chemically produced "drop-in" biofuels. These cases maintain the ICEVs with improved technology but involve a transition to large scale production and use of cellulosic biofuels. A final case also includes adoption of all the pricing policies described above. All cases include the IHUF, which increases from $0.42/gge in 2010 to $1.27/gge in 2050, and feebates that reflect the assumed societal willingness to pay for reductions in GHG emissions and oil use.

The technological progress enabling increased energy efficiency described in Chapter 2 (Table 2.11) will be devoted to improving vehicle fuel economy only if strong policies, such as increasingly stringent fuel economy and emissions standards, are put in place beyond 2025. The approximately revenue-neutral feebates, which are phased over 5 years beginning in 2017, amount to a tax of $60 per ICEV in 2021, with rebates of $770 per HEV, $1,650 per PHEV, $2,900 for each BEV, and $2,575 per FCEV. The feebates change over time as energy efficiencies, fuel properties, and the social willingness to pay for GHG and oil use reductions change. Assuming such standards are implemented, the midrange estimates of efficiency improvements and their costs result in estimated reductions in GHG emissions of 29 percent by 2030 and 52 percent by 2050 (Figure 5.15). For the same dates, petroleum consumption is estimated to be reduced by 33 and 64 percent, respectively. The reductions are due in part to the continued reduction in rates of fuel consumption for both ICEVs and HEVs (Figure 5.16) and by a steady shift from ICEVs to HEVs and BEVs (Figure 5.17).

If technology progresses as envisioned in the midrange scenario, the economic benefits of the efficiency improvements versus the Business as Usual case could be very large.

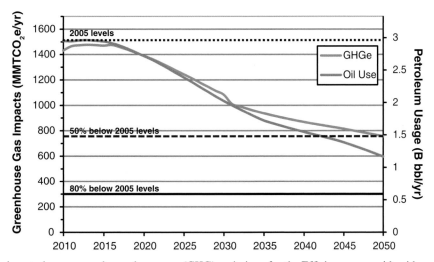

FIGURE 5.15 Changes in petroleum use and greenhouse gas (GHG) emissions for the Efficiency case with midrange technology estimates as compared to 2005 levels.

The key components of economic costs and benefits are shown in Figure 5.18 as annual costs, discounted to present value at the rate of 2.3 percent per year. The sum of the individual components grows to an estimated $130 billion per year by 2050. The largest component is "uncounted fuel savings," the future fuel savings not considered by consumers at the time of purchasing a new car but realized later over the life of the vehicle. Consumers' surplus, their net satisfaction with their vehicle purchases, decreases slightly after 2030 due to the increased cost of ICEVs over time. The net present social value of the transition to much higher efficiency vehicles is estimated to be on the order of $3.5 trillion.

Increasing the quantity of thermochemically produced, drop-in biofuels from 13.5 billion gge to 19.2 billion gge in 2030 increases the estimated reduction in petroleum use from 33 to 37 percent in that year. The 2030 reduction in GHG emissions is 32 percent versus 28 percent. In 2050, when the biofuels industry has expanded to produce 45 billion gge, the estimated impact is much greater: petroleum use is down 86 percent (compared with 64 percent) and GHG emissions are 66 percent lower than in 2005 (compared with 52 percent without advanced biofuels) (Figure 5.19).

If carbon emissions from the production of 20 percent of thermochemical biofuels were captured and stored, an estimated 78 percent reduction in GHG emissions versus

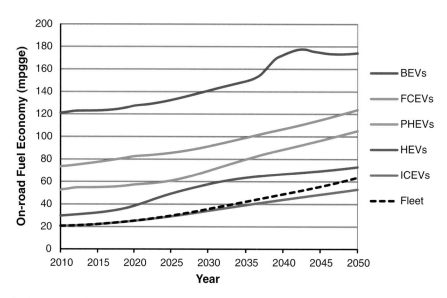

FIGURE 5.16 Average fuel economy of on-road vehicles for the Efficiency case with midrange technology estimates. The upturn in battery electric vehicle (BEV) fuel economy after 2040 reflects the rapidly increasing share of new BEVs on the road (and thus a larger fraction of the BEV fleet is the newest, most efficient BEVs). The downturn that follows is representative of an increasing number of battery electric trucks in the fleet.

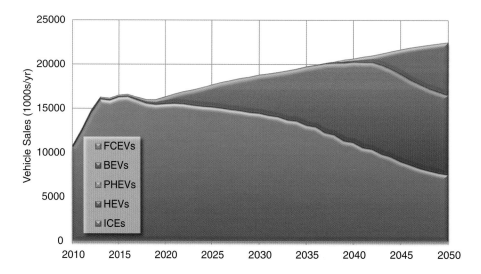

FIGURE 5.17 Vehicles sales by vehicle technology for the Efficiency case with midrange technology estimates.

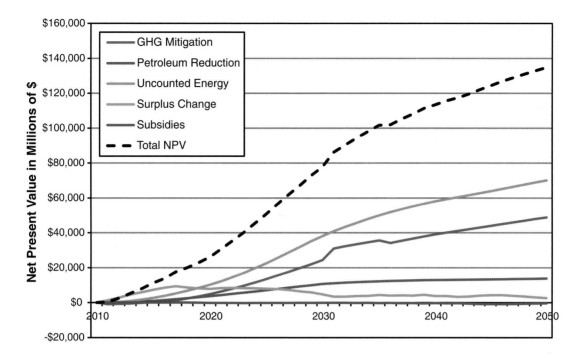

FIGURE 5.18 Estimated costs and benefits for the Efficiency case with midrange technology estimates.

2005 could be achieved by 2050. Given the uncertainty in the analysis, the 2050 goals would then be met for all practical purposes. The 2030 goal of a 50 percent reduction in petroleum use is still missed because of the low initial ramp-up in production, however; the estimated reduction is 37 percent. The cost of 20 percent CO_2 removal for biofuels blended into gasoline adds about $0.20 per gallon to the average price of gasoline in 2050. If CCS is applied to all biofuels, then the net GHG emissions from the LDV fleet could be slightly negative.

5.4.4 Emphasis on Pricing Policies

A great deal can be accomplished by means of policies that change the prices of fuels and vehicles and harness market forces to reduce GHG emissions. This scenario, like the others based on the midrange technology scenario, assumes that fuel economy standards are inducing manufacturers to produce increasingly efficient vehicles. However, it also introduces stronger feebates and adds to the cost of fuels the social willingness to pay for GHG and oil reduction. The additional feebate system capitalizes in vehicles' prices the uncounted energy savings due to consumers' assumed under-

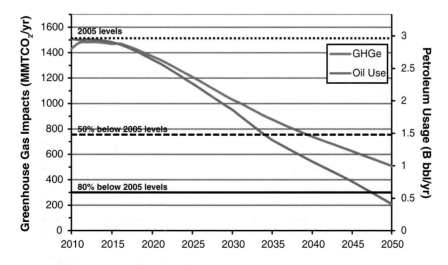

FIGURE 5.19 Changes in petroleum use and greenhouse gas (GHG) emissions for the Efficiency + Biofuels case as compared to 2005 levels.

valuing of future fuel savings.[12] Production of electricity and hydrogen via processes with low-GHG impacts is assumed, but not intensive use of drop-in biofuels.

The fully taxed price of gasoline increases from $2.70 per gallon in 2010 to $4.90 per gallon in 2020 and $5.50 per gallon in 2030. Gasoline prices continue to increase, reaching $6.60 per gallon in 2050, as a result of $2.70 per gallon in combined taxes. The price of electricity in 2050 is roughly equal to that of gasoline on an energy basis, but BEVs are more than three times more energy efficient than comparable ICEVs in 2050. Feebates also strongly encourage purchases of BEVs. The rebate for a BEV in 2020 is almost $14,000, while ICEVs are taxed at $300 each. The difference decreases as vehicles and fuels improve so that by 2050, BEVs receive a $1,300 per vehicle rebate, whereas ICEVs are taxed at $2,500 per vehicle (the incidence also shifts to approximate revenue neutrality).

The result is a massive shift to battery electric and hybrid electric vehicles. By 2050, an estimated 59 percent of new-vehicle sales are BEVs and 33 percent are HEVs. In 2050 almost 40 percent of the vehicles on the road are BEVs. In the absence of policies to put a hydrogen refueling infrastructure in place, fuel cell vehicles never achieve any significant share of the market. Battery electric vehicles are far less dependent on early infrastructure development, which gives them a decisive advantage over FCEVs in this scenario.

Light-duty vehicle petroleum use is estimated to be 38 percent lower than the 2005 level by 2030, and 87 percent below 2005 in 2050. Greenhouse gas emissions are reduced 74 percent by 2050. Vehicle miles of travel in 2050 are also more than 15 percent lower than in the Efficiency Case (identical assumptions but without the additional pricing policies). In part this is due to the higher energy prices, but it is also due to 7 percent fewer vehicles on the road and lower annual miles for the 39 percent of vehicles that are BEVs.

5.4.5 Plug-in Electric Vehicles

Plug-in electric vehicles (PEVs) possess some attributes that are substantially different from those of the other vehicle types. Battery electric vehicles (BEVs) not only have limited range but also have long recharging times. The combination of these two attributes limits the ability of BEVs to satisfy all the daily travel demands of most drivers. This reduces the total annual mileage of BEVs to two thirds of that of an ICEV, HEV, CNGV, or FCEV and detracts from their utility to most households. In the LAVE-Trans model, most but not all of the vehicle travel demand that cannot be satisfied by BEVs is shifted to other vehicles in the vehicle stock. To some degree the BEV's travel range limitations will be offset by its lower energy costs. In the midrange technology

scenario, the long-run, fully learned cost of BEVs is $20,000 more than that of ICEVs in 2010, although BEVs eventually become $600 less expensive by 2050 (see Figure 5.8). PHEVs, on the other hand, suffer no such limitations on use and can take energy from the grid or from the gas pump. However, their initial cost is higher and remains higher through 2050 in the midrange scenario. PHEVs start out with a high-volume, learned cost that is $10,000 more than that of an ICEV and remains at least $2,000 more expensive through 2050. The PHEV's higher price will be partly offset by lower energy costs, yet its price remains a significant barrier to full market success.

Two PEV transition policy scenarios are reported below. Both include feebates reflecting social willingness to pay for GHG and petroleum reduction plus the IHUF. In the first, PHEVs achieve a modest market share of 5 percent whereas BEVs account for 35 percent of new-car sales by 2050. The scenario continues the current levels of PHEV and BEV sales, which requires substantial, sustained subsidies: total subsidies per BEV decrease from $25,000 per vehicle in 2012 to just over $10,000 per vehicle in 2020.[13] When long-run PEV costs approach the prices of other technologies the transitional subsidies are removed (2028 for BEVs and 2033 for PHEVs) but the feebates and IHUF continue. By 2050, PEVs constitute 40 percent of the market, HEVs 34 percent, and advanced ICEVs 26 percent (Figure 5.20). In this case, petroleum use is 35 percent lower than the 2005 level in 2030 and 73 percent lower in 2050. GHG emissions are 31 and 63 percent below the 2005 level in 2030 and 2050, respectively. If the AEO 2011 reference grid assumptions are used, the GHG reductions are 31 percent in 2030 and 56 percent in 2050.

Despite the cost of vehicle subsidies (over $50 billion present value) this scenario still has a substantial positive net present value of over $500 billion. Most of the benefits (about 50 percent) are due to uncounted energy savings from PEVs, which have substantially lower energy costs than ICEVs or HEVs (Figure 5.21).

Adding greater volumes of advanced biofuels (45 billion gge in 2050) to the PEV Transition Policy case reduces petroleum use relative to 2005 by 40 percent in 2030 and 94 percent in 2050. The GHG reductions in those years are estimated to be 34 and 75 percent assuming electricity generated by a low-carbon grid.

As described above, the initial costs for PEVs are substantially higher than for other technologies, primarily due to battery costs. If subsidies are not applied until battery costs have come down significantly, there is still opportunity for significant benefits. If the current advanced vehicle tax credits are allowed to expire in 2020, it is possible to induce a transition

---

[12]It is likely that the feebates alone would induce manufacturers to realize fuel economy and emissions improvements similar to those assumed to result from standards, but that possibility has not been tested here (Greene et al., 2005).

[13]These estimated total subsidies may appear too high given a federal tax credit of only $7,500 for BEVs and $5,000 for PHEV-30s. However, states also offer incentives of up to $7,500, and manufacturers are very likely also subsidizing initial sales, partly to induce market success and partly to gain credits under the CAFE regulations and ZEV mandates.

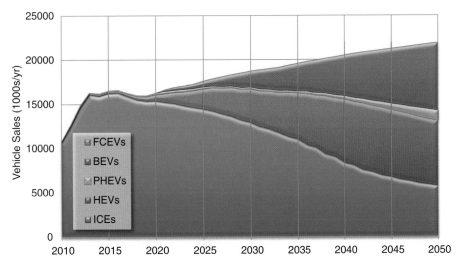

FIGURE 5.20 Vehicle sales by vehicle technology assuming midrange technologies and plug-in electric vehicle subsidies and additional incentives.

to PEVs by 2050 while waiting to apply technology-specific subsidies to PEVs until 2023. These subsidies are complemented by the usual IHUF and feebates. In this case, the total subsidy to BEVs is $13,000 per passenger car in 2023. However, it is reduced to $6,000 per vehicle by 2028, and by 2034 only the feebate remains. A similar subsidy trajectory is followed for PHEVs but is delayed by 6 years, beginning instead in 2029 after vehicle costs have been further reduced. By 2050, BEVs make up 35 percent of new-vehicle sales, while PHEVs are 6 percent, both shares similar to the cases

above. Likewise, petroleum usage in 2030 is reduced by 34 percent, and petroleum usage and GHG emissions in 2050 are reduced by 73 and 63 percent, respectively, compared to 2005 levels; these are almost identical to the PEV transition case without biofuels but with the low-carbon grid, discussed above. The net present value is nearly identical ($520 billion compared to $540 billion), and the total cost of the subsidies necessary to produce the transition are essentially the same, as well, $50 billion.

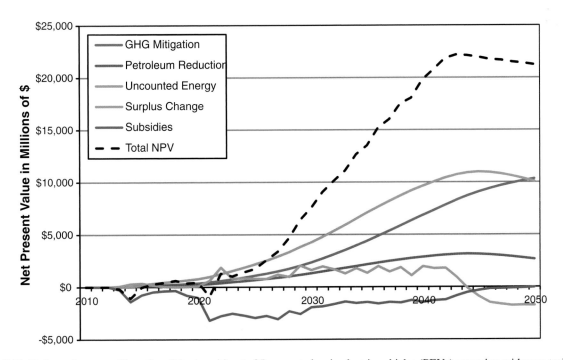

FIGURE 5.21 Estimated costs and benefits of the transition to 25 percent plug-in electric vehicles (PEVs) assuming midrange technologies and PEV subsidies and additional incentives.

5.4.6 Hydrogen Fuel Cell Electric Vehicle Cases

Given the midrange technology assumptions, if no early hydrogen infrastructure is provided and the existing tax credits are allowed to expire in 2020, a transition to FCEVs does not occur. An early transition to hydrogen fuel cell vehicles can be induced by ensuring that an adequate amount of hydrogen refueling infrastructure to support early vehicle sales is in place at least in some regions ahead of vehicle sales and that vehicle subsidies or mandates support early sales. All the FCEV transition cases include feebates reflecting social willingness to pay for GHG and oil reduction and the IHUF. The three carbon-intensity cases described in Chapter 3 were tested, beginning with low-cost hydrogen produced mainly by steam methane reforming without carbon capture and storage.

The first FCEV transition case assumes that 200 subsidized or mandated hydrogen refueling stations are put in place in 2014, 200 in 2015, and 100 more in 2016. These stations are likely to be geographically clustered, for example, in California and other states where ZEV requirements and other supporting policies are in place. Increased hydrogen vehicle subsidies (or mandates inducing implicit subsidies) begin in 2015 at $17,500 per vehicle (including the existing tax credit). The initial, high subsidies decline gradually to $16,000 per vehicle in 2020 and $6,000 by 2025. This induces modest levels of FCEV sales: 9,000 in 2015, followed by annual sales of 16,000, 21,000, and 26,000 in 2016-2018 (Figure 5.22). The transitional vehicle subsidy is ended in 2027, but the feebate system that in 2027 provides a $1,400 rebate for FCEVs and imposes a $500 tax on ICEVs remains in effect. By 2050, almost half of the vehicles on the road are FCEVs or HEVs.

In the low-cost hydrogen case, petroleum consumption is estimated to be 41 percent below the 2005 level in 2030. In 2050 petroleum consumption is down an estimated 90 percent relative to 2005 and GHG emissions are 59 percent lower. Assuming CCS is used in the production of hydrogen, greenhouse gas emissions are estimated to be 74 percent lower in 2050 and petroleum use is 95 percent below the 2005 level (Figure 5.23). Using assumptions to produce hydrogen with the lowest GHG impacts, 2050 GHG emissions are estimated to be 80 percent lower than in 2005, and petroleum use 96 percent below the 2005 level. Petroleum use in 2030 is estimated to be 42 percent below 2005 in this case. With the feebates in place, the higher-cost but lower-GHG-impact hydrogen increases FCEV sales: in the low-carbon production case FCEVs take an estimated 57 percent of the market in 2050, and in the low-cost production hydrogen case they capture 48 percent.

Despite the initial cost of subsidies that reach $6 billion per year in the mid-2020s, the estimated net present value of the policy-induced transition to hydrogen FCEVs is on the order of $1 trillion (Figure 5.24). The benefits are roughly equally composed of social benefits (GHG and petroleum reduction) and private benefits (fuel savings and consumers' surplus gains).

Adding advanced biofuels to the FCEV policy case reduces the 2050 market share of FCEVs from 57 to 46 percent. Low-GHG gasoline reduces the cost penalty that feebates levy on ICEVs; vehicles consuming drop-in biofuel instead of petroleum gasoline become more cost-competitive with FCEVs and gain market share. This is an illustration of how policies may interact in ways that make the combined impact smaller than the sum of the individual effects. Still, total petroleum use and GHG emissions are lower. In 2030, petroleum use is estimated to be 46 percent below the 2005 level and GHG emissions are 37 percent below (versus 42 and 35 percent, respectively, without advanced biofuels). In 2050, adding advanced biofuels reduces GHG emissions to 86 percent below the 2005 level and petroleum use to 100 percent below.

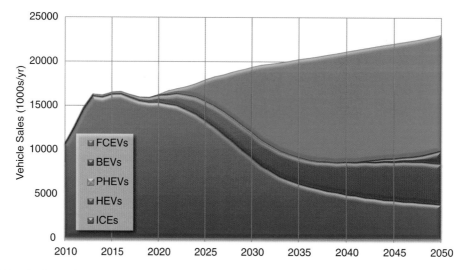

FIGURE 5.22 Vehicle sales by vehicle technology with midrange technology assumptions and low-carbon production of hydrogen, fuel cell vehicle subsidies, and additional incentives.

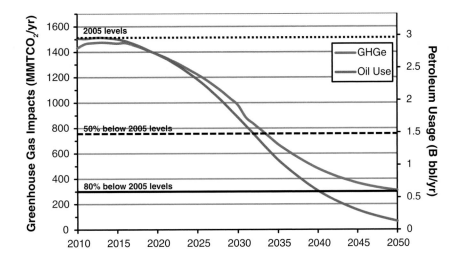

FIGURE 5.23 Changes in petroleum use and greenhouse gas (GHG) emissions with midrange technology assumptions, fuel cell vehicle subsidies and additional incentives, and a low-GHG infrastructure for the production of hydrogen.

5.4.7 Compressed Natural Gas Vehicles

Due to limitations of the LAVE model, CNGVs take the place of FCEVs; FCEVs are excluded from analyses in which CNGVs are included. Like FCEVs, only one type of CNGV is considered, CNG non-hybrid ICEVs. CNGVs have some advantages relative to other advanced technologies. Natural gas prices are lower than petroleum, biofuel, or hydrogen prices, and the infrastructure for natural gas production and distribution is nearly ubiquitous. This means that, unlike hydrogen, there is no initial phase of high prices at low volumes. Natural gas refueling stations are still required,

however, and natural gas vehicles have lower range than gasoline vehicles due to the lower energy density of CNG.

The CNG policy case includes feebates and the IHUF, both commencing in 2015. Also in that year there is a transitional subsidy/mandate of $10,000 per CNGV. Like the vehicle subsidies in other cases, this could be borne by the manufacturer or government or shared between the two. The transitional subsidy is reduced each year and ended by 2025. In addition, 100 subsidized refueling stations are opened in 2014, 200 in 2015, and 100 more in 2016. CNGV sales peak at 49 percent from 2031-2034, then decline to 33 percent in

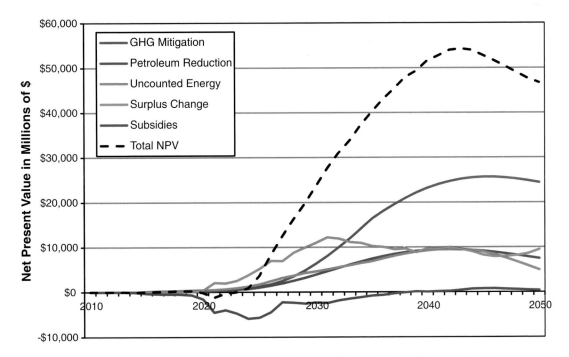

FIGURE 5.24 Present value cost and benefits of a transition to hydrogen fuel cell vehicles using midrange technology assumptions, fuel cell vehicle subsidies and additional incentives, and a low-GHG infrastructure for the production of hydrogen.

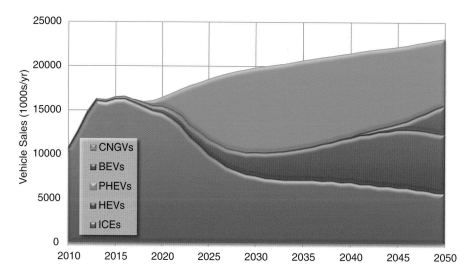

FIGURE 5.25 Vehicle sales by vehicle technology for midrange technology estimates and policies promoting compressed natural gas vehicles.

2050 (Figure 5.25), chiefly due to the feebates which favor BEVs and even HEVs and ICEVs over CNGVs.

In the CNG transition policy case, petroleum consumption is estimated to be 52 percent below the 2005 level in 2030. GHG emissions are 28 percent lower. In 2050, estimated petroleum use and GHG emissions are, respectively, 86 and 47 percent lower than 2005 levels (Figure 5.26). Adding advanced biofuels to the CNG transition case eliminates petroleum use in 2050 and reduces GHG emissions to an estimated 62 percent below the 2005 level. If it is assumed that some CNGVs will be hybrid vehicles, the model would suggest no more than a few additional percent reductions in GHG emissions because these CNG HEVs would not be further displacing gasoline-powered vehicles but rather less efficient CNG ICEVs. All greenhouse gas emissions for natural gas vehicles are strongly predicated on the methane

leakage rates outlined in Chapter 3 due to methane's large global warming potential.

5.4.8 Plug-in Electric Vehicles and Hydrogen Fuel Cell Electric Vehicles

Combining subsidies to PEVs and FCEVs with advanced biofuels and also including the usual feebates and IHUF on energy eliminates petroleum use in 2050 and reduces GHG emissions by 88 percent versus 2005 levels. In 2030, a 56 percent reduction in petroleum use is achieved. The implied subsidies required to achieve this result are substantial. In 2015, a BEV gets a total subsidy of $20,500; a FCEV, $27,500; and a PHEV, $13,000. The implied subsidies decrease to about $3,000 per BEV and FCEV and half of that for PHEVs in 2025, including feebates. After

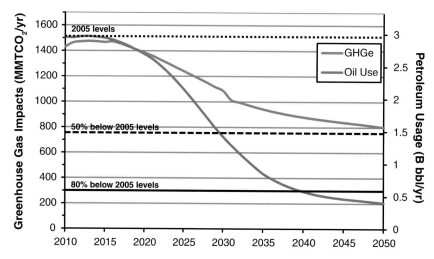

FIGURE 5.26 Changes in petroleum use and greenhouse gas emissions for midrange technology estimates with policies promoting compressed natural gas vehicles.

2030 the transitional subsidies are ended but the feebates remain. The total NPV of subsidies is approximately $140 billion. Although both technologies attain sustainable market shares, they compete with one another as well as with ICEVs and HEVs, which reduces their combined impact (Figure 5.27). The presence of several competing technologies in the marketplace tends to limit diversity of choice (fewer makes and models for any given technology) and to a lesser extent reduces fuel availability (due to fewer vehicles of any one type on the road), in comparison to a case dominated by one or two technologies. Nonetheless, this case achieves a NPV gain of $1.7 trillion versus the same technology assumptions without policy interventions to induce transitions. In this case, all of the liquid fuel used by vehicles with internal combustion engines is biofuel. In addition, the grid is low-carbon, as is hydrogen production. As a result, by 2050, there is almost no difference in the social costs (GHGs and petroleum use) of the different powertrain technologies. No vehicle receives a fee or a rebate that exceeds $50.

5.4.9 Optimistic Technology Scenarios

Optimistic technology scenarios imply breakthrough advancement of a given technology. These are taken to represent roughly a 20 percent likelihood occurrence in technological development for the respective technology. Such advancement is less likely than the midrange assumptions, although if it occurs, it changes the landscape for adoption of a technology, both in its costs and its benefits. In brief, the optimistic technology cases show that better-than-anticipated progress for plug-in vehicle technology combined with a decarbonized grid and assuming the same policies spelled out for the midrange cases above could come close to achieving the GHG and petroleum reduction goals by 2050 but fall short of the 2030 petroleum use goal. A parallel case for fuel

cell vehicles could achieve or exceed all of the goals. A full explanation of the optimistic cases is contained in Section H.4 in Appendix H.

5.4.10 Summary of Policy Modeling Results

The results of all the cases are summarized and compared in Tables 5.3 and 5.4 and in Figures 5.28 and 5.29. The fuel infrastructure investment costs modeled to achieve each scenario can be found in Section H.3 in Appendix H. The cases in Tables 5.3 and 5.4 are grouped in the same order as the case descriptions above. Abbreviations used in the table are explained in Table 5.5. For each group, there is a Base Case using the identical vehicle technology assumptions but without energy transition policies. All cases, including the Base Cases, assume policies requiring continued improvements in vehicle energy efficiencies and, therefore, all cases also include feebates consistent with the fuel economy and emissions standards. The inherent uncertainties in the model estimates should be kept in mind.

Only three cases are estimated to meet or exceed the 50 percent petroleum reduction goal in 2030 and the 80 percent petroleum and GHG reduction goals in 2050. One is based on optimistic assumptions for FCEV technology; the other two require both plug-in and hydrogen fuel cell market success, plus the low-carbon production of electricity and hydrogen, and the supporting policies of fuel economy/emissions standards and the IHUF on energy.

Two additional cases meet the 2030 petroleum goal but miss the 2050 GHG goal by wide margins. Both cases are based on a substantial transition to CNGVs, and such a transition may result in even more greenhouse gas emissions than modeled due to uncertainty in methane leakage rate and methane's substantial greenhouse warming potential. Eight cases are in the range of 40 to 50 percent petroleum reduction in 2030. Given the uncertainty inherent in the

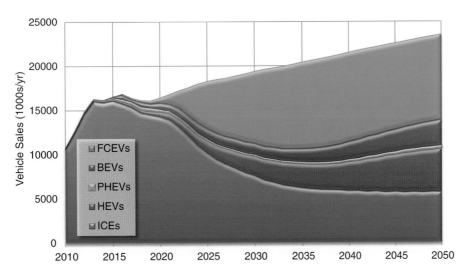

FIGURE 5.27 Vehicle sales by vehicle technology for midrange technologies and policies promoting the adoption and use of plug-in electric vehicles, hydrogen fuel cell electric vehicles, and biofuels.

TABLE 5.3 Summary of Estimated Petroleum and GHG Reductions in the Policy Cases

Scenario[a]	Oil Reduction (% reduction below 2005 level)		GHG Reduction (% reduction below 2005 level)		Oil Consumption (billion bbl/yr) 2005 = 2.96		GHG Emissions (MMTCO$_2$e/yr) 2005 = 1514	
	2030	2050	2030	2050	2030	2050	2030	2050
BAU	−5%	4%	−3%	9%	2.8	3.1	1,467	1,645
Reference	−29%	−27%	−26%	−22%	2.1	2.2	1,118	1,184
Eff+FBSC	−32%	−61%	−29%	−50%	2.0	1.2	1,082	755
Eff+FBSC+IHUF	−33%	−64%	−29%	−52%	2.0	1.1	1,071	721
Eff+Bio+FBSC+IHUF	−37%	−86%	−32%	−66%	1.9	0.4	1,030	508
Eff+Bio w/CCS+FBSC+IHUF	−37%	−86%	−35%	−78%	1.9	0.4	979	335
Eff+Intensive Pricing+LCe	−38%	−87%	−35%	−74%	1.8	0.4	990	389
PEV+ FBSC+IHUF+Trans+AEOe	−35%	−71%	−31%	−56%	1.9	0.9	1,049	662
PEV+ FBSC+IHUF+Trans+LCe	−35%	−73%	−31%	−63%	1.9	0.8	1,046	567
PEV(later)+FBSC+IHUF+Trans+LCe	−34%	−73%	−30%	−63%	2.0	0.8	1,055	563
PEV+Bio+FBSC+IHUF+Trans+LCe	−40%	−94%	−34%	−75%	1.8	0.2	1,005	381
FCEV+FBSC+IHUF+Trans+LH_2	−41%	−91%	−32%	−59%	1.7	0.3	1,025	621
FCEV+FBSC+IHUF+Trans+H$_2$CCS	−42%	−95%	−34%	−74%	1.7	0.1	993	391
FCEV+FBSC+IHUF+Trans+LCH$_2$	−42%	−96%	−35%	−80%	1.7	0.1	982	310
FCEV+Bio+FBSC+IHUF+Trans+LCH$_2$	−46%	−100%	−37%	−86%	1.6	0.0	949	209
CNGV+FBSC	−32%	−61%	−29%	−50%	2.0	1.2	1,082	755
CNGV+FBSC+IHUF+Trans	−52%	−86%	−28%	−47%	1.4	0.4	1,086	801
CNGV+Bio+FBSC+IHUF+Trans	−56%	−100%	−31%	−62%	1.3	0.0	1,045	568
Eff (Opt)+FBSC	−38%	−68%	−34%	−59%	1.8	0.9	1,000	620
Eff (Opt)+Bio+FBSC+IHUF	−43%	−95%	−37%	−76%	1.7	0.2	947	367
PEV (Opt)+FBSC+AEOe	−32%	−78%	−29%	−60%	2.0	0.7	1,082	607
PEV (Opt)+FBSC+IHUF+Trans+LCe	−35%	−89%	−31%	−76%	1.9	0.3	1,048	368
FCEV (Opt)+FBSC+LH_2	−32%	−61%	−29%	−50%	2.0	1.2	1,082	755
FCEV (Opt)+FBSC+IHUF+Trans+LCH$_2$	−50%	−100%	−41%	−90%	1.5	0.0	888	150
PEV+FCEV+FBSC+IHUF+Trans+LCe+LCH$_2$	−52%	−99%	−42%	−82%	1.4	0.0	872	267
PEV+FCEV+Bio+FBSC+IHUF+Trans+LCe+LCH$_2$	−56%	−100%	−45%	−87%	1.3	0.0	839	190

[a]Base Cases are indicated in boldface. Eff+FBSC serves as a Base Case for the four groups below it: Eff, Intensive Pricing, PEV, and FCEV, as well as for the mixed cases in the final grouping including both PEVs and FCEVs. See Table 5.5 for explanation of scenario components.

modeling analysis, that may be close enough. All of these cases meet the 2050 petroleum reduction goal. Six of the eight cases achieving an estimated 40 percent or greater reduction in petroleum use by 2030 also achieve a 70 percent or greater reduction in GHG emissions by 2050. Three are based on hydrogen fuel cell market success; one combines plug-in vehicle market success with biofuels. One relies on efficiency plus greater use of pricing policies, but this also induces a massive shift to plug-in vehicles by 2050. The final case combines optimistic efficiency improvements with biofuels to achieve an estimated 76 percent reduction in GHG emissions in 2050.

All five cases that meet the 2050 GHG reduction goal also imply near elimination of petroleum use. This is likely to be more difficult than the modeling analysis makes it appear. Near elimination of U.S. petroleum use, if it is also accomplished by other petroleum using countries, would cause world petroleum prices to fall. Falling petroleum prices have not been included in the modeling analysis but would make it more difficult for alternative technologies to succeed. This

effect could be countered by a policy setting a price floor on petroleum, as discussed in Chapter 6.

As uncertain as these estimates are, they provide several important insights. First, reaching the 2030 and 2050 goals will be difficult. It will require strong and sustained policies to continuously improve the energy efficiency of LDVs and to de-carbonize the systems supplying energy for the vehicles, and very likely it will also require strong policies to induce a transition to one or more of the advanced power-train technologies. Second, continued improvement in vehicle and fuel technologies is essential. Although the committee considers the technological progress assumed in the committee's scenarios to be reasonably likely, it is not guaranteed. Given that several technological advances are necessary to come close to meeting the goals, research and development of all the technologies considered in this report is a high priority.

If the alternative technologies develop and are deployed according to the committee's technological and market assumptions, the scenario modeling indicates that the additional costs of any transition may be much smaller

TABLE 5.4 Total Net Present Value in 2050 for Various Cases

Scenario[a]	Net Present Value (billions $)					
	Surplus Change	Subsidies	GHG Mitigation	Petroleum Reduction	Uncounted Energy	Total NPV
BAU	0.0	0.0	0.0	0.0	0.0	0.0
Reference	0.0	0.0	0.0	0.0	0.0	0.0
Eff+FBSC	0.0	0.0	0.0	0.0	0.0	0.0
Eff+FBSC+IHUF	−7.0	−0.8	44.6	16.6	78.2	131.7
Eff+Bio+FBSC+IHUF	−20.0	4.8	285.6	151.0	49.0	470.3
Eff+Bio w/CCS+FBSC+IHUF	−40.7	6.5	507.8	150.5	48.0	672.2
Eff+Intensive Pricing+LCe	−361.3	−128.0	520.1	253.4	1103.4	1387.5
PEV+ FBSC+IHUF+Trans+AEOe	43.0	−53.3	136.3	72.6	285.5	484.2
PEV+ FBSC+IHUF+Trans+LCe	23.8	−52.3	218.9	76.6	273.8	540.7
PEV(later)+FBSC+IHUF+Trans+LCe	29.2	−50.6	212.6	72.3	259.6	523.2
PEV+Bio+FBSC+IHUF+Trans+LCe	24.7	−45.7	437.1	205.6	226.8	848.6
FCEV+FBSC+IHUF+Trans+LH_2$	307.0	−38.6	210.3	200.4	287.0	965.9
FCEV+FBSC+IHUF+Trans+H$_2$CCS	279.7	−44.0	493.0	222.6	289.0	1240.3
FCEV+FBSC+IHUF+Trans+LCH$_2$	252.9	−45.7	591.2	225.3	198.0	1221.7
FCEV+Bio+FBSC+IHUF+Trans+LCH$_2$	276.2	−38.5	725.1	275.4	56.9	1295.1
CNGV+FBSC	0.0	0.0	0.0	0.0	0.0	0.0
CNGV+FBSC+IHUF+Trans	414.3	−32.7	−40.1	248.1	251.1	840.7
CNGV+Bio+FBSC+IHUF+Trans	396.0	−31.1	222.4	346.9	210.6	1144.7
Eff (Opt)+FBSC	0.0	0.0	0.0	0.0	0.0	0.0
Eff (Opt)+Bio+FBSC+IHUF	−20.1	2.5	291.7	153.5	60.2	487.8
PEV (Opt)+FBSC+AEOe	0.0	0.0	0.0	0.0	0.0	0.0
PEV (Opt)+FBSC+IHUF+Trans+LCe	18.0	−54.1	280.7	79.2	299.6	623.4
FCEV (Opt)+FBSC+LH_2$	0.0	0.0	0.0	0.0	0.0	0.0
FCEV (Opt)+FBSC+IHUF+Trans+LCH$_2$	596.0	−47.0	869.4	294.2	590.3	2302.9
PEV+FCEV+FBSC+IHUF+Trans+LCe+LCH$_2$	345.0	−145.3	739.8	300.8	442.0	1682.3
PEV+FCEV+Bio+FBSC+IHUF+Trans+LCe+LCH$_2$	371.7	−137.4	855.8	340.6	264.8	1695.5

[a]Base Cases are indicated in boldface. Eff+FBSC serves as a Base Case for the four groups below it: Eff, Intensive Pricing, PEV, and FCEV, as well as for the mixed cases in the final grouping including both PEVs and FCEVs. See Table 5.5 for explanation of scenario components.

than the sum of private and public benefits. The benefits considered in the model include both public and private benefits such as benefits to the owners of the vehicles (i.e., the uncounted energy savings), benefits to the owners of all vehicles (i.e., the increased consumer surplus), and benefits to society at large (the benefits of GHG emissions reduction and reduced petroleum use). Costs refer to the additional costs of the transition over and above what the market is willing to do voluntarily, as represented by the respective Base Case. These include any increases in vehicle or fuel costs not included in consumers' surplus, net subsidies, and consumers' surplus losses. If there were increases in GHG emissions or petroleum use, these would also be included in transition costs.

5.5 COMPARISON TO PREVIOUS WORK

The LAVE-Trans modeling represents a significant step toward more completely modeling scenarios by which the LDV fleet could be drastically changed. However, because of the uncertainties concerning the behavior of consumers

and firms that underpins these modeling results as well as the uncertainty in projected costs and available technologies, it is important to consider these results in the context of the large body of literature on transitions to alternative vehicles and fuels. Here the focus is on some of the key findings and assumptions of the modeling compared to the available literature; a detailed summary of key reports on the matter is given in Appendix D.

A number of policy scenarios modeled above include the use of a large volume of biofuels (up to 45 billion gge/yr) in order to meet the goals for petroleum reduction and/or GHG emissions reductions from the LDV fleet. As is described in Chapter 3, the biomass required for such volumes of cellulosic drop-in fuel is plausible; many previous studies indicate a similar level of available biomass resource (UCD, 2011; NRC, 2008, 2009b; DOE, 2011; Greene and Plotkin, 2011; Pacala and Socolow, 2004). However, there is still uncertainty about the levels of production necessary to contribute significantly to the fueling of the LDV fleet due to the unknown future cost of fuels produced from biomass relative to gasoline produced from petroleum (DOE, 2011).

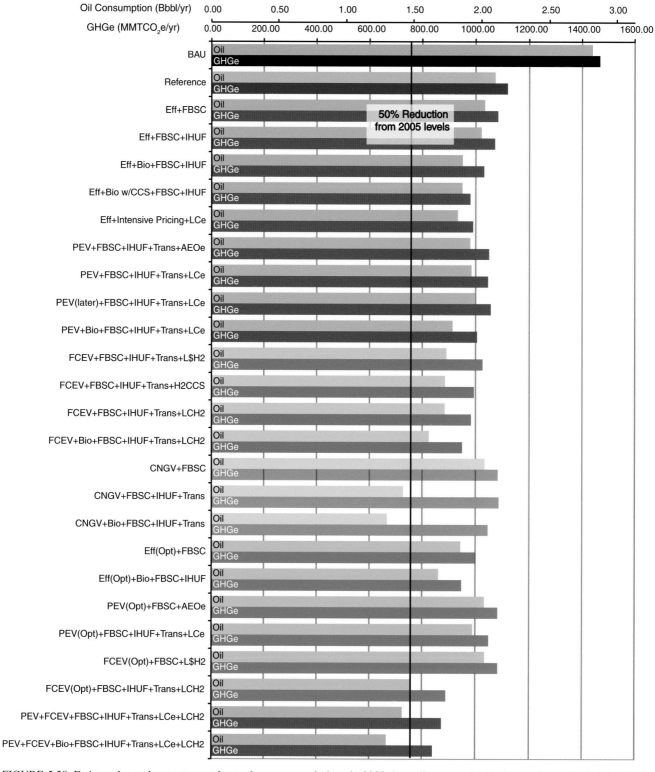

FIGURE 5.28 Estimated petroleum usage and greenhouse gas emissions in 2030, by policy scenario. Hydrogen fuel cell electric vehicle (FCEV) and plug-in electric vehicle (PEV) scenarios utilize Low-Carbon production of alternative fuel unless otherwise specified. See Table 5.5 for explanation of scenario components.

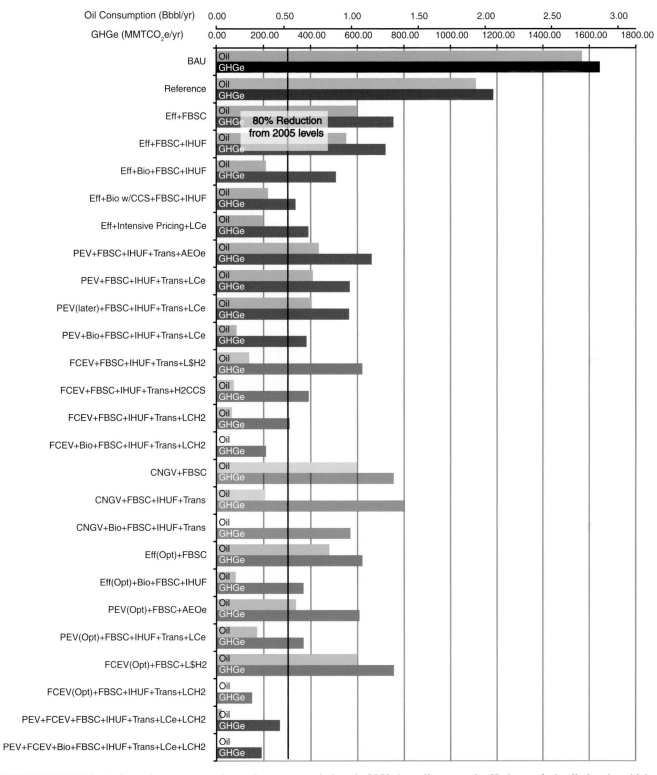

FIGURE 5.29 Estimated petroleum usage and greenhouse gas emissions in 2050, by policy scenario. Hydrogen fuel cell electric vehicle (FCEV) and plug-in electric vehicle (PEV) scenarios utilize Low-Carbon production of alternative fuel unless otherwise specified. See Table 5.5 for explanation of scenario components.

TABLE 5.5 Abbreviations for Policies Considered in the LAVE-Trans Model

Eff	Improved vehicle efficiency—midrange technology assumptions
FBSC	Feebates based on societal willingness to pay for GHG and petroleum reduction
IHUF	User fee on energy indexed to the average energy efficiency of all on-road vehicles
Bio	Assumes increased use of thermochemical biofuels up to 45 billion gge in 2050
CCS	Includes the use of carbon capture and storage
Intensive pricing	Includes IHUF, FBSC, carbon/oil tax, and feebates based on fuel savings (see Sections 5.4.2 and 5.4.4)
Trans	Transition policies consisting of vehicle and fuel infrastructure subsidies or mandates
AEOe	Reference Case electricity grid based on *Annual Energy Outlook 2011*
LCe	Low-Carbon electricity grid
L\$H$_2$	Low-Cost production of hydrogen
H$_2$CCS	Production of hydrogen with Carbon Capture and Storage
LCH$_2$	Low-Carbon production of hydrogen
(Opt)	Optimistic technology assumptions for the indicated technology (see Section H.4 in Appendix H for details)

NOTE: For more details on fuels production, see Table 5.1 and Chapter 3. Vehicle technology assumptions are described in Chapter 2. Policies are defined in Box 5.2.

A further limit on the availability of biofuels is likely to be competition from other uses, such as in aircraft or heavy-duty vehicles (UCD, 2011). Such limited availability would prevent achievement of an 80 percent reduction in GHG emissions without advanced progress in hydrogen fuel cell technology. A UC-Davis report noted that biofuels would play a pivotal role in any policy scenario designed to reduce GHG emissions from the transportation sector, particularly in the next two decades while deployment of advanced powertrain vehicles is still in its infancy (Yeh et al., 2008). The committee accepted as a premise in its modeling the achievement by 2030 of the production of volumes of biofuel specified in RFS2 and did not examine scenarios in which biofuel deployment did not achieve these levels.

As can be seen from the midrange cases in Figure 5.29, improvements in vehicle efficiency, particularly when combined with policies to drive consumers to purchase efficient vehicles, offer the possibility of large reductions in petroleum consumption and GHG emissions. These improvements in efficiency are dependent on the availability of the highly efficient vehicles described in Chapter 2. Based on the CAFE standards out to model year 2025 (EPA and NHTSA, 2011) as well as a number of studies looking out the next 20 years or more (ANL, 2011; DOT, 2010; UCD, 2011; NRC, 2010a,b; Bandivadekar et al.,

2008; Bastani et al., 2012), the committee's assessment of potential fuel consumption reductions in the near-future (to 2030) are largely in line with much of this literature, particularly given the committee's charge to assess the *potential* for future improvements. However, there is substantial uncertainty about vehicle efficiencies out to 2050. The committee chose to attack this problem of uncertainty by directly addressing the potential for reducing the losses in the vehicles' powertrains without prescribing particular technological solutions. It is worth noting that some of the technologies likely to be applied over the next few years (e.g., cooled exhaust gas recirculation) were not known to be viable 10 years ago, and continued improvement in materials and design has enabled load reductions in areas such as tire rolling resistance and weight reduction beyond what many would have thought practical just a decade or so ago. Although some of the known technologies may not pan out as planned, it is also plausible that there will be improvements beyond what is now known. The committee's analysis of the potential for technological improvement to LDVs has tried to balance these judgment issues. Based on these assumptions, the committee's projections for 2050 exceed those of many prior studies, particularly those that relied upon full-system simulation (UCD, 2011; ANL, 2009, 2011). Studies that are less optimistic about the possibility of significant load reductions yield little improvement in fuel consumption between the mid-term (2030) and long-term (2050) (DOT, 2010; NRC, 2010a, 2010b; EPRI and NRDC, 2007). If the committee's assessment of the long-term potential for highly efficient vehicles is proved incorrect, this will significantly hamper the effectiveness of all scenarios to reduce petroleum consumption and GHG emissions, since all alternative vehicles share the same basic load reductions enabling their high efficiencies. Recent efforts by Bastani et al. (2012) attempt to describe the most likely trajectory of the LDV fleet and show precisely this. Notably, the resultant *likely* efficiencies are far less than the committee's own assertions, as might be expected. Furthermore, this work shows the significance of meeting future fuel economy standards. As noted in the committee's own work, fuel economy standards will have to be an important driver in reducing vehicle energy consumption.

One of the major implications of the committee's modeling results is the difficulty in attaining the goals for reductions in GHG emissions and petroleum consumption chiefly through a transition to PEVs. The limited utility of BEVs and the higher costs of PHEVs remain a significant barrier in any scenario. The committee's assumptions on costs, however, agree with the majority of the literature on the topic; each report indicated a lower long-run cost for FCEVs and substantially elevated costs for both BEVs and long-range PHEVs (30+ mile all-electric range) (DOT, 2010; UCD, 2011; NRC, 2010a, 2010b; ANL, 2011). PHEVs with a lower range do show reduced cost barriers because of their smaller batteries, but they also offer significantly less potential for fuel displacement and reduced GHG emissions. Furthermore, as electricity prices increase over time with the increasingly clean electric grid and gasoline use by comparable ICE and hybrid vehicles decreases, the price advantage of fuel/electricity consumption of a BEV or a PHEV diminishes. These are factors not considered in any of the other reports on the transition to alternatively fueled LDVs.

Ultimately, the committee found numerous pathways to attain significant reductions in GHG emissions and petroleum consumption. The levels of GHG reductions are of similar magnitude to those described in previous studies (Figure 5.30); however, the specifics of the pathways themselves are often very different. For example, although the proposed UC-Davis scenarios for LDV GHG emissions reductions appear to be of a comparable magnitude, a large fraction of the reductions in the scenarios with the lowest GHG emissions come from a 25 percent decrease in VMT per capita, resulting in a 324 MMTCO$_2$ decrease in emissions from LDV transportation. There is also a notable difference in the Davis results for the FCEV scenario. Here, McCollum and Yang (2009) have limited penetration of FCEVs to 60 percent of new-car sales, whereas the NRC modeling results show the potential for much greater penetration of FCEVs, spurred on both through low future costs and policy action. Figure 5.30 indicates a sizable disparity between the efficiency cases of the VISION and LAVE-Trans models and previous NRC studies (NRC, 2008; 2010a). This difference is primarily a result of the more optimistic vehicle efficien-

cies presumed in the current work. A similar disparity is seen in a comparison with results of the HyTrans model (Greene and Leiby, 2007), although fuel production pathways, market analysis, and policies applied in the HyTrans analysis also deviate from those used in the committee's work. Small differences between the VISION and LAVE-Trans models are also observed for reasons outlined in Section 5.4.1.

The committee's modeling results are generally consistent with the available literature in both assumptions and results; however, the LAVE-Trans model has allowed the committee to build on this previous body of work to examine the transition costs associated with a shift to alternative vehicles and fuels in the LDV fleet. Moreover, the committee has examined several different policy options for achieving this transition, including multiple carbon pricing options, feebates, fuel taxes, and vehicle subsidies, leading to a number of pathways exhibiting sizable reductions in petroleum consumption and greenhouse gas emissions.

5.6 ADAPTING POLICY TO CHANGES IN TECHNOLOGY

Uncertainty is inherent in policy making for a transition to vehicles fueled by energy sources with reduced carbon impacts. The future path of technological development is uncertain. Future market conditions are also uncertain; indeed many economists have concluded that gasoline prices over the past several decades are best predicted by a

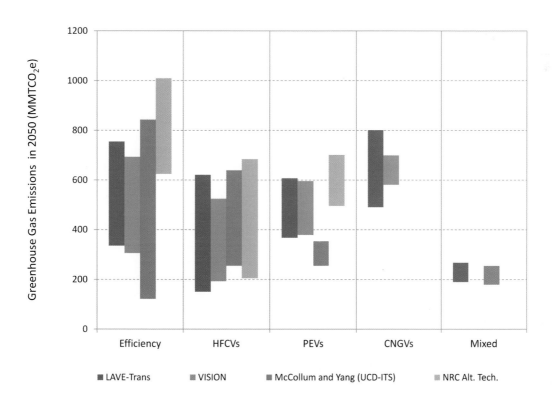

FIGURE 5.30 Comparison of greenhouse gas emissions scenarios.

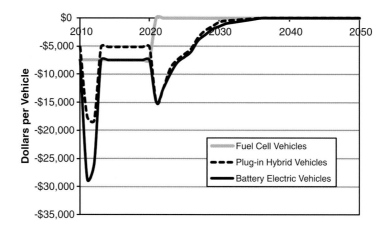

FIGURE 5.31 Assumed battery electric vehicle and plug-in hybrid electric vehicle subsidies in Optimistic EV Technology Scenario.

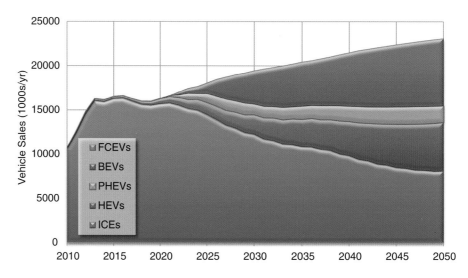

FIGURE 5.32 Vehicle sales by technology: Optimistic Plug-in Electric Vehicle Scenario.

random walk[14] (e.g., Hamilton, 2009; Anderson et al., 2011; Alquist et al., 2012). And as emphasized above, many of the parameters that drive the committee's modeling results are uncertain because knowledge of consumers' evaluation of limited-range vehicles, limited fuel availability, and other key factors is poor for present circumstances and worse for 30 to 40 years in the future. And, of course, the future will present opportunities and challenges that were not anticipated. In this section, the LAVE model is used to illustrate some of the challenges these uncertainties present to policy makers.

Policies that would work well if technologies advance as in the committee's midrange or optimistic cases may fail if technological progress stalls or is more expensive than anticipated. One technology may be expected to advance rapidly and yet a different technology turns out to exceed

expectations. To illustrate these points, the LAVE model was used to construct three hypothetical scenarios. These scenarios are not predictions, nor do they reflect the committee's judgments about the likelihood of success of the technologies used to illustrate the role of uncertainty. The choice of technologies that succeed or fail in the scenarios below is arbitrary.

The first scenario includes a policy of subsidies for PHEVs and BEVs that works well assuming optimistic technological progress for these two technologies and midrange progress for all others. The scenario also assumes high biofuel intensity and low-carbon production of electricity and hydrogen. The vehicle subsidies for 2010-2012 were chosen to match actual sales of BEV and PHEV vehicles in the United States and include the federal tax credit of $7,500 per vehicle, as well as state subsidies and implicit subsidies by manufacturers introducing these vehicles. Only the federal subsidy is assumed to continue until 2020 and then end. In 2021 a new subsidy of $15,000 per vehicle is assumed for

_____
[14]A random walk is a mathematical formalism for a stochastic process defined as a series of random steps. In this case, oil prices are considered as a Gaussian random walk, meaning that the size of the step follows a Gaussian probability distribution.

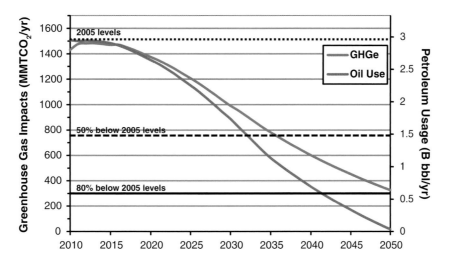

FIGURE 5.33 Changes in petroleum use and greenhouse gas emissions versus 2005: Optimistic Plug-in Electric Vehicle Scenario.

both vehicle types, decreasing each year until all subsidies are ended after 2035 (Figure 5.31).

The result is a successful, sustainable market penetration of PEVs. In 2050 BEVs attain a market share of 33 percent, PHEVs have an 8 percent share, and largely biofuel-powered HEVs and ICEVs claim 24 and 35 percent of the new-vehicle market, respectively (Figure 5.32).

The improvements in fuel economy, high penetration of drop-in biofuels (45 billion gallons in 2050), and market success of grid-connected vehicles powered by electricity produced by a low-carbon grid essentially eliminate oil use by LDVs and reduce GHG emissions by 78 percent, for all practical purposes meeting both 2050 goals. The 2030 goal

is almost met by a 41 percent reduction in petroleum use versus the 2005 level (Figure 5.33).

The cost of subsidies to induce the transition is substantial, $130 billion NPV discounted to 2010 at 2.3 percent per year. The subsidies together with the lower energy costs of plug-in vehicles generate consumers' surplus benefits that exceed the subsidy costs (Figure 5.34). When uncounted energy savings over the full life of the vehicles and the societal values of reduced GHG emissions and oil use are added to the other costs and benefits, the NPV of the transition policies is over $600 billion. The subsidies must be paid before most of the benefits are received, however, putting a large amount of capital at risk.

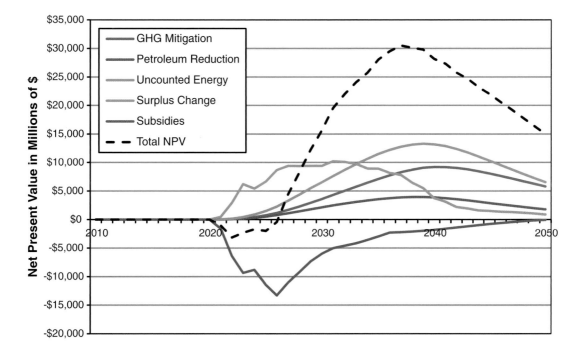

FIGURE 5.34 Net present value of the costs and benefits of the transition: Optimistic Plug-in Electric Vehicle Scenario.

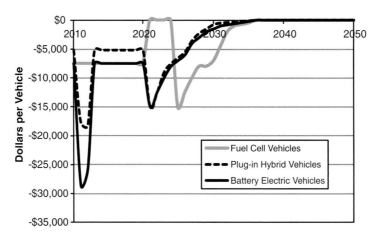

FIGURE 5.35 Assumed battery electric and plug-in hybrid electric vehicle subsidies in Pessimistic Plug-in Electric Vehicle Technology Scenario with Adaptation.

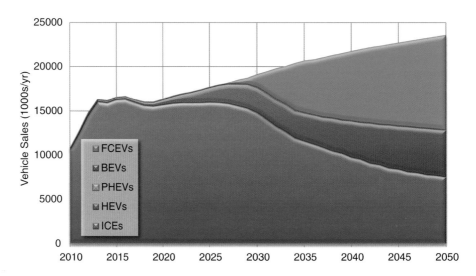

FIGURE 5.36 Vehicle sales by vehicle technology in Pessimistic Plug-in Electric Vehicle Technology Scenario with Adaptation.

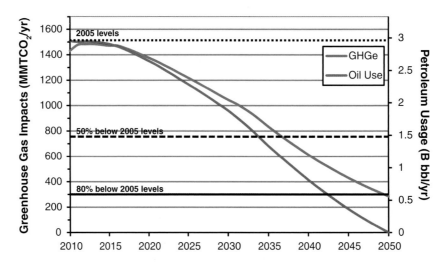

FIGURE 5.37 Changes in petroleum use and greenhouse gas emissions versus 2005 in Pessimistic Plug-in Electric Vehicle Technology Scenario with Adaptation to promote hydrogen fuel cell vehicles after 2024.

If the extreme assumption is made that the two technologies do not progress beyond their status today (BAU assumptions), the same subsidies that induced a sustainable transition in other cases are unsuccessful in achieving any sustainable market penetration. However, far less is spent on subsidies in this pessimistic PEV technology scenario, since the vehicles remain too expensive to attract many buyers. The total expenditures on the unsuccessful attempt to induce a transition amount to somewhat more than $1 billion. Costs exceed benefits, however, and the NPV of the scenario is on the order of –$250 million. Not surprisingly, the goals are not met in 2050, but petroleum use is still 75 percent lower than in 2005 and GHG emissions are 60 percent lower due to the much greater energy efficiency of ICEVs and HEVs and the extensive use of drop-in biofuels.

Suppose that the hypothetical failure of PEV technology to advance is quickly recognized, and that it is observed that FCEV technology is advancing more rapidly than expected. Further, assume that a decision is made to change course 3 to 4 years after the higher PEV subsidies are offered in 2021. Two hundred subsidized hydrogen refueling stations are built in 2024 followed by another 200 in 2025. Subsidies nearly identical to those previously offered for the plug-in vehicles are offered for FCEVs (Figure 5.35). Because it is assumed that the FCEV technology has progressed according to the midrange assumptions, this policy adaptation succeeds, resulting in nearly a 50 percent FCEV market share by 2050 (Figure 5.36). As a consequence, the 2050 goals for both oil and GHG reduction are met (Figure 5.37).

These scenarios are intended to illustrate the importance of uncertainty about future technology evolution and the value of adapting policies to the progress of technology. The choice of technologies for the illustration is entirely arbitrary. Which technology will succeed, if any, is uncertain. There will be costs to attempting to deploy technologies that do not progress to commercial competitiveness. However, if competitive alternatives emerge, and policies can be changed, it may still be possible to meet the long-term goals at a reasonable cost.

5.7 SIMULATING UNCERTAINTY ABOUT THE MARKET'S RESPONSE

In addition to uncertainty about the progress of alternative fuel and vehicle technologies (e.g., Bastani et al., 2012), there is also considerable uncertainty about how the market will respond to novel technologies. Many of the most important determinants of the market success of advanced technologies are poorly understood. These include the inconvenience cost of limited fuel availability for hydrogen and CNG, and limited range and long recharging times for BEVs. The number of innovators willing to pay a premium for novel technologies is largely unknown, as is the amount they would be willing to pay to get one of the first plug-in hybrid electric or hydrogen fuel cell vehicles. And while there are many estimates of consumers' willingness to pay for fuel economy, there is at present no consensus on the subject (Greene, 2010). There are dozens of studies providing estimates of the sensitivity of consumers' vehicle choices to price, yet little is known about the price sensitivity of choices among novel technologies. On the vehicle and fuels supply side, there is a great deal of uncertainty about learning rates, scale economies, and firms' aversion to risk. Furthermore, all these factors can and likely will change over a 40-year period.

It is possible to get a sense of how these uncertainties affect the committee's modeling results by means of simulation analysis. Table 5.6 lists 17 factors that determine market behavior in the LAVE model and provides mean values used in the model runs as well as uncertainty ranges based solely on the committee's judgment. Ten thousand simulations of the LAVE model were run to produce distributions for key model outputs, including the impacts on GHG emissions and

TABLE 5.6 Model Parameters Included in Simulation Analysis and Ranges of Values

Parameters	Distribution	Minimum	Mean	Maximum
Importance of diversity of makes and models to chose from	Triangle	0.50	0.67	0.9975932
Value of time ($/hr)	Triangle	$10.00	$20.00	$39.86
Maximum value of public recharging to typical PHEV buyer	Uniform	$500	$1,000	$1,500
Cost of 1 day on which driving exceeds BEV range	Uniform	$10,002	$20,000	$29,999
Maximum value of public recharging to typical BEV buyer	Uniform	$0	$500	$1,000
Importance of fuel availability relative to standard assumption	Triangle	0.67	1.00	1.67
Payback period for fuel costs (yr)	Triangle	2.0	3.0	5.0
Volume threshold for introduction of new models relative to standard assumptions	Uniform	0.80	1.00	1.20
Optimal production scale relative to standard assumptions	Uniform	0.75	1.00	1.25
Scale elasticity relative to standard assumptions	Uniform	0.50	1.00	1.50
Progress ratio relative to standard assumptions	Uniform	0.96	1.00	1.04
Price elasticities of vehicle choice relative to standard assumptions	Uniform	0.60	1.20	1.80
Percentage of new car buyers who are innovators	Triangle	5.0	15.0	20.0
Willingness of innovators to pay for novel technology ($/mo)	Uniform	$100	$200	$300
Cumulative production at which innovators' willingness to pay is reduced by half	Uniform	1,000,000	2,000,000	3,000,000
Majority's aversion to risk of new technology ($/mo)	Uniform	–$900	–$600	–$300
Cumulative production at which majority's risk is reduced by half	Uniform	$500,000	$1,000,000	$1,500,000

petroleum consumption, and the market shares of advanced vehicle technologies. Not all elements of market uncertainty are included in the simulations. In particular, the LAVE model does not include a representation of industry's likely aversion to risky investments. Nor do the simulation runs include uncertainty about future energy prices.

Two scenarios were simulated: policies to induce a transition to PEVs and policies to induce a transition to FCEVs. Both scenarios include 13.5 billion gallons of drop-in biofuel by 2050 and 10 billion gallons gasoline equivalent of ethanol, as well as the energy efficiency improvements of the midrange scenario.

The resulting uncertainty is strikingly large. The simulated distribution of GHG emissions reductions for the FCEV Policy Case ranges from 43 percent, corresponding to zero market penetration of fuel cell vehicles, to 83 percent at a 60 percent market share of FCEVs (Figure 5.38). It is bi-modal, reflecting the presence of tipping points that cause FCEVs to succeed to a greater or lesser degree, or fail to achieve any significant market share. The existence of tipping points reflects the many positive feedbacks in the transition process. The simulated distribution of greenhouse gas reductions due to plug-in vehicles has a similar bi-modal form and nearly as great a range: −42 to −71 percent (Figure 5.39). The modal separation is less because EVs do not have the strong dependence on fuel availability that hydrogen vehicles do.

The impacts are highly uncertain chiefly because the market response to electric-drive technology is uncertain. The simulated distribution of BEVs' share of the new LDV market in 2050 is shown in Figure 5.40. Although there is a peak in the vicinity of 30 percent, there is a reasonable probability of almost any market share between zero and 50 percent, and a nearly 30 percent probability of almost no market share. The situation for FCEVs is similar but there is a greater separation between market success and failure (Figure 5.41). The simulation analysis can also identify those parameters that have the greatest influence on market success. Both technologies are highly sensitive to assumptions about scale economies in the automotive industry, to the number of innovators and their willingness to pay for novel technology, and to the value to consumers of having a diverse array of vehicles to choose from. BEVs do better if consumers are more sensitive to energy costs and less sensitive to initial price. Consumers' concern about range and recharging-time limitations is also very important for BEVs. Fuel cell vehicles' market success is strongly dependent on the importance of fuel availability, but this factor is of much less importance for BEVs.

There are many reasons that these results should be interpreted cautiously, not the least of which is that a fixed policy strategy is assumed, regardless of the parameter values chosen. As is the case for uncertainty about technological progress, adapting policies to suit the realities of the marketplace would undoubtedly produce better results. All of the frequency distributions shown are conditional on a set of specific policy assumptions that are held constant for all simulations.

The uncertainties illustrated here can be reduced by research and analysis, and by learning from experience. Clearly, there is a great deal of benefit to be gained from a better understanding of both the technologies and the behavior of the market. Uncertainty analysis does not describe the future as it must be or as it will be; it is an attempt to describe

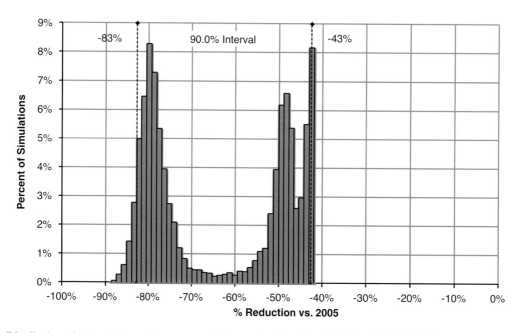

FIGURE 5.38 Distribution of estimated greenhouse gas emissions reductions from 2005 level: Fuel Cell Electric Vehicles Case.

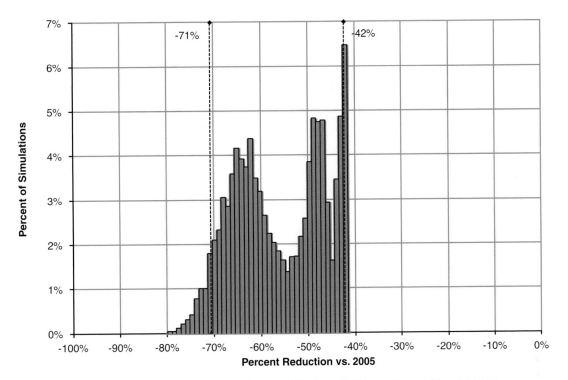

FIGURE 5.39 Distribution of estimated greenhouse gas reduction in 2050 from 2005 level: Plug-in Electric Vehicles.

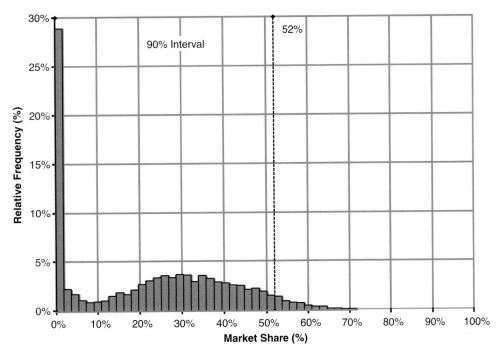

FIGURE 5.40 Distribution of battery electric vehicle market share in 2050: Plug-in EV Policy Case.

what we think we do and do not know about the distant future, as viewed from the present. Learning—increasing knowledge of the processes and behaviors that will affect a transition, as well as the costs and performance of the technologies that could enable one—is likely to be essential if the 2030 and 2050 goals are to be achieved efficiently. Inducing

a transition to non-petroleum energy sources with extremely low GHG emissions is an unprecedented challenge for public policy. To support effective policy making, a much better understanding of how markets and technology will interact is likely to be highly beneficial.

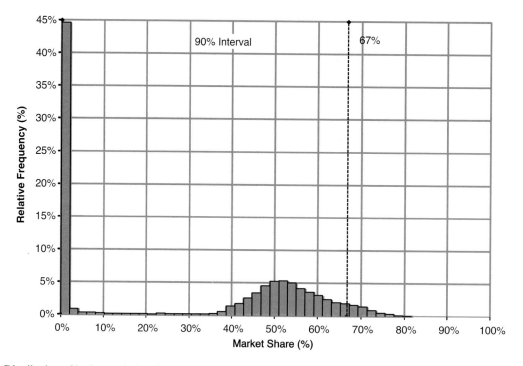

FIGURE 5.41 Distribution of hydrogen fuel cell vehicle market share in 2050: Fuel Cell Electric Vehicle Policy Case.

5.8 FINDINGS

Large and important reductions in petroleum use and greenhouse gas emissions can be achieved by increasing the fuel economy of light-duty vehicles in line with the CAFE standards for 2025 and embodied in the RFS2 (25-30 percent by 2030 and 30-40 percent by 2050). Even greater reductions will be possible if advances in vehicle and fuel technologies beyond those required to meet the 2025 CAFE standards and the RFS2 standards can be realized.

Achieving the 2030 and 2050 goals for reduction of oil use and greenhouse gas emissions will require a mix of strong public policies, market forces that encourage greater energy efficiency, and continued improvements in vehicle and fuels technologies. As the comparison of VISION and LAVE-Trans model estimates illustrates, reaching the goals is likely to be more difficult than previous "what if" analyses have concluded due to economic feedback effects and competition among technologies and fuels. These feedback effects include increased vehicle use with reduced energy costs, increased new-vehicle demand with improved technology, and competition for market share among advanced technologies. They are also almost certain to include lower petroleum prices as a consequence of reduced petroleum demand, although no attempt has been made to model that in these analyses. These feedback effects are much smaller in magnitude than the direct effects of energy efficiency improvement and displacement of petroleum with alternative energy sources; still, they increase the difficulty of achieving the 2050 goals.

Achieving a 40 percent reduction in petroleum use over 2005 levels by 2030 is a more realistic and achievable goal than a 50 percent reduction. Whether or not this level of reduction would be sufficient to achieve the objective of solving the nation's oil dependence problem given expected increases in domestic petroleum supply should be carefully evaluated.

Even if the nation should fall short of the 2050 goals, there are likely to be environmental, economic, and national security benefits resulting from the reductions that are achieved. The committee's modeling suggests that reductions in petroleum use on the order of 70 to 90 percent are possible given very strong policies and continued advances in the key technologies: electric-drive vehicles (hybrid, plug-in hybrid, battery, and fuel cell) and drop-in biofuels. In the committee's judgment, reductions in greenhouse gas emissions on the order of 60 to 80 percent are possible but will require effective and adaptive policies over time as well as continued advances in the technologies described in Chapters 2 and 3.

Including the social costs of GHG emissions and petroleum dependence in the cost of fuels (e.g., via a carbon tax) provides important signals to the market that will promote technological development and behavioral changes. Yet these pricing strategies alone are likely to be insufficient to induce a major transition to alternative, net-low-carbon vehicle technologies and/or energy sources. Additional strong, temporary policies may be required to break the lock-in of conventional technology and overcome the market barriers to alternative vehicles and fuels.

If two or more of the fuel and/or vehicle technologies evolve through policy and technology development as shown in a number of the committee's scenarios, the committee's model calculations indicate benefits of making a transition to a low-petroleum, low-GHG energy system for LDVs that exceed the costs by a wide margin. Benefits include energy cost savings, improved vehicle technologies, and reductions in petroleum use and GHG emissions. Costs refer to the additional costs of the transition over and above what the market is willing to do voluntarily. However, as noted above, modeling results should be viewed as approximations at best because there is by necessity in such predictions a great deal of uncertainty in estimates of both benefits and costs. Furthermore, the costs are likely to be very large early on with benefits occurring much later in time.

Depending on the readiness of technology and the timing of policy initiatives, subsidies or regulations for new vehicle energy efficiency and the provision of energy infrastructure may be required, especially in the case of a transition to a new vehicle and fuel system. In such cases, substantial subsidies might be required for at least 5 to 10 years, and possibly as long as 20 years if technological progress is slow (e.g., starting grid-connected vehicles now is likely to require 20 years of subsidy to stabilize them at a significant market share). And, as shown above, there is likely to be a high degree of risk in policies targeted to a particular technology. For these reasons, it is important to consider carefully when and if such policies are necessary and to make policy adaptable to changing evidence about technology and market conditions. It is also very important that policy makers obtain objective, expert advice on the readiness of both fuel and vehicle technologies and markets. Scenario analysis has identified strong tipping points for the transition to new vehicle technologies. If sufficiently large subsidies are not applied to overcome the early cost differentials, then the transition will not occur and the subsidies will have been wasted. In pursuing these goals, the rate of cost decline and the subsidies applied at each stage must be carefully weighed to establish that the program is effective.

Advance placement of refueling infrastructure is critical to the market acceptance of hydrogen fuel cell and CNG vehicles. It is likely to be less critical to the market acceptance of grid-connected vehicles, since many consumers will have the option of home recharging. However, the absence of an outside-the-home refueling infrastructure for grid-connected vehicles is likely to depress demand for these vehicles. Infrastructure changes will not be needed if the most cost-effective solution evolves in the direction of more efficient ICEVs and HEVs combined with drop-in low-carbon biofuels.

Empirical knowledge of the barriers to major energy transitions is currently inadequate to make robust assessments of public policies. The modeling analysis presented in this chapter is intended to be an initial step in the right direction rather than a definitive assessment of alternatives.

Research is needed to better understand key factors for transitions to new vehicle fuel systems such as the costs of limited fuel availability, the disutility of vehicles with short ranges and long recharge times, the numbers of innovators and early adopters among the car-buying public, as well as their willingness to pay for novel technologies and the risk aversion of the majority, and much more. More information is also need on the transition costs and barriers to production of alternative drop-in fuels, especially on the type of incentives necessary for biofuels. The models that the committee and others have used to analyze the transition to alternative vehicles and/or fuels are first-generation efforts, more useful for understanding processes and their interactions than producing definitive results.

5.9 REFERENCES

Alquist, R., L. Kilian, and R.J. Vigfusson. 2012. Forecasting the price of oil. In *Handbook of Economic Forecasting, Volume 2* (G. Elliott and A. Timmermann, eds.). Amsterdam: Elsevier-North-Holland.

Anderson, S.T., R. Kellogg, and J.M. Sallee. 2011. What Do Consumers Believe about Future Gasoline Prices? Working Paper 16974. *National Bureau of Economic Research Working Paper Series.* Cambridge, Mass.: National Bureau of Economic Research. April.

ANL (Argonne National Laboratory). 2009. *Multi-Path Transportation Futures Study: Vehicle Characterization and Scenario Analyses.* ANL/ESD/09-5. Prepared for the U.S. Department of Energy. Argonne, Ill.: Energy Systems Division, Argonne National Laboratory. July.

———. 2011. *Light-Duty Vehicle Fuel Consumption Displacement Potential Up to 2045.* ANL/ESD/11-4. Argonne, Ill.: Energy Systems Division, Argonne National Laboratory. July.

Bandivadekar, A., K. Bodek, L. Cheah, C. Evans, T. Groode, J. Heywood, E. Kasseris, M. Kromer, and M. Weiss. 2008. *On the Road in 2035: Reducing Transportation's Petroleum Consumption and GHG Emissions.* Laboratory for Energy and the Environment Report No. LFEE 2008-05 RP. Cambridge, Mass.: Massachusetts Institute of Technology. July.

Bastani, P., J.B. Heywood, and C. Hope. 2012. The effect of uncertainty on U.S. transport-related GHG emissions and fuel consumption out to 2050. *Transportation Research, Part A: Policy and Practice* 46(3):517-548.

Boardman, A.E., D.H. Greenberg, A.R. Vining, and D.L. Weimer. 2011. *Cost-Benefit Analysis: Concepts and Practice.* Fourth Edition. Upper Saddle River, N.J.: Prentice Hall.

Brown, S.P.A., and H.G. Huntington. 2010. Reassessing the Oil Security Premium. Discussion Paper RFF DP 10-05. Washington, D.C.: Resources for the Future.

Copulos, M.R., 2003. America's Achilles Heel: The Hidden Costs of Imported Oil. Washington, D.C.: National Defense Council Foundation.

Daly, A., and S. Zachary. 1979. Improved multiple choice models. In *Identifying and Measuring Transportation Mode Choice* (D. Hensher and Q. Dalvi, eds.). London: Teakfield.

Delucchi, M.A., and J.J. Murphy. 2008. US military expenditures to protect the use of Persian Gulf oil for motor vehicles. *Energy Policy* 36:2253-2264.

DOE (U.S. Department of Energy). 2011. *U.S. Billion Ton Update: Biomass Supply for a Bioenergy and Bioproducts Industry.* R.D. Perlack and B.J. Stokes, Leads. ORNL/TM-2011/224. Oak Ridge, Tenn.: Oak Ridge National Laboratory. August.

DOT (U.S. Department of Transportation). 2010. *Transportation's Role in Reducing U.S. Greenhouse Gas Emissions.* Prepared by the U.S. DOT Center for Climate Change and Environmental Forecasting, Washington, D.C. April.

EIA (Energy Information Administration). 2011a. *Annual Energy Outlook 2011.* DOE/EIA-0383(2011). Washington, D.C.: Department of Energy, Energy Information Administration. March.

———. 2011b. *Annual Energy Review 2010.* DOE/EIA-0384(2010). Washington, D.C.: Department of Energy, Energy Information Administration. October.

———. 2012. *Annual Energy Outlook 2012.* DOE/EIA-0383ER(2012). Washington, D.C.: Department of Energy, Energy Information Administration. Available at http://www.eia.gov/forecasts/aeo/er/.

EPA and NHTSA (U.S. Environmental Protection Agency and National Highway Transportation Safety Administration). 2011. *Draft Joint Technical Support Document: Proposed Rulemaking for 2017-2025 Light-Duty Vehicle Greenhouse Gas Emission Standards and Corporate Average Fuel Economy Standards.* EPA-420-D-11-901. Washington, D.C. November.

EPRI (Electric Power Research Institute) and NRDC (Natural Resource Defense Council). 2007. *Environmental Assessment of Plug-In Hybrid Electric Vehicles. Volume 1: Nationwide Greenhouse Gas Emissions.* Palo Alto, Calif.: EPRI. July.

GAO (U.S. Government Accountability Office). 2006. *Strategic Petroleum Reserve: Available Oil Can Provide Significant Benefits, But Many Factors Should Influence Future Decisions About Fill, Use and Expansion.* GAO-06-872. Washington, D.C.: GAO. August.

Greene, D.L. 2009. Measuring energy security: Can the United States achieve oil independence? *Energy Policy* 38(4):1614-1621.

———. 2010. *How Consumers Value Fuel Economy: A Literature Review.* EPA-420-R-10-008. Washington, D.C.: U.S. Environmental Protection Agency.

Greene, D.L., and P.N. Leiby. 1993. *The Social Costs to the U.S. of Monopolization of the World Oil Market, 1972-1991.* ORNL-6744. Oak Ridge, Tenn.: Oak Ridge National Laboratory. March.

———. 2007. *Integrated Analysis of Market Transformation Scenarios with HyTrans.* ORNL/TM-2007/094. Oak Ridge, Tenn.: Oak Ridge National Laboratory. June.

Greene, D.L., and S.E. Plotkin. 2011. *Reducing Greenhouse Gas Emissions from U.S. Transportation.* Prepared for the Pew Center on Global Climate Change. Arlington, Va. January.

Greene, D.L., P.D. Patterson, M. Singh, and J. Li. 2005. Feebates, rebates and gas-guzzler taxes: A study of incentives for increased fuel economy. *Energy Policy* 33(6):757-775.

Hamilton, J.D. 2009. Understanding crude oil prices. *The Energy Journal* 30(2):179-206.

Huntington, H.G. 2007. Oil Shocks and Real U.S. Income. *The Energy Journal* 28(4):31-46.

Interagency Working Group on Social Cost of Carbon. 2010. *Technical Support Document: Social Cost of Carbon for Regulatory Impact Analysis Under Executive Order 12866.* Washington, D.C. February.

Jones, D.W., P.N. Leiby, and I.K. Paik, 2004. Oil price shocks and the macroeconomy: What has been learned since 1996? *The Energy Journal* 25(2):1-32.

Kaufmann, W.W., and J.D. Steinbruner, 1991. *Decisions for Defense: Prospects for a New Order.* Washington, D.C.: Brookings Institution. April.

Leiby, P.N. 2008. Estimating the Energy Security Benefits of Reduced U.S. Oil Imports, Final Report. ORNL/TM-2007/028. Oak Ridge, Tenn.: Oak Ridge National Laboratory, Revised March 14.

McCollum, D.L., and C. Yang. 2009. Achieving deep reductions in U.S. transport greenhouse gas emissions: Scenario analysis and policy implications. *Energy Policy* 37(12):5580-5596.

McConnell, J.M., 2008. Annual Threat Assessment of the Director of National Intelligence for the Senate Select Committee on Intelligence. SSCI ATA FEB 2008-DNI. Unclassified Statement for the Record. Washington, D.C.: Office of the Director of National Intelligence. February 5.

McFadden, D.L. 1978. Modeling the choice of residential location. In *Spatial Interaction Theory and Planning Models* (A. Karlqvist, L. Lundqvist, F. Snickars, and J. Weibull, eds.). Amsterdam: North Holland.

Military Advisory Board, 2011. *Ensuring America's Freedom of Movement: A National Security Imperative to Reduce Oil Dependence.* Alexandria, Va.: Center for Naval Analyses.

Moreland, H., 1985. A Few Billion for Defense Plus $250 Billion for Overseas Military Intervention. New Policy Papers #1. Washington, D.C.: Coalition for a New Foreign and Military Policy.

NRC (National Research Council). 2008. *Transitions to Alternative Transportation Technologies: A Focus on Hydrogen.* Washington, D.C.: The National Academies Press.

———. 2009a. *Hidden Costs of Energy: Unpriced Consequences of Energy Production and Use.* Washington, D.C.: The National Academies Press.

———. 2009b. *America's Energy Future: Liquid Transportation Fuels from Coal and Biomass.* Washington, D.C.: The National Academies Press.

———. 2010a. *Transitions to Alternative Transportation Technologies: Plug-in Hybrid Electric Vehicles.* Washington, D.C.: The National Academies Press.

———. 2010b. *America's Energy Future: Real Prospects for Energy Efficiency in the United States.* Washington, D.C.: The National Academies Press.

OMB (Office of Management and Budget). 2012. Appendix C: Discount Rates for Cost-Effectiveness, Lease, Purchase, and Related Analyses. OMB Circular No A-94. Memorandum from the Executive Office of the President of the United States. Washington, D.C. January 3.

Pacala, S., and R. Socolow. 2004. Stabilization wedges: Solving the climate problem for the next 50 years with current technologies. *Science* 305(5686):968-972.

Parry, I.W.H., and J. Darmstadter. 2004. The Costs of US Oil Dependency. Washington, D.C.: Resources for the Future. November 17.

Ravenal, E.C. 1991. Designing Defense for a New World Order: The Military Budget in 1992 and Beyond. Washington, D.C.: Cato Institute.

Singh, M., A. Vyas, and E. Steiner. 2003. VISION Model: Description of Model Used to Estimate the Impact of Highway Vehicle Technologies and Fuels on Energy Use and Carbon Emissions to 2050. Prepared by Argonne National Laboratory for the Department of Energy under Contract No. W-31-109-ENG-38. Report#ANL-ESD-04-1. Argonne, Ill.: Argonne National Laboratory. December.

UCD (University of California, Davis). 2011. *Sustainable Transportation Energy Pathways: A Research Summary for Decision Makers* (J. Ogden and L. Anderson, eds.). Davis, Calif.: Institute of Transportation Studies.

Vyas, A.D., D.J. Santini, and L.R. Johnson. 2009. Potential of plug-in hybrid electric vehicles' to reduce petroleum use: Issues involved in developing reliable estimates. *Journal of the Transportation Research Board* 2139:55-63.

Williams, H.C.W.L. 1977. On the formation of travel demand models and economic measures of user benefit. Pp. 285-344 in *Environment and Planning, Part A,* Volume 9. London: Pion, Ltd.

Yeh, S., A.E. Farrell, R.J. Plevin, A. Sanstad, and J. Weyant. 2008. Optimizing U.S. mitigation strategies for the light-duty transportation sector: What we learn from a bottom-up model. *Environmental Science and Technology* 42(22):8202-8210.

6

Policies for Reducing GHG Emissions from and Petroleum Use by Light-Duty Vehicles

To reach the twin goals addressed in this study, significant changes in policy will be needed to induce a move toward vehicle-fuel systems whose petroleum demand and greenhouse gas (GHG) emissions are very different from those of today. The modeling and results from Chapter 5 suggest a range of possible policy and technology pathways by which these goals might be met. This chapter reviews policy options, including those analyzed in Chapter 5 (for example, vehicle fuel economy and GHG standards and renewable fuel standards), that may offer promise. Each policy is described and assessed based on evidence about its use, effectiveness, and any shortcomings. Policy suggestions based on these assessments are provided in Chapter 7.

The policies needed to reach the goals for reductions in petroleum use and GHG emissions will have to differ dramatically from those of the past and could incur a high up-front cost. However, as the modeling results in Chapter 5 illustrate, these costs may be more than recouped in later years.

Policies are needed that can promote major changes in direction in the extensive private investments associated with vehicle manufacturing, fuel production and related infrastructure—changes that in turn will affect the market decisions made by consumers and businesses which ultimately shape such investments. The extent to which the resulting transition to a low-petroleum light-duty vehicle (LDV) system with low net GHG emissions will require displacing the incumbent internal combustion, liquid fueled vehicle technology is not known. However, major changes clearly will be needed in the use of natural resources and in the impacts of GHG emissions associated with supplying LDV fuels. Given the inherent uncertainties, an adaptive policy framework is needed that will be responsive to markets, technologies, and progress toward achieving the goals.

6.1 POLICIES INFLUENCING AUTOMOTIVE ENERGY USE AND GREENHOUSE GAS EMISSIONS

Several arenas of policy are relevant as means of influencing automotive energy use and GHG emissions: land-use, transportation, energy, environmental protection, and technology. These arenas are interrelated and the relationships are sometimes implicit. Failure to recognize the interrelationships between policy arenas could result in poor coordination or even contradictions among policy signals. Some of the relationships have been made explicit as policy makers have realized, for example, the interactions between land-use planning and transportation planning. The challenge of achieving deep reductions in petroleum use and GHG emissions requires an even greater degree of coordination among the policy arenas influencing the LDV sector.

Figure 6.1 shows on a normalized scale the total nationwide levels of several key LDV-related impacts that have been a subject of public policy. From 1970 through 2005, light-duty vehicle miles traveled (VMT) increased by 160 percent. Over the same period, gains in fuel efficiency held LDV petroleum demand and CO_2 emissions to a 74 percent increase. Modest absolute declines were achieved for traffic fatalities. The greatest improvement was seen in vehicle conventional air pollution, which achieved an absolute reduction of 65 percent by 2005 relative to its 1970 level.

6.1.1 Land-Use Policy

Land-use policies are perhaps the deepest foundation of the automotive system, helping, along with geography, to shape transportation patterns through the ages. U.S. land-use governance remains highly localized, and many levels of administration are involved in the planning, permitting, and

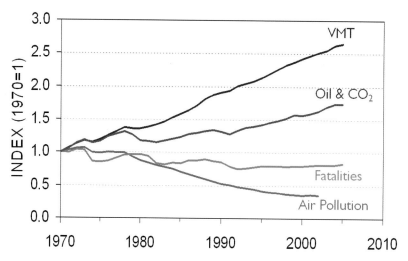

FIGURE 6.1 Trends in impacts of U.S. light-duty vehicles.
SOURCE: DOT, DOE, and EPA statistics.

zoning of land use. Higher levels of government traditionally show substantial deference to local prerogatives.

Academic understanding of the links between land use and transportation has translated only slowly into policies that might restrain travel demand growth tied to land use. Researchers have identified five land-use features, the "five Ds," influencing demand for automobile travel: population **d**ensity, land-use **d**iversity, neighborhood **d**esign, major **d**estination accessibility, and transit stop **d**istance from departure and arrival points of transit stops (TRB 2009, p. 52).

Although only recently considered in the context of transportation-related petroleum demand and GHG emissions, programs that support or constrain the expansion of croplands and managed forests used for sourcing biofuel feedstocks or for carbon sequestration through afforestation and grassland restoration are another important aspect of land-use policy. Determining the optimal use of land with respect to climate protection raises issues that may require rethinking of such policies (Righelato and Spracklen, 2007; Wise et al., 2009; Zhang et al., 2008).

6.1.2 Transportation Policy

Transportation policies center on the provision and operation of the infrastructure needed for mobility. For the automobile, they have focused on building, maintaining, and supporting roadways. In urban areas, transportation policy also supports mass transit, as well as sidewalks and bike paths, and so affects the availability and affordability of alternatives to auto travel. There is a clear emphasis in the U.S. Department of Transportation's (DOT's) official mission statement on ensuring speed of conveyance as well as safety and efficiency. With the automobile being by far the dominant mode of transportation for most Americans, facilitating auto travel has been a major part of DOT's mission. Much

of the necessary investment for highways and major roads is accomplished through a federal-state partnership approach, while most local roads are handled by municipalities with varying degrees of state involvement.

A key financial reason for the success of automobiles is that the vehicles themselves are purchased by individual consumers, who also pay for operating costs, notably fuel. That leaves to government the provision and maintenance of infrastructure. This contrasts with public transit modes, which require a public or public-private partnership to acquire and operate the vehicles and their supporting infrastructure. Consumers ultimately pay for all aspects of any transport system, with taxes or other user fees supporting the publicly provided elements.

6.1.3 Energy Policy

U.S. energy policies have roots in natural resource policy. Most pertinent to the auto sector are policies that have facilitated the development of petroleum resources over the years and those related to ensuring access to overseas supplies and securing them vis-à-vis geopolitical considerations. On the domestic front, policies supporting the economic development of oil and gas resources confronted environmental considerations and the need to balance competing players' demands for use of lands and offshore locations. Thus, the U.S. Department of the Interior long has been involved in petroleum-related activity. The Energy Research and Development Administration was created in 1974, and its successor, the U.S. Department of Energy (DOE), was formed in 1977, following the 1970s petroleum crisis.

In recognition of the importance of petroleum for military operations and as a critical resource for the entire economy, efforts to secure and expand the supply of petroleum have long been and continue to be a key part of U.S. energy policy.

The 1973-1974 energy crisis prompted the development of policies to encourage energy conservation and promote alternatives to petroleum. The LDV fleet became a key target, and vehicle efficiency standards known as the Corporate Average Fuel Economy, or CAFE, standards, were enacted as part of the Energy Policy and Conservation Act of 1975 (P.L. 94-163). A "gas-guzzler" tax followed in 1978 with passage of the Energy Tax Act (ETA; P.L. 95-618).

The 1970s also saw the development of policies to support alternatives to petroleum ranging from synthetic fossil fuels to biofuels. The ETA also introduced an excise tax exemption for gasohol,[1] which subsequently was extended and transformed into a tax credit for ethanol, the volumetric ethanol excise tax credit (VEETC), which until recently stood at $0.45 per gallon of ethanol. A tariff was imposed on imported ethanol to foster domestic biofuel production. Both the tax credit and the tariff expired at the end of 2011.

The CAFE credits program for alternative fuel vehicles (AFVs) was created by the Alternative Motor Fuels Act of 1988 (P.L. 100-94). It provided credit incentives for the manufacture of vehicles that used alcohol or natural gas fuels, either exclusively or as an alternative to gasoline or diesel fuel. This program induced automakers to sell a large number of dual-fuel vehicles capable of running on E85. However, for a variety of reasons including limited availability of E85 retail outlets, the program has not fostered significant use of alternative fuels (DOT-DOE-EPA, 2002). The Energy Policy Act of 1992 (EPAct; P.L. 102-486) established an expanded set of incentives and programs to promote alternative fuels and AFVs. They include mandates for AFV use in the federal fleet and certain state and utility fleets, and authorization for federal support of voluntary AFV deployment programs, which were subsequently implemented by DOE through the Clean Cities program.

Among the most recent developments in U.S. energy policy with respect to the LDV sector is the Renewable Fuel Standard (RFS) instituted as part of the 2005 EPAct (P.L. 109-58). The RFS put in place for the first time a nationwide mandate for use of a fuel other than petroleum. The 2005 EPAct also included expanded incentives for the production and commercialization of a range of AFV technologies. It included tax incentives for AFVs and infrastructure for alternative fuels that are not drop-in fuels. Incentives were provided on a graduated scale to encourage the production of different AFVs (DOE-EERE, 2011a; TIAP, 2012). Metrics related to technical fuel efficiency were used to determine the level of incentives. The incentives were limited to the first 60,000 qualifying vehicles produced by any one automaker.

Tax credits initially were for hybrid-electric, battery-electric, and fuel-cell electric vehicles and for qualified diesel and natural gas LDVs. Numerous modifications have occurred over the years and Congress allowed the tax credits for hybrid electric vehicles, and diesel and natural gas vehicles

to expire at the end of 2010. Tax credits for battery-electric motorcycles, three-wheeled electric vehicles, and low-speed neighborhood-electric vehicles expired at the end of 2011, as did a credit for converting conventional gasoline and diesel vehicles to plug-in hybrid or all-electric propulsion systems.

Currently, under the American Recovery and Reinvestment Tax Act and the Emergency Economic Stabilization Act of 2008, the United States uses a program that extends a federal tax credit of up to $7,500 to buyers of qualified plug-in hybrid and battery-electric LDVs. The credit is applicable in the year of the vehicle's purchase. Subsequent legislation limits the credit to the first 200,000 eligible vehicles from each qualified automaker. When that threshold is reached, the tax credit for subsequent vehicles sold is reduced in stages, disappearing completely after six calendar quarters.[2] Fuel-cell electric vehicles remain eligible for a federal tax credit of $4,000-$8,000, depending on their fuel economy ratings, but it is scheduled to expire in 2014.

In December 2007, the Energy Independence and Security Act (EISA; P.L. 110-140) expanded the RFS to target 35 billion gallons of ethanol-equivalent biofuels plus 1 billion gallons of biomass-based diesel by 2022, with life-cycle GHG emissions stipulations designed to foster cellulosic and other advanced biofuels. The same legislation raised the combined light-duty fleet CAFE standard to a 35 mpg level by 2020 while authorizing other structural reforms in the standards. The EISA also established a loan guarantee program for construction of manufacturing facilities for advanced-vehicle batteries and battery systems and requires a phase-out of the dual-fuel vehicle CAFE credit program by 2020.

6.1.4 Environmental Policy

Because automobiles and their supporting infrastructure impact the environment in numerous ways, many aspects of environmental policy come into play. However, it is control of the direct emissions from motor vehicles that is most relevant.

The history of Los Angeles smog, the pioneering work of Arie Haagen-Smit in linking smog to tailpipe pollution, and the subsequent development of emissions regulations first in California and then federally with the broad authority established by the Clean Air Act (CAA 1970) all are elements of one of the iconic stories of U.S. environmental policy (Mondt, 2000; CARB, 2011). At the beginning of this process in the 1960s, air pollution science was in its infancy and controls were rudimentary. As development continued, progressively tighter standards were set for restricting tailpipe emissions, prescribing fuel formulations, and limiting fuel evaporation from vehicles and fuel pumps.

The most stringent regulations for combustion-based vehicles, such as California's partial zero emission vehicle

---

[1] A fuel consisting of a blend of gasoline and ethanol.

[2] This tax credit is described in greater detail at http://www.fueleconomy. gov/feg/taxevb.shtml, accessed February 6, 2012.

standard, cut emissions per vehicle-mile by over two orders of magnitude, reducing an LDV's direct conventional air pollution impacts to nearly negligible levels. Quantitatively, the CAA policies addressing emissions have been by far the most effective areas of policy, resulting in a substantial absolute reduction of conventional pollution from LDVs even in the face of rising VMT (see Figure 6.1).

The CAA's overarching requirement for healthy air, embodied in the National Ambient Air Quality Standards (NAAQS), is what ultimately anchors the policy. The law obligates the U.S. Environmental Protection Agency (EPA) to pursue fact-based assessments of air pollutants' impacts on public health and welfare and to promulgate NAAQS solely on that basis. Economic considerations can enter in only when EPA develops the regulations that determine how the NAAQS will be met.

The Supreme Court (2007) interpreted the CAA's definition of air pollutants to include greenhouse gases and said that they could be subject to regulation if found to endanger public health or welfare. The EPA subsequently made such an "endangerment" finding (2009), setting in motion a regulatory process that started with GHG emissions standards for motor vehicles and is being extended to other sources.

6.1.5 Technology Policy

A large number of policy measures have the potential to influence technical innovation in LDVs and fuels. The federal department involved most actively in directly promoting new automotive technology has been the DOE. Its role primarily has been one of funding basic science and engineering research related to vehicles and fuels and pursuing demonstration and deployment programs that might foster market adoption of the technologies developed. Many National Research Council (NRC) studies reviewed this research, development, demonstration, and deployment approach while suggesting refinements and highlighting the challenges and obstacles involved. Examples of such energy technology policy programs include the Advanced Battery Consortium, the Partnership for a New Generation of Vehicles, the hydrogen-oriented FreedomCAR program, and the present US DRIVE program that emphasizes electric vehicles and plug-in hybrids. From the 1970s forward, parallel efforts have been aimed at developing renewable fuels.

6.1.6 Decision Making Through the Matrix of Policy Arenas

Based on methods of technology assessment and economic analysis as discussed below in this chapter, policy measures are established through a matrix of policy arenas such as those outlined above. The preceding overview of the different arenas of public policy that influence the LDV sector—transportation, land use, environmental protection, energy, and technology—underscores the complexity of the

challenge from a practical policy-making perspective. A national decision to reach goals such as those given in this committee's statement of task will likely need to involve all of these different policy arenas and the associated diversity of congressional committees, federal agencies, and stakeholder interests, along with an analogous range of interests at state and local levels of government. Reaching a national decision to achieve the goals will be a complicated undertaking that requires an adaptive policy framework as discussed below in this chapter.

6.2 WAYS TO INFLUENCE PETROLEUM USE AND GHG EMISSIONS EFFECTS IN THE LDV SECTOR

Policies that affect petroleum use, GHG emissions, or both ultimately exert their influence through a few key parameters:

- Vehicle energy intensity—typically, the energy required to move the average vehicle of the on-road LDV fleet 1 mile;
- Petroleum share of the energy used to power LDV fleets (when energy security and dependence on petroleum is the issue) or net GHG emissions balance of the fuel system (when climate disruption is the issue); the latter is often described as the average well-to-wheels GHG emissions of the energy used to power the vehicle fleet;[3] and
- Volume of travel—typically, the VMT by the on-road LDV fleet.

It sometimes is argued that system efficiency constitutes an independent fourth parameter, but that is not the case. Policies that affect system efficiency influence GHG emissions or petroleum use only through one or more of the three parameters listed above.[4]

A common analytic framework for transportation energy and climate-change analysis involves factoring emissions based on the three key parameters, which interact multiplicatively. Addressing all three (vehicle energy intensity, petroleum share of energy use in LDVs, and travel activity) is important because a policy that focuses only on a single parameter is likely to require it to be pushed to extraordinary lengths.[5]

Whether the policies target one or more of the parameters, they operate by influencing market actors whose decisions determine the values of the parameters, which in turn determine LDV petroleum use and GHG emissions. Policies that

---

[3]For advantages and disadvantages of the use of well-to-wheels approaches to regulating GHG emissions related to fuels, see below.

[4]However, in view of the interest in policies promoting system efficiency, they are discussed below in this chapter.

[5]For example, the average on-road fleet fuel economy would have to exceed 180 mpg if vehicle energy intensity were the only parameter targeted for reducing LDV petroleum use; see footnote 2 in Chapter 2.

target one parameter may influence others. A well-known example is the difference in impact on vehicle GHG emissions and energy use produced by motor fuel taxes versus efficiency standards. Motor fuel taxes stimulate demand for more fuel-efficient vehicles. They also raise the variable cost of driving, which in turn reduces VMT. In contrast, fuel economy standards require the sale of more fuel-efficient vehicles, *reducing* the cost of driving, thereby *increasing* VMT. CAFE is effective at pushing new technology into the fleet but is unlikely to affect the size of vehicles that consumers purchase (at least with the current footprint-based system). Taxes discourage people from driving more and encourage consumers to purchase smaller vehicles. The benefits of CAFE and taxes are largely independent of one another. Both policies have been found to reduce LDV fuel use overall, but the *amount* by which each policy reduces total LDV petroleum use or GHG emissions differs.

Finally, the cost of reducing emissions by changing any single parameter is likely to rise as the magnitude of required change increases or the time over which the required change is to be accomplished decreases.

Policies such as carbon pricing that affect more than a single parameter are generally considered by economists to be most cost-effective.

Vehicle energy intensity, petroleum share of the fuel market, and travel demand each is an outcome of market decisions. Thus, the market actors whose decisions affect each of the parameters (and whose decisions on one parameter can affect their decisions on another) have to be examined in assessing policy options.

In general, the ultimate actor is the consumer—the owner or other end user of LDVs who purchases vehicles and fuel and, through tax dollars, user fees, and bundled transactions, also pays for roads and other parts of the transportation infrastructure. Through factors including their choices of where to live, work, and shop, consumers determine the urban-regional forms and broader built environment that automotive transportation shapes and serves.

The markets that influence LDV petroleum use and GHG emissions involve cash flow from consumers or other end users to the suppliers of transportation-related products and services, most notably, the automobile industry and the motor fuels industry.[6] In most cases, policies designed to influence decisions about motor vehicle purchase and use are directed at these entities rather than at the consumer.[7] For example, motor vehicle fuel economy standards are imposed on vehicle manufacturers, not vehicle purchasers. The penalties for not meeting such standards are directly imposed on these firms and, along with the costs of meeting the standards, may be wholly or in part passed onto consumers. It is therefore left to vehicle manufacturers and fuel producers not only to develop and produce the products required to meet the regulations to which they are subject but also to generate the economic signals that induce the purchase of their products in the required quantities by consumers.

6.3 POLICIES AIMED AT REDUCING VEHICLE ENERGY INTENSITY

The ultimate aim of policies to reduce vehicle energy intensity is to lower the average actual on-road fuel consumption of the total LDV fleet. There are two broad approaches available for achieving this. The first is to reduce the average fuel consumption of the typical new vehicle, in all size classes, largely through incorporating technologies that reduce fuel consumption. The second is reduce or eliminate the heaviest and thus least efficient vehicles in the LDV fleet by encouraging the purchase and use of lighter vehicles, which can lead to reduced performance or utility (e.g., reduced load-carrying ability or acceleration). Individual policies can emphasize one of these two approaches, can encourage one while discouraging the other, or can be neutral.

6.3.1 Vehicle Energy Efficiency and GHG Emissions Standards

Several countries have enacted standards that mandate the level of energy efficiency or the level of CO_2 emissions that the average newly produced vehicle must achieve by a certain date. Anderson et al. (2011), Eads (2011), and An et al. (2007) describe the vehicle efficiency and GHG emissions standards programs that are in place or under development around the world. The CAFE standards have been in effect the longest, have been studied extensively, and are most pertinent to the committee's task.

6.3.2 U.S. CAFE Standards

The initial CAFE standards were enacted as part of the 1975 Energy Policy and Conservation Act (see Figure 2.1 for historical and projected LDV vehicle fuel economy). Although the U.S. standards are considered to be regulatory rather than economic, they are enforced through economic penalties. Manufacturers whose annual factory sales of vehicles do not meet the CAFE standards for each of their fleets (domestic and imported cars and domestic and imported trucks) must pay a civil penalty.[8] For the 2011 model year,

---

[6]Although the complex interactions and transactions that determine the provision of transportation infrastructure, associated land-use patterns, and related services are difficult to characterize as a distinct "market," they also involve a set of actors whose decisions can be viewed through an economic lens.

[7]There are exceptions. As is discussed below, "feebates"—subsidies for more fuel-efficient vehicles and taxes on less fuel-efficient ones—cause resources to flow directly between the government and consumers. The same thing is true of direct tax credits.

[8]"Factory sales" are sales by the manufacturer to the dealer. Therefore, the number of vehicles of a certain model year actually reaching the con-

the penalty was $5.50 for each tenth of a mile per gallon that the manufacturer's average fuel economy fell short of the standard, multiplied by the total volume of vehicles in the affected fleet (EPA, 2009). As of July 2011, NHTSA had collected a total of $795 million in civil penalties over the life of the CAFE program (NHTSA, 2011).

6.3.2.1 *The Lag Between the Fuel Economy of New Vehicles and That of the On-Road Fleet*

There is a significant difference between the fuel economy of the average new vehicle and that of the average on-road vehicle. The average LDV's lifetime has been increasing and, according to R.L. Polk, is now about 10.8 years (R.L. Polk and Company, 2011). Vehicles are driven less as they age, and so it takes about 15 years for the age- and travel-weighted average fuel economy of the on-road fleet to reach 90 percent of the average level of new vehicles in a given year, based on the most recently published vehicle survivability statistics (NHTSA, 2006). The CAFE standards apply only to the new-car fleet, and rarely has an effort been made to impact the pace of fleet turnover.[9]

6.3.2.2 *Recent Changes in the U.S. CAFE Standards*

In 2007, new legislation set a fuel economy target of 35 mpg (2.9 gal/100 mi) for the combined LDV fleet of cars and trucks, to be achieved by model year 2020.[10] The legislation authorized NHTSA to set standards on the basis of vehicle attributes. The agency settled on vehicle "footprint," defined as the track width times the wheelbase, as a basis for all LDV standards, building on the similar approach adopted in the 2006 CAFE reform rule for light trucks. Therefore, CAFE standards now vary with the size mix of an automaker's fleet (Box 6.1). Pursuant to the Obama Administration's agreement with automakers and other parties to develop a single national program for CAFE standards in coordination with federal and California LDV GHG emissions standards, a more ambitious target date was set, requiring that a 35.5 mpg CAFE-equivalent (counting non-fuel-economy-related GHG emissions) new fleet average be met by model year 2016 (EPA and NHTSA, 2010). This target implies an annual rate of improvement in average new LDV fleet fuel economy of 5 percent.

In November 2011, NHTSA and EPA jointly published a Notice of Proposed Rulemaking to further strengthen CAFE

---

sumer differs somewhat from that model year's factory sales. Manufacturers can carry forward or backward excess CAFE credits for 3 model years in order to offset any shortfalls to a given fleet. Manufacturers cannot transfer credits between fleets or between manufacturers. Penalties are assessed for a given model year and fleet if any shortfall in CAFE during that model year is not offset by these credits (NHTSA, 2012).

[9]The most notable exception was the "Cash for Clunkers" program adopted by the Obama Administration in 2009. This is discussed in more detail below.

[10]This legislation was the Energy Independence and Security Act.

> ### BOX 6.1
> ### The "Footprint" Approach
>
> According to the "footprint" formula now used in computing CAFE, in model year 2016 a compact car such as the Honda Fit, with a model footprint of 40 square feet, would have a fuel economy target of 41.4 mpg (2.42 gal/100 mi), while a full size car, such as the Chrysler 300, with a model footprint of 53 square feet, would have a fuel economy target of 32.8 mpg (3.05 gal/100 mi). A large pickup truck such as the Chevrolet Silverado, with a model footprint of 67 square feet, would have a fuel economy target of 24.7 mpg (4.05 gal/100 mi).
>
> ---
>
> SOURCE: Davis et al. (2011), Table 4-19.

standards and GHG emissions standards for LDVs for the model year (MY) 2017-MY2025 period. The agencies proposed an increase in the standards to a MY2025 target of 54.5 mpg, with GHG emissions reductions (CO_2 equivalent) corresponding to a fleet average of 163 g/mi (EPA and NHTSA, 2011). In fuel economy terms, the agencies project LDV fleet average compliance levels of 40.9 mpg (2.4 gal/100 mi) in 2021 and 49.6 mpg (2.0 gal/100 mi) in 2025. The agreement would provide CAFE credits for the production of vehicles employing certain advanced technologies. The initial 5-year phase, for MY2017-MY2021, provides for a slower rate of increase for light trucks, averaging 2.9 percent per year, compared to a 4.1 percent increase in passenger car standards for the same period. The program also provides for a comprehensive mid-term evaluation prior to finalization of the MY2022-MY2025 standards. Although subject to revision under the mid-term review, the rates of increase proposed for the second phase of the program, covering MY2022-MY2025, are 4.7 percent per year for light trucks and 4.2 percent per year for passenger cars. The projected average annual rate of fuel economy increase for the recently finalized and currently proposed CAFE regulations is 3.6 percent per year over the 2010-2025 period, rising from an achieved MY2010 compliance level of 29.3 mpg (NHTSA, 2012)

These two rulemakings reflect a significant change in the way CAFE standards are developed and issued. Previously, the task had been solely the responsibility of NHTSA, in consultation with other agencies such as the EPA. The standards applied only to the fuel economy of new vehicles. However, NHTSA and the EPA issued the final MY2010-MY2016 standards jointly, and the MY2017-MY2025 standards are being developed and proposed by both agencies in order to address the fuel economy of vehicles and the GHGs they emit.

NHTSA's authority for issuing fuel economy standards remains the 1975 Energy Policy and Conservation Act, as amended by the 2007 EISA. EPA's authority for GHG emissions standards is the CAA. The factors NHTSA may consider in developing fuel economy standards are not precisely the same as the factors that EPA may use in developing GHG emissions standards, and so the promulgation of a rule covering both fuel economy and GHG emissions requires a considerable amount of interagency coordination to ensure a consistent set of requirements. An additional level of coordination is involved because the state of California, subject to EPA waiver of the CAA preemption provision, has authority to set its own motor vehicle emissions standards. California has agreed to harmonize its standards with the EPA and NHTSA under the single national program terms.

6.3.3 Subsidies for More Fuel-Efficient Vehicles and Fees on Less Fuel-Efficient Vehicles

Another policy for encouraging the production and sale of vehicles that are more fuel-efficient and/or emit less CO_2 is to use subsidies, taxes, or both, based on fuel use, CO_2 emissions, or a combination. In the United States, a gas guzzler tax was established by the Energy Tax Act of 1978. Phased in over 1981-1985, this program now involves a graduated level of taxation on passenger cars having a fuel economy below 22.5 mpg (regulatory level, as used for CAFE standards).[11] The gas guzzler tax is proportional to the increase in fuel consumption rate above that of a 22.5 mpg car and the current maximum is $7,700 on cars rated at less than 12.5 mpg. The gas guzzler tax does not apply to light trucks and for at least the past two decades has applied to only a small fraction of vehicles, typically high-performance sports and luxury cars. In its early years, the gas guzzler tax was effective in helping to motivate fuel economy improvements in the least efficient cars in the fleet (Khazzoom, 1994; DeCicco and Gordon, 1995). Japan, many countries in Western Europe, and a few others have had graduated vehicle taxation schedules based on fuel consumption, engine displacement, or some other metric defined for tax purposes. Some of these programs have been recast in recent years to be based on vehicle CO_2 emissions rate.

When subsidies for efficient vehicles are added to a vehicle taxation program, it becomes what is referred to as a "feebate" program. Such a program was under discussion as part of the response to the 1973 energy crisis (Difiglio, 1976), but only the gas guzzler tax portion was implemented. Over the years, feebate programs were proposed in a number of states but were never enacted. In 1991, the Canadian Province of Ontario enacted a tax for fuel conservation that levied modest graduated taxes on inefficient vehicles and provided subsidies for a subset of efficient vehicles.

In recent years, France has pursued a feebate-type program, known as the "bonus-malus" system, applied at the time of purchase of a vehicle (Bastard, 2010). The amount charged ("malus") or rebated ("bonus") depends on the vehicle's CO_2 type approval test emissions figure.[12] Originally, the amounts ranged from a bonus payment of €1,000 for cars rated under 100 g/km to a fee of €2,600 for cars rated above 250 g/km. A bonus payment of €5,000 applied for vehicles with a CO_2 emissions value below 60 g/km. The incentive provided by these bonus-malus values has been estimated to be broadly equivalent to €150/metric ton of CO_2 (Bastard, 2010).

According to Bastard (2010), "the system demonstrated high effectiveness: in 2008, CO_2 emissions from new vehicles in France fell by 9 g/km compared to 2007, falling from 149 g/km to 140 g/km, most of the decrease resulting from the bonus-malus system." The decrease resulted from three separate impacts: (1) a downsizing in the segment mix, (2) a downsizing in power, and (3) a move to diesel in certain segments. The measure was intended to be revenue neutral, but has turned out to have a net cost for the French state, as the shift in the market to smaller vehicles was higher than anticipated. Bastard estimates that the net budgetary cost was approximately €200 million in 2008 and €500 million in 2009.[13]

6.3.4 Motor Fuel Taxes as an Incentive to Purchase More Fuel-Efficient Vehicles

A third type of policy to incentivize the purchase of more fuel-efficient vehicles is motor fuel taxes. Nearly every country levies taxes on motor fuel, but the level of tax varies widely. Table 6.1 shows the variance in motor fuel taxes for several major developed countries, in 1990 and in 2010.

Fuel prices impact both vehicle purchase decisions with respect to fuel economy and how much vehicles are driven. The sum of these impacts is measured by the elasticity of demand for fuel—defined as the percentage change in fuel purchased divided by the percentage change in fuel price. This elasticity has been estimated by many studies, which generally differentiate between short-term (2 years or less) and long-term (more than 2 years) elasticity. Short-term elasticity generally is interpreted as reflecting changes in VMT. Long-term elasticity is interpreted as reflecting changes in the fuel economy of vehicles purchased and the long-term VMT changes generated by changes in where people live and work.

In January 2008 the Congressional Budget Office (CBO) reviewed the literature on fuel price elasticity and concluded:

> Estimates of the long-run elasticity of demand for gasoline indicate that a sustained increase of 10 percent in price eventually would reduce gasoline consumption by about 4 percent. That effect is as much as seven times larger than the

---

[11]See http://www.epa.gov/fueleconomy/guzzler/index.htm for a gas guzzler tax program overview and lists of vehicles subject to the tax.

[12]This is similar to the "as tested" CAFE standard.

[13]Bastard (2010), p. 25. The tax and subsidy values have been adjusted in an effort to make the system more nearly revenue-neutral.

TABLE 6.1 Gasoline and Diesel Prices, Tax, and Percent Tax in 1990 and 2010

	France		Germany		Japan		United Kingdom		United States	
	1990	2010	1990	2010	1990	2010	1990	2010	1990	2010
Gasoline										
Total price	$5.60	$6.72	$4.09	$6.86	$4.87	$5.93	$4.35	$6.81	$2.08	$2.71
Tax	$3.97	$4.16	$2.56	$4.30	$2.29	$2.75	$2.63	$4.37	$0.56	$0.50
Percent tax	70.9%	61.9%	62.7%	62.7%	47.1%	46.3%	60.4%	64.2%	26.7%	18.2%
Diesel										
Total price	$2.66	$5.59	$4.06	$5.94	$2.61	$5.04	$3.05	$6.95	$1.48	$2.94
Tax	$1.67	$3.01	$2.30	$3.25	n/a	$1.66	$1.80	$4.39	$0.41	$0.53
Percent tax	62.8%	53.8%	56.6%	54.7%	n/a	32.9%	59.0%	63.1%	27.9%	17.9%

SOURCE: Data from Davis et al. (2011), Figures 10.2 and 10.3.

estimated short-run response, but it would not be fully realized unless prices remained high long enough for the entire stock of passenger vehicles to be replaced by new vehicles purchased under the effect of higher gasoline prices—or about 15 years . . . consumers also might adjust to higher gasoline prices by moving or by changing jobs to reduce their commutes—actions they might take if the savings in transportation costs were sufficiently compelling. Those long-term effects would be in addition to consumption savings from short-run behavioral adjustments attributable to higher fuel prices. CBO (2008)

A 2009 study titled *Moving Cooler: An Analysis of Transportation Strategies for Reducing Greenhouse Gas Emissions* (Collaborative Strategies Group, 2009) modeled how much lower LDV fleet GHG emissions would be in 2050 (relative to 2005) if fuel prices or carbon taxes were used to boost U.S. motor fuel prices to West European levels. The elasticities used in the analysis were comparable to those cited in the 2008 CBO study. The reduction as a result of the improved fuel economy portion of the fuel price impact was 19 percent. The reduction as a result of the VMT impact portion was 8 percent (Collaborative Strategies Group, 2009 pp. B-15 and D-11).

Fuel taxes can also differ by fuel type, thereby influencing the choice of engine used to power a vehicle. In Europe, most vehicle models are available in both gasoline and diesel versions. The diesel versions cost more but deliver better fuel economy. France, in particular, taxes diesel at a much lower rate than gasoline—in 2010, the tax on diesel was $3.01/gal whereas the tax on gasoline was $4.16/gal. That differential has been credited with being an important factor in causing a rise in the diesel share of new automobiles in France from 2 percent in 1973 to 74 percent in 2007.

Fuel economy improvements reduce motor fuel tax revenues, all else equal, because under current law the amount of tax per gallon of fuel is constant. Inflation also erodes the real value of fuel tax revenues. Finally, substitution of hydrogen or electric vehicles for conventional vehicles would further diminish tax revenues unless those fuels were brought within the purview of the tax law. One solution would be to tax all

forms of energy used by vehicles and index the motor fuel tax to inflation and also to the average energy efficiency of all vehicles on the road. For example, if total vehicle miles of travel per unit of energy increased by 3 percent from one year to the next, the tax in the following year would be increased by 3 percent. Such an indexed highway user fee on energy would maintain a constant tax rate per vehicle mile of travel while encouraging car buyers to purchase energy-efficient vehicles.

6.3.5 A Price Floor Target for Motor Fuels

A major impediment to investment in new alternative technologies, even when petroleum prices are high, is uncertainty about the future price path. Investors and consumers are less likely to invest in fuel-efficient technologies that require substantial up-front costs when they are uncertain about the payoffs from those investments. Prices of crude oil have been volatile in the past (Figure 6.2). In the late 1970s and early 1980s, the private and public sectors invested heavily in alternative fuels and AFVs, but many of the alternatives became uneconomic when prices of crude oil fell in the mid-1980s and remained low until the early 1990s.

One policy to stabilize the prices of petroleum-based fuels at a level that will help ensure a transition to more energy efficiency is through the use of a tax or surcharge on the price of oil that is applied only when oil prices fall below a specified target price. This surcharge would then be inversely related to the price of oil. For example, if the target price of oil with existing taxes is $90/bbl, and the price falls to $85/bbl over a specified period, the surtax would be $5/bbl, ensuring that the market price remains at $90/bbl. If the market price fell below $85/bbl, the surtax would increase, and if the market price rose above $90, then the surtax would be zero. The setting of the target price would be a policy choice made by Congress and the President and implemented in ways similar to other taxes on oil sales with the goal of stabilizing prices of petroleum-based fuels above a minimum price.

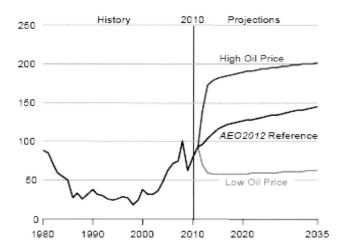

FIGURE 6.2 Actual average annual world oil prices from 1980 to 2010 and projected annual world oil prices from 2010 to 2035 under three different scenarios (in 2010 dollars per barrel). SOURCE: *Annual Energy Outlook 2012* (EIA, 2012).

Such a price floor or fuel price stabilization policy could be implemented on crude-oil sales in the United States as suggested above, or it could apply only to imported crude oil (Hubbard and Navarro, 2010). Borenstein (2008) shows how the concept of the oil price floor could be tied only to gasoline or other specific fuels that are derived from crude oil.

Revenues from any such surcharge would vary over time. They could be earmarked for use in current and proposed subsidies for alternative vehicles and AFVs, or used more broadly for tax or deficit reduction.

6.3.6 Policies to Change the Size and Weight Composition of the LDV Fleet

The average on-road fuel economy of the LDV fleet can also be changed by altering the fleet's composition. One example of the impact that such a change can have is the decline in U.S. on-road fleet average fuel economy due to the increase of trucks in the fleet mix between MY1980 and MY2004.[14] Policies could be designed to encourage or discourage such a shift.

An example of a policy change that discourages a shift in fleet mix is the 2007 legislation updating the CAFE program. Before then, CAFE standards were set for and had to be met by each of four fleets (U.S. cars, imported cars, U.S. trucks, and imported trucks) of each manufacturer selling vehicles in the United States. This approach permitted manufacturers to "downsize" or "upsize" their fleets as part of their CAFE fulfillment strategy and helped lead to the proliferation of trucks and SUVs in the fleet mix. The reform rule of 2006 restructured light-duty truck standards, basing them on a

vehicle attribute, for which NHTSA selected footprint as discussed above. The EISA 2007 legislation authorized similar restructuring for all LDVs, and footprint-based standards have subsequently been promulgated for passenger cars and light trucks. Although the exact effect depends on the shape of the curve that maps vehicle footprint to regulatory targets for fuel economy, a general intent of this structure is that similarly sized vehicles would be required to achieve a similar increase in fuel economy. The regulatory curves are flattened at the extremes, to avoid standards that are too stringent for the smallest vehicles or standards that are too weak for the largest vehicles.

Designing and estimating the effects of such standards involve complex evaluations of many factors that influence vehicle design and engineering, customers' preferences, and automakers' product strategies. The analysis given in the EPA and NHTSA rulemaking studies concludes that the adopted footprint-based standards appropriately balance the many considerations that the agencies were required to weigh and does not provide any motivation for automakers to change their fleet mixes for CAFE purposes. Some have argued, though, that the chosen footprint curves inhibit downsizing as a cost-effective compliance strategy and may create an incentive to upsize the LDV fleet in a way that reduces fuel savings and GHG emissions reductions attributable to the standards (Whitefoot and Skerlos, 2012).

Other policies, such as the "Cash for Clunkers" program undertaken in 2009, have been designed to encourage consumers to dispose of lower-efficiency vehicles (which were then rendered inoperable) and replace them with higher-efficiency new vehicles, providing a stimulus to new-car sales. While the program was operating, it encouraged the purchase of fuel-efficient vehicles (Yacobucci and Canis, 2010). Because the program was temporary, most observers believe that it operated primarily to shift vehicle purchases in time rather than achieve any long-term impact on fleet composition. In a report published in October 2009, Edmunds.com estimated that of the nearly 690,000 new vehicles sold during the period the program was operating, only 125,000 of the sales were incremental (Edmunds.com, 2009).

6.3.7 Assessment of Vehicle Fuel Economy Improvement Strategies

The various policies described above each have demonstrated a potential to reduce the LDV fleet's average fuel consumption. It is generally agreed that the U.S. CAFE standards have been effective in stimulating the production and sale of more fuel-efficient vehicles (NRC, 2002). According to the EPA, the composite average LDV new-vehicle fuel economy (laboratory rated at 55 percent city driving and 45 percent highway driving) increased from 15.3 mpg (6.5 gal/100 mi) in MY1975 to 28.6 mpg (3.5 gal/100 mi) in MY2011, the latest year for which data have been published (EPA, 2010). Most of this increase occurred between

---

[14]In 1980, "trucks" accounted for 16.5 percent of LDV production; by 2004 the truck share had reached 52.0 percent (EPA, 2010, Table 1, p. 7).

MY1978 and MY1988. The political acceptability and track record of the CAFE program have established it as a leading option among policies for meeting LDV petroleum use and GHG reduction goals. However, as discussed elsewhere in this chapter, a strict CAFE standard alone is not sufficient for meeting ambitious petroleum and GHG reduction goals because it fails to address issues of consumer motivation, travel demand, and other factors that shape the on-road fuel consumption of the LDV fleet.

Although there is less experience with their use, subsidies and taxes based on projected vehicle fuel consumption and imposed at the time of vehicle acquisition (feebates) could supplement (or, in principle, even substitute for) CAFE standards. So also could higher fuel taxes. Both types of policies have been shown to be effective in encouraging the purchase (or lease) of more fuel-efficient vehicles. However, the reluctance in the United States to raise taxes of any kind, consumers' undervaluation of fuel economy, and the level to which taxes would have to be raised to achieve results comparable to those seen with fuel economy standards, especially if supplemented by feebates, make their use problematic.

6.4 POLICIES TO REDUCE THE PETROLEUM USE IN OR GHG EMISSIONS IMPACTS OF FUEL

The second major factor influencing the LDV sector is petroleum's share of fuel use or the overall GHG emissions impact of supplying and using the fuel. Although numerous policies intended to reduce petroleum use by LDVs have been pursued over the years (e.g., the Energy Security Act of 1992 [96 P.L. 294], the Alternative Motor Fuels Act of 1988 [100 P.L. 494), the EPAct of 1992 and 2005 [109 P.L. 58], and the EISA of 2007 [110 P.L. 140] (DOE-EERE, 2011b), to date they have had little impact on the overwhelming dominance of petroleum-derived gasoline and diesel fuel. Nevertheless, many policy makers still show considerable interest in pursuing similar strategies for encouraging or mandating the use of biofuels, natural gas, hydrogen, electricity, or other non-petroleum fuels to power LDVs. Regulations, subsidies, various forms of tax incentives, and loan guarantees are now being used both in the United States and other countries to encourage the use of non-petroleum-based fuels and fuels that are expected to emit fewer GHGs. Fuel taxes and price floors on petroleum-based fuels also would discourage petroleum use.

Although the goals of reducing petroleum use and GHG emissions commonly are treated together (as in the case of the statement of task for this study; see Appendix A), the scientific, economic, and technical issues associated with these two goals are not identical. Each goal has its distinctive challenges associated with the design of fuels policies. Implementing any such policies requires appropriate metrics and the ability to track and measure effects throughout the fuel supply, distribution, and end-use systems that the policies seek to influence.

Measuring and tracking petroleum reduction require that the feedstocks used for producing fuel be quantified and reported. Given a legal definition of what qualifies as "non-petroleum" (e.g., as specified by AMFA [1988] and subsequent energy legislation), determination of the extent of petroleum reduction is conceptually straightforward. However, determining various fuels' net GHG emissions impacts is difficult, for several reasons:

- At least some of the GHG emissions or CO_2 uptake occurs upstream from the use of any fuel. For example, battery electric vehicles do not have tailpipe emissions, but the production of electricity to fuel the batteries may emit GHGs.
- The quantification of net CO_2 uptake, sequestration, and related emissions is uncertain in some cases. For example, the storage of carbon in soil by perennial bioenergy feedstocks depends on prior land condition and is difficult to estimate with high certainty.
- A significant portion of the GHG emissions impacts of all alternative fuels occurs outside the LDV sector. For example, the GHG emissions from electricity generation for powering battery electric vehicles or for producing hydrogen to power hydrogen fuel-cell vehicles occur in the power generation or hydrogen production sector. Therefore, the GHG emissions must be tracked in multiple sectors beyond the LDV sector.
- Biofuel-induced land-use changes and nitrous oxide flux from nitrogen fertilization could affect the net GHG emissions effects. Yet, the quantification of those net GHG emissions effects could be difficult.

6.4.1 Tax Incentives for Fuels and Their Infrastructure

During the energy crisis in the 1970s, policies were developed to support alternatives to petroleum ranging from synthetic fossil fuels to biofuels. The Energy Tax Act of 1978 (P.L. 95-618) introduced an excise tax exemption for gasohol. The exemption subsequently was extended and transformed into a tax credit for ethanol called the VEETC, a $0.45/gal tax credit. Congress also approved a tariff on imported ethanol to foster domestic biofuel production. Both the VTEEC and the tariff expired at the end of 2011.

The EPAct of 2005 also established a tax credit of up to $30,000 for the cost of fueling equipment for alternative fuels including hydrogen, natural gas, propane, electricity, E85, and diesel fuel blends containing at least 20 percent biodiesel.[15] Residential fueling equipment was eligible for a credit of up to $1,000. The tax credits for hydrogen run through 2014; they expired at the end of 2011 for all other fueling equipment (DOE-EERE, 2011a).

---

[15]California and a number of other states have policies for subsidizing a variety of alternative fuel vehicles (AFVs) and related infrastructure.

6.4.2 Fuel-Related Regulations

Traditional fuel regulations, as authorized under the original language of the CAA's Section 211, addressed fuel composition and its physical and chemical properties. These fuel-performance standards were based in principle on measurable fuel properties. Fuel suppliers could certify their products through laboratory testing or analytic methods based on physiochemical characteristics. Regulators could readily verify that standards were met by directly sampling fuel products, although this was rarely done. Because fuel additives and formulation requirements may not be finally incorporated into a consumer fuel until they are blended in at a distribution terminal, tanker truck, or even a fuel pump, the regulated entity may vary in fuel standards (40 CFR 80.2, Definitions).[16] The point of regulation is the point of finished fuel product distribution, which is where most fuel properties are determined.

Compliance with the complex model for gasoline emissions was a departure from this standard. There are a large number of fuel parameter combinations that could meet the requirements for compliance, and compliance was referenced to the base fuel of each individual fuel supplier. Compliance was determined before the fuel left the production facility. Once the fuel was distributed and comingled with other fuels that complied at production, it was no longer possible to determine compliance at the final point of distribution.

As energy policy considerations came into play, regulations were designed to stipulate the use of certain fuels derived from specified non-petroleum feedstocks. Thus, fuel regulations developed for energy policy take the form of a legal requirement to supply a certain amount of a fuel manufactured from particular resources. Others take the form of a requirement to supply a minimum percentage of a group of fuels derived from desired sources or to supply a mix of fuels that on average meet requirements for being derived from desired sources. Such is the case for the amended Renewable Fuel Standard (RFS2) in the EISA. An approach generalized to require a mix of unspecified fuels that meet specified average net GHG emissions over their life cycle is known as a Low-Carbon Fuel Standard (LCFS), and such a standard has been established in California.

6.4.3 Renewable Fuel Standard[17]

RFS2 was intended to move the United States "toward greater energy independence and security" and to "increase the production of clean renewable fuels" (110 P.L. 140). RFS2 is actually a collection of mandates for fuel providers to supply categories of renewable fuels defined by their feedstock type and life-cycle GHG emissions (Box 6.2). The

volume mandate for the "renewable fuel" category has been met by corn-grain ethanol and is expected to be met up to 2022 (NRC, 2011). Production capacity is available for meeting the volume mandate for biomass-based diesel. However, commercial production of cellulosic biofuels has fallen far short of the volume for that category mandated by EISA. Indeed no compliance-tracking renewable identification numbers (RINs) had been generated for cellulosic biofuels as of April 2012 (EPA, 2012a).[18]

EISA gives EPA the right to waive or defer enforcement of RFS2 under a variety of circumstances. For example, RFS2 can be waived if sufficient biofuels are not likely to be produced for blending or if its enforcement has been deemed to cause economic dislocation (NRC, 2011). For example, the governors of nine states, 26 members of the U.S. Senate, and 156 members of the U.S. House of Representatives petitioned EPA to grant the RFS waiver, citing the effects of the 2012 drought on U.S. food and feed prices as the reason for potential economic dislocation. The EPA has been exercising its discretion to reduce the level of cellulosic biofuels required in RFS2. Specifically, the EPA reduced the mandate for cellulosic biofuels by 93 percent in 2010 (from 100 million to 6.5 million gallons), by 97 percent in 2011 (from 250 million to 6.6 million gallons), and by 98 percent for 2012 (from 500 million to 10.5 million gallons) (EPA, 2012b). When there is a waiver, blenders are permitted to buy RINs from EPA instead of actually purchasing cellulosic biofuels. There also is a clause that allows blenders to buy RINs from EPA even if cellulosic biofuels are available but substantially more expensive than petroleum-based fuels (Thompson et al., 2010; NRC, 2011). Although the intent was to protect consumers from high prices relative to gasoline, the clause effectively eliminates a guaranteed demand for cellulosic biofuels. The potential waiver and clause regarding the purchase of RINs reduce the incentive for the major fuel producers to develop and deploy technology for producing cellulosic biofuels, particularly when large financial investments and risks are involved. But without the waiver, blenders are required to purchase fuel that is not being made; their only option is to buy RINs. The cost of cellulosic fuels has not come down as some had hoped. The combination of high cost, the potential waiver, and the clause described above have undermined the effectiveness of RFS2 in driving an increase in cellulosic biofuels.

---

[16]Code of Federal Regulations, Title 40, Part 80, "Regulation of Fuels and Fuel Additives," Definitions section; available at www.gpoaccess.gov/cfr/index.html.

[17]This description is taken from EPA's website.

[18]"The Renewable Identification Number (RIN) system was developed by the EPA to ensure compliance with RFS2 mandates. A RIN is a 38-character numeric code that corresponds to a volume of renewable fuel produced in or imported to the United States. RINs are generated by the producer or importer of the renewable fuel. RINs must remain with the renewable fuel as the renewable fuel moves through the distribution system and as ownership changes. Once the renewable fuel is blended into motor vehicle fuel, the RIN is no longer required to remain with the renewable fuel. Instead, the RIN may be separated from the renewable fuel and then can be used for compliance, held for future compliance, or traded" (McPhail et al., 2011, p. 5).

**BOX 6.2
Life-Cycle Assessment for
Greenhouse Gas Emissions**

Life-cycle assessment (LCA) is a tool available for the accounting of net greenhouse gas (GHG) emissions effects of different fuel pathways. However, the use of LCA to determine policy compliance is a marked departure from traditional approaches to fuels regulation, which prior to the RFS had always been based on physiochemical fuel properties. Standards based on a fuel's physiochemical properties are enforceable through measurement or measurement-based analytic methods that allow verifiable assurance of fuel providers' compliance. However, fuel property standards are not adequate for regulating the GHG emissions associated with both production and use of a fuel. Fuel property standards cannot account for upstream emissions associated with any fuel. Therefore, LCA is used to assess the GHG emissions impacts of fuels. However, GHG emissions occur in multiple sectors in geographically dispersed locations and over multiple periods of time. For example, for biofuels, CO_2 uptake by biomass and sequestration in soil or GHG emissions from indirect land-use changes occur remotely from locations of fuel use in the transportation sector. Thus, accounting for life-cycle emissions is more complicated and uncertain than it is for direct emissions (NRC, 2011).

Some members of the committee believe that a problem with using LCA in policy regulation is a misplaced burden of proof (DeCicco, 2012) because some of the CO_2 sequestration and emissions occur outside the LDV sector and are not under the control of fuel producers, fuel retailers, or fuel users. Others believe that it is appropriate to hold fuel providers responsible for the upstream emissions of their products. The parties that are responsible for the direct and indirect emissions from all the different parts of the biofuel supply chain have not been clearly established. If the United States is to limit the GHG emissions impacts of LDVs and their associated fuel supply systems (as opposed to their direct tailpipe emissions only), then policies are needed to address the GHG emissions from other sectors upstream from fuel use. Although GHG emissions from the transportation sector could be reduced in the United States by RFS2, the policy may not contribute to reducing global GHG emissions.

RFS2 requires EPA to determine whether the four types of renewable fuel meet their respective GHG thresholds. Although the intent was to ensure that biofuels have lower GHG emissions impacts compared to petroleum-based fuels, whether the policy will actually contribute to a reduction in GHG emissions is uncertain. The NRC report *Renewable Fuel Standard: Potential Economic and Environmental*

Effects of U.S. Biofuel Policy concluded that "RFS2 may be an ineffective policy for reducing GHG emissions because the effect of biofuels on GHG emissions depends on how the biofuels are produced and what land-use or land-cover changes occur in the process" (NRC, 2011; p. 2-4). The same physical fuel can have widely different life-cycle GHG emissions depending on numerous factors, including the feedstock used (e.g., corn stover or switchgrass), the management practices used to produce the feedstock (e.g., nitrogen fertilization during biomass growth), the energy source used in the biorefinery (e.g., coal or renewable electricity), and whether any indirect land-use changes were incurred as a result of feedstock production. For example, the use of crop or forest residues for feedstock is less likely to cause indirect land-use changes than is the use of planted crops. Moreover, indirect land-use changes as a result of bioenergy feedstock production and the associated GHG impacts are difficult to ascertain.

6.4.4 Possible Alternative to RFS2

Because GHG sources and sinks are dispersed across sectors (agricultural, forestry, and industrial) and international borders, some committee members believe that policies that target them at the location where they occur are likely to be much more effective than RFS2 in reducing GHG emissions impacts. RFS2 includes a GHG accounting system that can account for upstream emissions. This system requires an elaborate tracking mechanism and a combination of real-world measurement and estimation of GHG emissions at each source and sink along the supply chain to verify overall claimed benefits from the production and transport of the biomass through conversion and distribution of the final products.

6.4.5 California's Low Carbon Fuel Standard

A regulatory effort to encourage the use of alternative fuels, with the specific intent of lowering GHG emissions, is California's LCFS. On January 18, 2007, California's then-governor issued Executive Order S-1-07 that called for a reduction of at least 10 percent in the carbon intensity of California's transportation fuels by 2020. The California Air Resources Board developed regulations to implement the order, approving them in April 2009. After delays due to litigation the regulations were promulgated on June 4, 2012, under an April 2012 court order permitting the promulgation to occur pending the results of an appeal that was still underway at the time of this writing.

The LCFS uses life-cycle assessment (LCA) rather than direct measurement of fuel properties to determine compliance. It applies to essentially all transportation fuel used in the state. Regulated parties are defined broadly as fuel producers and importers and some owners of alternative fuels or alternative fuel sources. The regulation defines a carbon intensity (CI)[19] metric based on LCA. Fuel suppliers are required to progressively lower the average CI of the fuel they supply. The targeted GHG-emission reduction is 10 percent in 2020 compared to the average baseline of transportation fuels in 2010. The LCFS assigns a CI to different types of biofuel (e.g., corn-grain ethanol produced via different pathways with different types of energy input gets different scores) and a CI for land-use change and other indirect GHG effects. However, the actual GHG effects from land-use change and other indirect effects could span a wide range (Mullins et al., 2010; Plevin et al., 2010). Given the large uncertainties, the extent to which LCFS actually contributes to reducing net GHG emissions is unclear. One committee member considers that the uncertainties in LCA are such that one cannot have confidence in the efficacy of an LCFS or other policies using LCA to ensure reductions of GHG emissions from fuels. As is the case with RFS2, fuel providers are held accountable for upstream GHG emissions and GHG emissions from indirect effects. They do not control these effects but can mitigate them by their choice of the source of their fuel supply.[20]

LCFS allows fuel providers to petition for individualized CI score. LCFS proponents view such provisions as beneficial for fostering innovation in low-emission fuel production. For example, some ethanol producers could sequester their CO_2 emissions, account for them, and seek credit for these reductions under the LCFS. Similarly, oil companies practicing enhanced oil recovery could seek credits for the portion of the injected CO_2 that remains in the water phase of the oil well and the portion that is dissolved in the unrecovered oil.

6.5 POLICIES TO IMPACT VEHICLE MILES TRAVELED

Since 1970, increases in U.S. LDV vehicle miles traveled have more than offset improvements in LDV on-road fleet fuel economy (see Figure 6.1). As a result, LDV petroleum use and CO_2 emissions have increased over the period. With VMT being such a driver of increased petroleum use and CO_2 emissions, it is natural that attention has been devoted to finding ways of reducing its rate of growth—or even its absolute total. This section reviews the principal policies that have been examined and what is known about their likely impact.

6.5.1 Historical and Projected Future Growth in LDV VMT

Between 1970 and 2005, VMT in the U.S. LDV fleet grew by an average annual rate of 2.8 percent. This rate of growth is not expected to continue. Indeed, the average annual rate of VMT growth from LDVs projected by the Energy Information Administration's (EIA) *Annual Energy Outlook* (EIA, 2011) for the period 2010 to 2035 is only about 60 percent of the average rate experienced between 1970 and 2007, the peak year prior to the recent economic recession.

But VMT is still projected to grow. Indeed, if the 1.49 percent annual growth rate of VMT over the last 5 years of EIA's projection period is assumed to be realized as well during the 2035-2050 period, VMT in 2050 will be 5.0 trillion—an 85 percent increase relative to its 2010 level.[21]

6.5.2 Reducing the Rate of Growth of VMT by Increasing Urban Residential Density

The relationships among household location, workplace location, trip-making activity, and LDV travel have been subjects of research and policy debate for many years. These relationships have been difficult to establish for many reasons, including the problem of controlling for variables such as self-selection bias as households locate in places that best suit their travel needs, preferences, and capabilities. However, there is general agreement that higher urban density is associated with less driving. The important issues are (1) the magnitude of this relationship and (2) the extent to which VMT might be altered by changes in urban density.

An NRC study analyzed in great detail the impact of compact development (another term for increased urban density) on motorized travel, energy use, and CO_2 emissions. The principal findings of the 2009 study can be summarized as follows:

- Developing more compactly, that is, at higher residential and employment densities, is likely to reduce VMT.
- Doubling residential density across an individual metropolitan area might lower household VMT by about 5 to 12 percent, and perhaps by as much as 25 percent, if coupled with higher employment concentrations, significant public transit improvements, mixed uses, and other supportive demand management measures[22] (NRC, 2009, pp. 2-6).

The 2009 analysis suggests that reductions in national VMT resulting from compact, mixed-use development

---

[19]Carbon intensity as defined in the LCFS is equivalent to life-cycle greenhouse gas emissions.

[20]In 2012, California was enjoined from enforcing the Low Carbon Fuel Standard because of a December 29, 2011, decision by the Federal Eastern District Court of California in the case of *Rocky Mountain Farmers Union vs. Goldstene*. The state is appealing the decision to the Ninth Circuit Court of Appeals, which lifted the lower court's injunction on April 23 and thereby allowed the state to proceed with LCFS implementation pending appeal (CARB, 2012).

[21]This is the number used in the business-as-usual and reference cases and in most of the policy simulations in this report.

[22]The 2009 committee commented on its second conclusion as follows: "Doubling residential density alone without also increasing other variables, such as the amount of mixed uses and the quality and accessibility of transit, will not bring about a significant change in travel" (NRC, 2009, p. 89).

might range from less than 1 percent to 11 percent. The high estimate would require 75 percent of new development to be built at double the density of existing development, a significant departure from the declining densities recorded in most urban areas over the past 30 years. The study emphasizes that increasing densities and mixing land uses may be more achievable in some metropolitan areas than others. Metropolitan areas differ a great deal in their geographic characteristics, land area, historical growth patterns, economic conditions, and local zoning and land-use controls. Policies that affect land use are local in the United States and in some areas in the past have led to decreasing density as urban areas have expanded. In others, strong regional authority with a commitment to more compact land use has increased density through land-use policy.

The present committee concluded that the likely changes in VMT as a result of changes in residential density would be small in the aggregate.

6.5.3 Reducing the Rate of Growth of VMT Through the Use of Pricing Strategies

Many strategies in addition to those encouraging increased residential density have been suggested as having the potential to reduce the rate of growth of VMT. The 2009 *Moving Cooler* study (Collaborative Strategies Group, LLC. 2009) mentioned above examined a number of pricing strategies, including congestion pricing, intercity tolls, pay as you drive (PAYD) insurance, a VMT tax, and a gas or carbon tax. Each of these pricing measures produced a reduction of 1 percent or greater in 2050 urban VMT under all levels of policy intensity studied—extended current practice, aggressive implementation, and maximum implementation. Indeed, the VMT impact of a fee per mile traveled at maximum implementation was estimated to reduce 2050 urban VMT by about 8 percent.[23]

6.5.4 Reducing the Rate of Growth of VMT Through Other Policies

Moving Cooler (Collaborative Strategies Group, LLC. 2009) also examined a range of additional policies deemed to have the potential to reduce the future rate of VMT growth. As in the case with pricing strategies, each of these other policies was evaluated at three levels of implementation. Three of the non-pricing strategies were estimated to have a 1 percent or greater impact on 2050 urban VMT with expanded current practice; four had a 1 percent or greater impact with practice more aggressive than current practice; and five had an impact of 1 percent or greater with maximum implemen-

tation. Although some of these strategies may be additive, many are not. Also, some strategies (such as the transit strategies, pedestrian strategies, and certain of the employer-based commute strategies) may already be reflected in the density-based VMT impacts reported earlier in this chapter. Indeed, the 25 percent reduction in VMT cited in NRC (2009) as a possible upper bound due to higher density was generated by a combination of VMT-related policies, not merely increased density.

6.5.5 Summary of the Impact of Policies to Reduce the Rate of Growth of VMT

Policies designed to reduce the rate of growth of VMT are likely to have limited impact compared with policies targeting vehicle efficiency and new energy sources. Even the extreme reorganization of national economic activity needed to produce the higher level of urban density examined in NRC (2009) would yield only an 11 percent reduction in VMT. And it should be remembered that the various VMT-related policies are not additive. Nevertheless, this limited VMT impact should not lead to the inference that such policies might not be valuable for other reasons.

6.5.6 Policies to Improve the Efficiency of Operation of the LDV Transport Network

As noted above, there has been considerable recent interest in the extent to which policies designed to improve the operating efficiency of the LDV transport network might also serve to reduce GHG emissions or petroleum use. Examples of such policies are eco-driving programs; ramp metering; variable message signs; active traffic, integrated corridor, incident, road weather, and signal control management; traveler information; and vehicle infrastructure integration. Many of these policies focus on reducing congestion to help even out vehicle speeds and reduce time spent stopped in traffic. Others provide drivers with the knowledge and information needed to learn to drive their existing vehicles using less fuel.

There is no dispute that drivers can, if they are careful and attentive, significantly improve the fuel economy they experience on the road. There also is no dispute that congested conditions waste fuel as well as drivers' time. The question is how widespread the use of eco-driving or the implementation of technologies that have a potential to reduce congestion (e.g., vehicle-to-vehicle communications, also known as telematics) become and how great the aggregate impact of such policies and technologies might be at the national level.

The challenge in developing such estimates is somewhat similar to the challenge of estimating the impact of increased urban density on VMT growth. In both cases, examples showing major potential, and sometimes actual, improvements in specific local situations can be cited. But how generalizable are these local results either to other localities

---

[23]"Maximum implementation" is a $0.12/mi fee, representing the increment needed to represent Western European motor fuel tax levels. It was derived based on an additional tax of approximately $4/gal on an approximate average on-road 33 mpg.

or, more importantly, to the national level? And are there factors that can be expected to offset these improvements to some degree over time?

Little research has been done to address these issues. Indeed, the only estimate of the possible national impact over time that the committee is aware of appears in the *Moving Cooler* report discussed above.

That report counts as "benefits" only the fuel savings and associated GHG emissions reductions resulting from the various measures to improve the operational efficiency of the road transport network.[24] It subtracts from these benefits an amount that reflects the VMT increase projected to result from reduced congestion.[25] *Moving Cooler* also takes into account the rate and extent of deployment of these strategies (Collaborative Strategies Group, 2009). Even at maximum deployment, the only strategy that reduces GHG emissions and fuel consumption by more than 0.5 percent as of 2050 is ecodriving, which yields a 4 percent reduction in GHG in that year.[26]

Moving Cooler acknowledges that these estimates are rather rough and might be greater if deployment of the strategies occurs sooner, of development is more widespread, or if the strategies themselves are more effective than they now appear to be. Clearly, there is much need for additional research on this topic.

6.6 POLICIES IMPACTING THE INNOVATION PROCESS

Identification, development, and commercialization of technologies that yield vehicles that are more efficient than current vehicles, AFVs, fuels from non-petroleum resources, fuel production systems with reduced GHG emissions, and, in some cases, even the means of reducing the rate of VMT growth often stem from research undertaken years before the technologies appear in the market. This section examines the different stages of the innovation process to address the questions about the role of government in this process.

There is no universally agreed-upon taxonomy for the stages of the innovation process, but one common framework divides the process into four stages (NSF, 2007):

1. *Research*, or "systematic study directed toward fuller knowledge or understanding." Research may be basic or applied. Basic research is directed toward the "fundamental aspects of phenomena and of observable facts without specific applications toward processes or products in mind." Applied research is directed toward "determining the means by which a recognized and specific need may be met" (NSF, 2007; p. 1).
2. *Development*, which takes the knowledge produced in research and systematically applies it toward the production of useful materials, devices, and systems or methods to meet specific requirements, often culminating in prototypes.
3. *Demonstration*, which tests the feasibility of the developed technology at an appropriate scale to identify all significant impediments to commercial success.
4. *Deployment*, in which the technology becomes widely used.

The need for government intervention is most widely accepted for the first two of these four stages: research and development (R&D). R&D builds the nation's intellectual capability to address energy problems. Even in the presence of strong intellectual property protection, private businesses generally cannot capture all of the benefits generated by their R&D investments, especially any investments that they might make in basic research. Because of this "spillover effect," private investment in R&D falls short of the socially optimal amount, thereby justifying public support.

There is an even stronger case to be made for publicly funded R&D to reduce greenhouse gas emissions and to displace petroleum use. The production and use of petroleum-derived energy generate negative environmental externalities and impose national security costs, neither of which is fully reflected in market prices. These social concerns compound the insufficient motivation for private firms to invest in R&D aimed at achieving these particular objectives.

Although this committee is not in a position to recommend specific levels of government R&D spending to advance vehicle and fuel technology, one insight from its analysis is that maintaining a diverse R&D portfolio is appropriate given the nature of the challenge. The committee's scenarios demonstrate that several pathways involving combinations of advanced vehicle and fuel technologies have the potential to achieve the goals of an 80 percent reduction in petroleum use and GHG emissions from LDVs. R&D critical to success for many key vehicle and fuel innovations includes:

---

[24]Estimates given in *Moving Cooler* (Collaborative Strategies Group, 2009) cover all road vehicle traffic, not merely LDVs. Other benefits that are not counted by *Moving Cooler* include time savings that may result from these measures.

[25]This "induced driving" effect is used by opponents of building more roads to argue that doing so only causes more driving. Using Federal Highway Administration models, *Moving Cooler* estimates that a systemwide average reduction in delay of 1 hr/1000 VMT in the absence of induced demand results in a systemwide increase in VMT of 2.13 percent. This increase in VMT results in a proportionate increase in fuel consumption and GHG emissions. This increase will be less in the short run than in the long run. *Moving Cooler* adjusts GHG emissions from increased VMT in the initial year of strategy deployment by (2.13 percent × 0.5), ramping this increase to the full 2.13 percent after 10 years (Collaborative Strategies Group, 2009, p. B-88).

[26]Some of these strategies might be somewhat additive, but it does not appear reasonable to claim that they are totally additive. Even if they were, the impact of the policies other than ecodriving would total only 1.4 percent (Collaborative Strategies Group, 2009, p. D-12).

- Low-cost, conductive, chemically stable plate materials for fuel cells;
- New, durable, low-cost membrane materials for the fuel cell stack and batteries;
- New catalyst structures that increase and maintain the effective surface area of chemically active materials and reduce the use of precious metals;
- New processing techniques for catalyst substrates, impregnation, and integration with layered materials;
- Energy storage beyond lithium-ion batteries;
- Reduced cost of carbon fiber and alternatives to polyacrolyonitrile as feedstock;
- Replacements for rare-earths in motors;
- Waste heat recovery; and
- "Smart car" technology.

Key fuel technologies include:

- "Drop-in" biofuels with low net GHG emissions;
- Carbon capture and storage; and
- Advanced hydrogen production technologies with low net GHG emissions.

These two lists may not be exhaustive over the time horizon examined; rather, they represent options already included in DOE's R&D portfolio. All of the fuel options entail combinations of new energy resources or carbon capture and storage technologies sufficient to deliver biofuels or other synthetic fuels, electricity, hydrogen, or combinations thereof with low net GHG emissions impacts. It is unclear which pathway is most likely to succeed, because each depends on technology success, cost reduction, consumer acceptance, and public policies.

6.6.1 Demonstration

Once a technology moves beyond research and development, the case for government support becomes more controversial. Consider the case of federal funding of demonstration projects. Suppose that R&D has yielded a new way of producing a fuel for LDVs. The R&D process may have shown that the technology works in a laboratory, but it does not demonstrate system integration in a production setting that might be scaled to a commercial level.[27] Before private industry will invest the large sums required to construct large numbers of commercial-scale plants employing such technology, someone must construct a first-of-a-kind commercial-scale plant. Prior to that, there may be a need to test and refine the workability of the technology in a production-like setting through the construction and operation of a pilot plant at less-than-commercial scale. Industry

also may be unwilling to shoulder those costs. The cost of producing fuel in a first-of-a-kind commercial-scale plant likely will not be as low as it might become, because the first-of-a-kind commercial-scale plant will not have the benefit of the "learning-by-doing" that can lower construction and operating costs. But without this step, full-scale commercialization will not occur.

If the technology is protected by strong patents or can be kept secret, and if the price that the firm constructing a first-of-a-kind commercial-scale plant can expect to receive for output from the plant reflects nearly all of the benefits that the technology creates, a private firm may run the risk of constructing this demonstration plant on its own. But if the developer of the technology cannot protect it from being easily appropriated by others, or if a significant share of the benefits cannot be captured in the price that the output of the plant will sell for, a private firm is not likely to be willing to take this step. In such cases, the first-of-a-kind commercial-scale plant may not be built without some form of government financial assistance. However, if the eventual business case for the technology relative to competing options (including potential progress in an incumbent technology) is weak, then even a government-financed first-of-a-kind demonstration may not lead to the scale-up to multiple or larger plants needed for commercial success. (However, some of the government policies discussed above in this chapter, such as a carbon tax or a price floor on oil, could improve the business case for certain alternative technologies.)

Vehicles pose a different demonstration issue. Manufacturers commonly create demonstration fleets to generate information on the in-use performance of vehicles using new powertrain systems. Examples of demonstration fleets include the General Motors EV1 and Equinox hydrogen fuel cell vehicles, Honda FCX Clarity, Audi A3 E-tron, and the Mini E. In these cases, only a limited number of vehicles have been produced. They were made available only to screened applicants, and ownership of the vehicle remains with the manufacturer.

Circumstances may dictate that government must directly participate in the demonstration project. In the case of the DOE's National Fuel Cell Vehicle Learning Demonstration Project, the primary goal was to validate vehicle and infrastructure systems using hydrogen as a transportation fuel for LDVs under real-world conditions using multiple sites, varying climates, and a variety of sources for hydrogen.[28] Specific objectives included validating hydrogen vehicles with more than a 250-mile range, 2,000-hour fuel cell durability, and hydrogen production costs of $3 per gallon of gasoline equivalent. The project was structured around a highly collaborative relationship with four industry teams—Chevron/Hyundai-Kia, Daimler/BP, Ford/BP, and GM/Shell—with the National Renewable Energy Laboratory (NREL) collecting and analyzing the data and publishing results. A

---

[27]A pilot facility is a form of demonstration that integrates technologies developed in the laboratory into a production system. The pilot demonstration is a step toward commercial design.

[28]This description is taken from Wipke et al. (2010).

total of 140 fuel cell vehicles, covering both Generation 1 and Generation 2 technology, were deployed over the course of the project. Twenty refueling stations, utilizing four different types of refueling technology, were deployed. The geographic regions covered were the San Francisco to Sacramento region, the Los Angeles metro area, the Detroit metro area, the Washington, D.C., to New York region, and the Orlando metro area.

The project established specific goals for many of the technical and operating questions and periodically reported progress toward meeting these goals. A detailed summary of the project's results through 2009, identifying which goals had been met and which goals still needed to be met, was published in 2010. Some teams ended their participation in 2009, but some continued at least through most of 2011. At an update made available in early 2012, NREL reported that through the third quarter of 2011, vehicles assigned to the project had accumulated a total of 154,000 operating hours and had traveled a total of 3.6 million vehicle-miles (NREL, 2012).

What seems to have made this demonstration project successful was its careful design that involved the coordination of a simultaneous demonstration of vehicles and fueling infrastructure, its focus on measurable goals that were critical to the eventual success of hydrogen-electric vehicles, mandatory reporting of detailed performance data including safety to establish expected baselines for commercialization, and its use of paired teams of vehicle manufacturer and fuel manufacturer, with the government playing a facilitating and coordinating role.

6.6.2 Deployment

The next step after demonstration is deployment—the roll out of a fully demonstrated technology with all of the technical and economic aspects as fully defined as possible. This is likely to be particularly challenging in the case of vehicles using non-liquid fuels, such as grid-connected-electric, natural gas, or hydrogen fuel cell vehicles, Even after successful completion of the demonstration phase, potential vehicle purchasers would need to be convinced that the technology is reliable and that the form of energy it requires will be available, while energy suppliers and vehicle manufacturers would need to be convinced that the vehicle/fuel system would be purchased by consumers in increasing volumes within timeframes relevant to major private investment planning. Cost reduction through learning-by-doing and by increasing sales volumes to achieve economy of scale likely would be necessary to ensure availability of a range of vehicle makes and models to consumers. Refueling infrastructure also would have to be widely enough available to sustain an expanding market.

The analysis in Chapter 5 illustrates that the timing of deployment is critical if the 2050 petroleum use and GHG reduction goals are to be achieved. Because of the long

lifetime of the LDV fleet, vehicles incorporating the sorts of technologies described in Chapter 2 would have to be in the market in substantial quantities by about 2035. If these vehicles require a new fuel infrastructure, enough of it would have to be in place even prior to this date to quell vehicle owners' anxieties about fuel availability. The Chapter 5 analyses suggest that transitions in energy resource and supply sectors or to alternative fuels, AFVs, or any combinations would have to be forced more rapidly than would occur through private market forces alone if the goals are to be achieved by 2050. Therefore, financial inducement from either private or public resources will be required.

The condition for private investment in deployment is that the technology is so promising that the potential investor would prefer it over other opportunities. Because of the long timeframe and uncertain outcome, potential private investors will require a high rate of return on that investment and will limit the amount they will invest to a level that does not endanger their long-term financial viability. The analyses in Chapter 5 suggest that deployment of alternative LDV and fuel technologies will in some cases be too large, last too long, and be too uncertain for a private entity to support financially. Further, the modeling shows that a substantial part of the return on the investment will accrue to society at large rather than to the private investor.

Policy-driven deployment is likely to be necessary to encourage and support a new technology through the early phases of market introduction, particularly if the success of the investment depends heavily on societal benefits. For AFV systems, publicly funded deployment encouraged by public policy might be especially important for addressing two major barriers:

- The scale-related cost problem associated with the fact that new vehicle-fuel systems lack sufficient economies of scale during the early stages of commercialization, and
- The coordination of commercial deployment of AFVs with the fueling infrastructure for those vehicles.

Nevertheless, given the uncertainties involved, technology-specific deployment programs may not be needed. If such programs are needed, several general principles should be followed:

- The deployment effort should be undergirded by and based on a long-term, substantial market signal to focus and drive reducing petroleum use and GHG emissions. An example would be a carbon tax or an equivalent means of setting costs for carbon GHG emissions.
- The cost of deployment would have to be known and the amount be acceptable.
- The time period over which any public investment is provided would have to be limited, and a technical

agency would be used to develop metrics to assess progress and guide adjustments by policy makers based on the achieved results, including building on effective activities and terminating activities that are ineffective or are overcome by events.

- A condition of public investment needs to be the presence of one or more legally committed private partners obligated to make a substantial investment so as to have a stake in the success of the technology deployment.

The government has tools in addition to direct investment that it can use to ease the investment hurdle for private capital. These could include loan guarantees to lower the rate of interest paid by the investor for the necessary private capital and direct loans to the investor at less than market rates. The government can also use mandates. Government mandates that set goals that are truly technology neutral tend to be attractive because they allow industry rather than the government to select the most promising means to meet the requirement. However, loan guarantees, loans and below-market interest rates, and mandates all share the disadvantage that they tend to hide the true cost of the government support.

A strongly mitigating factor against government involvement in technology-specific deployment is that there is little or no successful experience to guide the selection of policies and tactics for such actions for vehicles and fuel technologies. One relevant precedent is the successful reduction of air pollutants from LDVs. The government's effort there was generally directed at the outcome rather than a particular achievement path.

The route to achieving the 2050 goals is not clear, and so the government's approach to pursue these goals has to be flexible and adaptive. The government must be able to assess candidate activities, select only those with a high chance of success, accept some risk because success is not guaranteed in every case, and be robust enough to survive when approaches initially chosen fail. The government needs to make unbiased and prompt assessments of progress and act swiftly to modify ineffective efforts and terminate those that are failing.

6.7 POLICIES IMPACTING PUBLIC SUPPORT

Fostering public understanding of the rationale underpinning various policy decisions, regulatory actions, and vehicle and fuel technologies designed to achieve the nation's GHG and petroleum reduction goals for the LDV fleet is critical to achieving public support of same.

It has been demonstrated that proper dissemination of information that increases consumers' awareness of and knowledge about a particular policy or program—alone or in concert with incentives—can have a more permanent impact on consumers' behavior than do incentive programs alone (Hopper and Nielsen, 1991; Iyer and Kashhyap, 2007).

Regarding the adoption of hybrid vehicles, it has been demonstrated that consumers today have little knowledge of the technology and limited knowledge of the potential benefits of the technology (e.g., roominess, power, and quiet operation) beyond the financial and environmental benefits (Ozaki and Sevastyanova, 2011).

In addition, financial and other incentives unsupported by effective public information programs do little to increase the number of people performing the behavior being incentivized and typically have the most influence on those already disposed to accept the policies and goals being promulgated, whereas public information programs have the potential of helping consumers form positive opinions about the recommended goals and policies—especially those who held no opinion, and even those who were opposed to such goals and policies, before being exposed to the information campaign (Allen et al., 1993; Ditter et al., 2005).

Overcoming the lack of knowledge about the need to achieve the recommended goals also is critical. Although some number of consumers will respond solely to policies providing financial benefit for modifying their driving habits to lower VMT and/or facilitate their purchase of low-emitting and AFVs, it will be decades, if ever, before such vehicles demonstrate performance and cost-of-ownership characteristics that make them clearly competitive with conventional vehicles. In their 2004 study of the impact of energy-efficiency audits on the adoption by industry of efficiency technologies such as energy-efficient lighting, heating, and cooling systems, Anderson and Newell (2004, p. 2) found that "access to more accurate performance information can reduce the uncertainty and risk associated with adopting technologies that are new, or that receive differing reviews from equipment vendors, utilities, or consultants." The Washington State Department of Transportation, which has a long history of successful public transit and VMT reduction programs, has found that public education "is a vital element" in its transportation demand management projects (McBryan et al., 2000).

6.8 ADAPTIVE POLICIES

As discussed throughout this report, many uncertainties surround advanced LDV, fuel, and energy supply technologies. Today's knowledge of the feasibility, scalability, costs, and benefits associated with the options analyzed in this report is insufficient to craft policies framed around any specific vehicle-fuel systems. Analysis performed today can be suggestive but is never dispositive about what technologies will succeed in the future. Neither can the market responses of the diverse actors whose decisions determine both technology adoption and the real-world impacts of its use be predicted with much certainty. As Dwight D. Eisenhower remarked, "Plans are nothing; planning is everything." Policy makers face a need to design measures that can be modified as new information becomes available while maintaining a focus on meeting the goals of the policy. Although it addresses a different issue, a RAND (RAND Europe, 1997, p. 2) study summarizes this sensibility by saying that "a realistic approach to the formulation of policy should explicitly confront the fact that policy will be adjusted as the world changes and as new information becomes available." An example of such an adaptive policy is provided in Chapter 5 (see Section 5.6, "Adapting Policy to Changes in Technology"). In that example, a mid-course change in policy was made as a result of an unanticipated improvement in one vehicle technology or fuel type, and the study goals were met.

In considering what such an adaptive policy framework might look like, it is important that it not be trivialized to a mere exhortation that "policy makers should adapt." Policy makers adapt all the time. Although the criticism, "America lacks an energy policy," is often heard, the country in fact has as energy policy that has developed over time, including evolving measures to address transportation energy use. Congress and successive administrations have adapted laws, regulations, and other programs to new conditions and new information, satisfying different needs and interests to different degrees (perhaps leaving some unsatisfied). Vehicle efficiency standards have been modified over the years depending on the public priority placed on petroleum conservation and more recently coordinated with CAA-authorized GHG emissions standards in response to climate concerns.

The track record of the existing approach to transportation energy policy is decidedly mixed. CAFE standards have helped to limit growth in oil demand and GHG emissions, but at uneven rates over the years. Whatever learning may have been achieved, in the United States alternative fuel and vehicle technologies have had little impact on the sector's petroleum dependence and no measurable benefit on its net GHG emissions intensity (which may in fact have worsened). Corn ethanol has displaced a portion of petroleum gasoline, but there is no evidence for the beginning of a broader transition to non-petroleum resources beyond the levels mandated by the RFS. If changes in energy use and GHG emissions of the magnitude given in this committee's task statement are to be achieved, the country will need a policy framework that is much more effective in moving the LDV-fuel system toward specified goals. Although a formal adaptive paradigm has not been used for transportation and energy policy to date, some guidance can be obtained from other contexts where it has been used. Insights can also be found in the history of public policies that have resulted in varying degrees of progress on the impacts of LDVs.

One issue for which discussions of adaptive policy have been published is that of climate adaptation, i.e., measures for handling the impacts of climate change rather than mitigating its causes. This body of work builds on prior thinking about adaptive frameworks for natural resource and ecological systems management. Swanson and Bhadwal (2009) characterize adaptive policies as those that not only anticipate the range of conditions that lie ahead, but also have an up-front design that is robust in the face of unanticipated situations. Aspects of such design include integrated and forward-looking analysis, policy development deliberations that involve multiple stakeholders, and the definition of key performance indicators that are then monitored in order to trigger automatic adjustments in parameters of the policy. Adaptive policies ideally are able to navigate toward successful outcomes even while encountering developments (including lack of hoped-for outcomes) that cannot be anticipated in advance.

An example of such an adaptive framework for the transport-sector GHG emissions is the proposal contained in a 2009 consensus report by the U. S. Climate Action Partnership (USCAP), a group of 31 corporations and public-interest groups. The USCAP proposal states (2009, p. 23):

> Congress should require EPA, in collaboration with the Department of Transportation (DOT) and other federal and state and local agencies, to carry out a periodic in-depth assessment of current and projected progress in transportation sector GHG emissions reductions. . . . This assessment should examine the contributions to emissions reductions attributable to improvements in vehicle efficiency and GHG performance of transportation fuels, increased efficiency in utilizing the transportation infrastructure, as well as changes in consumer demand and use of transportation systems, and any other GHG-related transportation policies enacted by Congress.
>
> On the basis of such assessments EPA, DOT and other agencies with authorities and responsibilities for elements of the transportation sector should be required to promulgate updated programs and rules—including revisions to any authorized market incentives, performance standards, and other policies and measures—as needed to ensure that the transportation sector is making a reasonably commensurate contribution to the achievement of national GHG emissions targets.

Committee members hold a range of views on the merits of the USCAP proposal. This committee presents its own proposal for an adaptive framework in Chapter 7.

6.9 REFERENCES

Allen, J., D. Davis, and M. Soskin. 1993. Using coupon incentives in recycling aluminum: A market approach. *Journal of Consumer Affairs* 27(2):300-318.

An, F., D. Gordon, H. He, D. Kodjak, and D. Rutherford. 2007. *Passenger Vehicle Greenhouse Gas and Fuel Economy Standards: A Global Update.* Washington, D.C.: The International Council on Clean Transportation.

Anderson, S.T., and R.G. Newell. 2004. Information programs for technology adoption: The case of energy-efficiency audits. *Resource and Energy Economics* 26:27-50.

Anderson, S.T., I.W.H. Parry, J.M. Sallee, and C. Fischer. 2011. Automobile fuel economy standards: Impacts, efficiency, and alternatives. *Review of Environmental Economics and Policy* 5(1):89-108.

Bastard, L. 2010. The impact of economic instruments on the auto industry and the consequences of fragmenting markets—Focus on the EU case. In *Stimulating Low-Carbon Vehicle Technologies*. France: Organisation for Economic Cooperation and Development.

Borenstein, S. 2008. *The Implications of a Gasoline Price Floor for the California Budget and Greenhouse Gas Emissions*. CSEM WP 182. Berkeley, Calif.: University of California Energy Institute.

CARB (California Air Resources Board). 2011. Key Events in the History of Air Quality in California. Available at http://www.arb.ca.gov/html/brochure/history.htm. Accessed January 18, 2012.

———. 2012. LCFS Enforcement Injunction is Lifted, All Outstanding Reports Now Due April 30, 2012. Available at http://www.arb.ca.gov/fuels/lcfs/LCFS_Stay_Granted.pdf. Accessed May 1, 2012.

CBO (Congressional Budget Office). 2008. *Effects of Gasoline Prices on Driving Beharior and Vehicle Markets*. Washington, D.C.: Congressional Budget Office.

Collaborative Strategies Group, LLC. 2009. *Moving Cooler: An Analysis of Transportation Strategies for Reducing Greenhouse Gas Emissions*. Washington, D.C.: Collaborative Strategies Group, LLC.

Davis, S.C., S.W. Diegel, and R.G. Boundy. 2011. *Transportation Energy Data Book: Edition 30*. Oak Ridge, Tenn.: Oak Ridge National Laboratory.

DeCicco, J.M. 2012. Biofuels and carbon management. *Climatic Change* 111(3-4):627-640.

DeCicco, J.M., and D. Gordon. 1995. Steering with prices: Fuel and vehicle taxation as market incentives for higher fuel economy. In *Transportation and Energy: Strategies for a Sustainable Transportation System* (D. Sperling and S. Shaheen, eds.). Washington, D.C.: American Council for an Energy-Efficient Economy.

Difiglio, C. 1976. Analysis of fuel economy excise taxes and rebates. In *Strategies for Reducing Gasoline Consumption Through Improved Motor Vehicle Efficiency*. Washington, D.C.: Transportation Research Board.

Ditter, S.M., R.W. Elder, R.A. Shults, D.A. Sleet, R. Compton, and J.L. Nichols. 2005. Effectiveness of designated driver programs for reducing alcohol-impaired driving: A systematic review. *American Journal of Preventative Medicine* 28(5S):280-287.

DOE-EERE (U.S. Department of Energy-Energy Efficiency and Renewable Energy). 2011a. Energy Policy Act of 2005. Available at http://www.afdc.energy.gov/afdc/laws/epact_2005. Accessed April 12, 2012.

———. 2011b. Key Federal Legislation. Available at http://www.afdc.energy.gov/afdc/laws/index.php?p=key_legislation&print=y. Accessed April 12, 2012.

DOT-DOE-EPA (U.S. Department of Transportation, U.S. Department of Energy, and U.S. Environmental Protection Agency). 2002. *Report to Congress. Effects of the Alternative Motor Fuels Act CAFE Incentives Policy*. Washington, D.C.: U.S. Department of Transportation, U.S. Department of Energy, and U.S. Environmental Protection Agency.

Eads, G.C. 2011. *50by50: Prospects and Progress*. London: Global Fuel Economy Initiative.

Edmunds.com. 2009. Cash for Clunkers Results Finally In: Taxpayers Paid $24,000 per Vehicle Sold, Reports Edmunds.com. Available at http://www.edmunds.com/about/press/cash-for-clunkers-results-finally-in-taxpayers-paid-24000-per-vehicle-sold-reports-edmundscom.html?articleid=159446&. Accessed February 14, 2012.

EIA (Energy Information Administration). 2011. *Annual Energy Outlook 2011 With Projections to 2035*. Washington, D.C.: U.S. Department of Energy.

———. 2012. *Annual Energy Outlook 2012*. DOE/EIA-0383ER(2012). Washington, D.C.: Department of Energy, Energy Information Administration. Available at http://www.eia.gov/forecasts/aeo/er/.

EPA (U.S. Environmental Protection Agency). 2009. Endangerment and cause or contribute findings for greenhouse gases under Section 202(a) of the Clean Air Act: Final rule. *Federal Register* 74(239):66495-66546.

———. 2010. *Light-Duty Automotive Technology, Carbon Dioxide Emissions, and Fuel Economy Trends: 1975 Through 2010*. EPA-420-R-10-023. Washington, D.C.: U.S. Environmental Protection Agency.

———. 2012a. RFS2 EMTS Informational Data. Available at http://www.epa.gov/otaq/fuels/rfsdata/. Accessed May 23, 2012.

———. 2012b. Renewable Fuels: Regulations and Standards. Available at http://www.epa.gov/otaq/fuels/renewablefuels/regulations.htm. Accessed March 15, 2012.

EPA and NHTSA (U.S. Environmental Protection Agency and National Highway Traffic Safety Administration). 2010. Light-Duty Vehicle Greenhouse Gas Emission Standards and Corporate Average Fuel Economy Standards. Final Rule. *Federal Register* 75(88):25323-25378.

———. 2011. 2017 and later model year light-duty vehicle greenhouse gas emissions and Corporate Average Fuel Economy Standards. *Federal Register* 76(213):74854-75420.

Hopper, J.R., and J.M. Nielsen. 1991. Recycling as altruistic behavior—Normative and behavioral strategies to expand participation in a community recycling program. *Environment and Behavior* 23(2):195-220.

Hubbard, G., and P. Navarro. 2010. How Democrats and Republicans Can Break Our Foreign Oil Addiction Overnight. Available at http://money.cnn.com/2010/10/13/news/economy/Glenn_Hubbard_oil_excerpt.fortune/index.htm. Accessed November 12, 2011.

Iyer, E.S., and R.K. Kashhyap. 2007. Consumer recycling: Role of incentives, socialization, information and social class. *Journal of Consumer Behavior* 6:32-47.

Khazzoom, J.D. 1994. *An Econometric Model of Fuel Economy and Single-Vehicle Highway Fatalities*. Greenwich, Conn.: JAI Press.

McBryan, B., J. Shadoff, D. Pike, and D. Loseff. 2000. *TDM Guide for Planners: Including Transportation Demand Management (TDM) Strategies in the Planning Process*. Olympia, Wash.: Washington State Department of Transportation.

McPhail, L., P. Wescott, and H. Lutman. 2011. *The Renewable Identification Number System and U.S. Biofuel Mandates*. Washington, D.C.: U.S. Department of Agriculture-Economic Research Service.

Mondt, J.R. 2000. Cleaner Cars: The History and Technology of Emission Control Since the 1960s. Warrendale, Pa.: Society of Automotive Engineers.

Mullins, K.A., W.M. Griffin, and H.S. Matthews. 2010. Policy implications of uncertainty in modeled life-cycle greenhouse gas emissions of biofuels. *Environmental Science and Technology* 45(1):132-138.

NHTSA (National Highway Traffic Safety Administration). 2006. *Vehicle Survivability and Travel Mileage Schedules*. DOT HS 809 952. Springfiled, Va.: National Technical Information Service.

———. 2011. Summary of CAFE Fines Collected. Available at http://www.nhtsa.gov/staticfiles/rulemaking/pdf/cafe/CAFE_fines_collected_summary.pdf. Accessed January 30, 2013.

———. 2012. CAFE Overview. Frequently Asked Questions. Available at http://lobby.la.psu.edu/_107th/126_CAFE_Standards_2/Agency_Activities/NHTSA/NHTSA_Cafe_Overview_FAQ.htm. Accessed June 23, 2012.

NRC (National Research Council). 2002. *Effectiveness and Impact of Corporate Average Fuel Economy (CAFE) Standards*. Washington, D.C.: National Academy Press.

———. 2009. *Driving and the Built Environment: The Effects of Compact Development on Motorized Travel, Energy Use, and CO_2 Emissions*. Washington, D.C.: The National Academies Press.

———. 2011. *Renewable Fuel Standard: Potential Economic and Environmental Effects of U.S. Biofuel Policy*. Washington, D.C.: The National Academies Press.

NSF (National Science Foundation). 2007. *Federal Research and Development Funding by Budget Function: Fiscal Years 2005-2007*. Arlington, Va.: National Science Foundation.

Ozaki, R., and K. Sevastyanova. 2011. Going hybrid: An analysis of consumer purchase motivations. *Energy Policy* 39(5):2217-2227.

Plevin, R.J., M. O'Hare, A.D. Jones, M.S. Torn, and H.K. Gibbs. 2010. Greenhouse gas emissions from biofuels' indirect land use change are uncertain but may be much greater than previously estimated. *Environmental Science and Technology* 44(21):8015-8021.

RAND Europe. 1997. *Adaptive Policies, Policy Analysis, and Civil Aviation Policy-making.* DRU-1514-VW/VROM/EZ. Leiden, The Netherlands: RAND Europe.

Righelato, R., and D.V. Spracklen. 2007. Carbon mitigation by biofuels or by saving and restoring forests? *Science* 317(5840):902-902.

R.L. Polk and Company. 2011. Average Age of Vehicles Reaches Record High, According to Polk. Available at https://www.polk.com/company/news/average_age_of_vehicles_reaches_record_high_according_to_polk. Accessed May 5, 2012.

Swanson, D.A., and S. Bhadwal. 2009. *Creating Adaptive Policies: A Guide for Policy-Making in an Uncertain World.* Winnipeg, M.B., Canada: International Institute for Sustainable Development.

Thompson, W., S. Meyer, and P. Westhoff. 2010. The new markets for renewable identification numbers. *Applied Economic Perspectives and Policy* 32(4):588-603.

TIAP (Tax Incentives Assistance Project). 2012. Consumer Tax Incentives: Passenger Vehicles. Available at http://energytaxincentives.org/consumers/vehicles.php. Accessed April 12, 2012.

TRB (Transportation Research Board). 2009. *TRB Special Report 298: Driving and the Built Environment: The Effects of Compact Development on Motorized Travel, Energy Use, and CO_2 Emissions.* Washington, D.C.: Transportation Research Board of the National Academies.

USCAP (U.S. Climate Action Partnership). 2009. A Blueprint for Legislative Action. Consensus Recommendations for U.S. Climate Protection Legislation. Available at http://www.c2es.org/docUploads/USCAP-legislative-blueprint.pdf. Accessed February 15, 2012.

Whitefoot, K.S., and S.J. Skerlos. 2012. Design incentives to increase vehicle size created from the US footprint-based fuel economy standards. *Energy Policy* 41:402-411.

Wipke, K., S. Sprik, J. Kurtz, T. Ramsden, C. Ainscough, and G. Saur. 2010. VII.1 Controlled hydrogen fleet and infrastructure analysis. In *DOE Hydrogen and Fuel Cells Program FY 2010 Annual Progress Report.* Golden, Colo.: National Renewable Energy Laboratory.

———. 2012. *National Fuel Cell Electric Vehicle Learning Demonstration: Final Report.* Golden, Colo.: National Renewable Energy Laboratory.

Wise, M., K. Calvin, A. Thomson, L. Clarke, B. Bond-Lamberty, R. Sands, S.J. Smith, A. Janetos, and J. Edmonds. 2009. Implications of limiting CO_2 concentrations for land use and energy. *Science* 324(5931):1183-1186.

Yacobucci, B.D., and B. Canis. 2010. *Accelerated Vehicle Retirement for Fuel Economy: "Cash for Clunkers."* Washington, D.C.: Congressional Research Service.

Zhang, C., H.Q. Tian, S.F. Pan, M.L. Liu, G. Lockaby, E.B. Schilling, and J. Stanturf. 2008. Effects of forest regrowth and urbanization on ecosystem carbon storage in a rural-urban gradient in the Southeastern United States. *Ecosystems* 11(8):1211-1222.

7

Policy Options

Previous chapters demonstrate that achieving a 50 percent reduction in petroleum consumption by light-duty vehicles (LDVs) by 2030 and 80 percent reductions in both petroleum consumption and greenhouse gas (GHG) emissions by LDVs by 2050 will be extremely challenging. What likely will be required to achieve those goals is some combination of the following:

- Major improvements in existing LDV powertrains;
- Major reductions in the weight and other loads of all sizes and types of LDVs;
- Changes in the energy resources or fuels used to power LDVs, and the effective control of net GHG emissions in the sectors that supply fuels for LDVs; and
- The successful introduction and widespread use of one or more entirely new powertrain systems (e.g., electric vehicles and fuel-cell electric vehicles [FCEVs]).

Reaching the ambitious goals for 2050 will be made easier by any reductions in the rate of growth in vehicle miles traveled (VMT) that might be practical and by technological advances that increase the operating efficiencies of transportation systems. However, the primary focus of the findings and policy options identified in this chapter is on how to bring about changes in vehicles and fuel supply sectors, and in consumer demand, necessary to meet the goals addressed in this study.

If the increases in new LDV fuel economy reflected in the standards finalized by the National Highway Traffic Safety Administration (NHTSA) and the U.S. Environmental Protection Agency (EPA) are attained by 2025, as noted in Chapter 5, considerable progress will have been made in moving the new LDV fleet toward lower levels of energy use and GHG emissions. This progress will have been achieved primarily by production and sale of LDVs with improved efficiency employing existing powertrain concepts, including

conventional hybrid electric vehicles. Despite such progress, however, this strategy alone is insufficient to decrease LDV petroleum consumption by 50 percent by 2030.

To meet the goals addressed in this study, vehicle and fuel-supply advances will be needed in the period from 2025 through 2050. One possible pathway to meet the 2050 petroleum use and GHG emission reduction goals could be combining high LDV fuel economy with high levels of drop-in biofuels produced using processes with low net GHG emissions. Another possible pathway could be a transition to other alternative fuel and alternative powertrain technologies (e.g., plug-in hybrid electric vehicles [PHEVs], battery electric vehicles [BEVS], and FCEVs) to constitute a significant share of the on-road fleet by 2050. The time required for fleet turnover means that vehicles incorporating these technologies will need to begin to enter the new LDV fleet in significant numbers by the 2030s. The technical, economic, and consumer acceptance barriers currently faced by these technologies may have been largely overcome by then. The uncertainties about technology improvements and costs are such that the committee cannot rule out either pathway for meeting the goals addressed in this study.

If new fuels are required to enable use of alternative powertrain technologies, these fuels will have to be available widely enough by 2025 to enable early adopters not to be overly concerned about fuel availability. Because physical stock changes in major energy supply systems occur more slowly than LDV stock turnover, enough measurable progress in this regard must be seen by 2030 so that it is clear that the 2050 reductions of in-sector LDV GHG emissions enabled by the advanced powertrain technologies will not be largely offset by the emissions generated by the production and distribution of the fuels themselves.

The objective of the policy actions suggested in this chapter is to substantially increase the probability of achieving the goals specified in the statement of task. The policy options identified in this chapter as most promising by the committee are based on its review of the past experience with and potential effectiveness of the possible policies described in Chapter 6, and on the committee's own evaluation of policies and policy combinations in Chapter 5. Regulatory policies such as Corporate Average Fuel Economy (CAFE) standards, pricing policies (either economy-wide or directed at fuel supply sectors) such as feebates for vehicles, and regulatory or pricing policies directed at fuel supply sectors will likely be essential to attaining the 2050 goals for reducing LDV petroleum consumption and GHG emissions. Additional policies may also be required if a transition to alternative vehicle and fuel systems turns out to be the best way to attain the goals. Such transition policies include infrastructure investments and possible subsidies. Because of uncertainties and unforeseen circumstances in the future, policies must be adaptive in response to technology and to market conditions over time to ensure that the goals are met in a cost-effective way.

7.1 POLICIES TO ENCOURAGE THE CONTINUED IMPROVEMENT OF THE FUEL EFFICIENCY OF THE LIGHT-DUTY VEHICLE FLEET

Even if the fuel economy and CO_2 reduction standards for new LDVs currently being implemented by NHTSA and the EPA are met, further improvement in the fuel efficiency of vehicles could be made in and after model year (MY) 2025. Although the committee believes that it is premature to suggest a specific fuel economy target for new LDVs by MY2050, a "ballpark" estimate is that a further doubling (that is, a doubling beyond the doubling that is scheduled to occur between 2005 and 2025) of the average new LDV fleet fuel economy standard by 2050[1] will be technically feasible but costly. The modeling results in Chapter 5 indicate that such an increase in the CAFE standard could reduce GHG emissions by about 50 percent in 2050 compared to the 2005 level. Reaching such ambitious fuel economy targets will require a mix of policies that affect the decisions of vehicle manufacturers to produce fuel-efficient vehicles and the decisions of consumers to purchase them.

FINDING. The CAFE standard has been effective in reducing vehicle energy intensity, and further reductions can be realized through even higher standards if combined with policies to ensure that they can be achieved.

POLICY OPTION. The committee suggests that LDV fuel economy and GHG emissions standards continue to be strengthened to play a significant role after model year

---

[1]Such a further doubling of on-road fleet fuel economy between 2025 and 2050 cannot, by itself, achieve the goals set forth in the charge to this committee. Additional changes involving fuels and VMT also will be needed. See the committee's scenarios in Chapter 4 for details.

2025 as part of this country's efforts to improve LDV fuel economy and reduce GHG emissions.

FINDING. "Feebates," rebates to purchasers of high-fuel-economy (i.e., miles per gallon [mpg]) vehicles balanced by a tax on low-mpg vehicles is a complementary policy that would assist manufacturers in selling the more-efficient vehicles produced to meet fuel economy standards.

POLICY OPTION. The committee recognizes that U.S. government "feebates" based on the fuel consumption of LDVs could have a role as a complement to LDV fuel economy and GHG emissions standards to facilitate and accelerate the introduction of significantly more efficient vehicles into the market to the meet the 2050 timing of the goals. The committee suggests that the U.S. government include "feebates" as part of a policy package to reduce LDV fuel use.

7.2 POLICIES TARGETING PETROLEUM USE

Petroleum consumption can be reduced by a variety of policies. Placing a quantity constraint on petroleum consumption (also known as rationing) would reduce its use directly and increase its price. A tax on petroleum would directly increase the price of petroleum, providing a signal to both producers and consumers to find ways to reduce use of petroleum-based fuels, redesign vehicles, or replace petroleum-based fuels with other fuels. Other approaches include requiring quantities of alternative fuels to be sold (such as through application of the Renewable Fuel Standard) or using subsidies to reduce the prices of alternative fuels to make their cost lower than the cost of petroleum-based fuels. As discussed in Chapter 6, it can be difficult to design a policy that successfully mandates the sale of certain fuels when they are more expensive than petroleum-based fuels. Subsidies require government revenue to fund, whereas taxes raise revenue that either can be used to fund programs related to energy and GHG emissions reduction or can be refunded to the taxpayer.

Placing a quantity limit on oil consumption (or use of petroleum fuels by LDVs specifically) has rarely been proposed and would be expected to have significant adverse social impacts.

What has been widely discussed for many years is taxation that would directly target petroleum demand or petroleum imports. Existing U.S. motor fuel taxes were adopted to raise revenues for funding roads. Historically, these taxes have helped support petroleum demand by facilitating vehicle use while remaining low enough to avoid significantly affecting fuel demand. A small exception to the historical rationale was the $0.043 per gallon gasoline tax increase of 1993 (the last time U.S. fuel taxes were raised), which had been proposed originally as a "Btu tax" to foster energy conservation and reduce the federal deficit. However, the funds from that levy were redirected back to the Highway Trust Fund in 1997.

To be used extensively, alternative fuels, together with the vehicles that they power, would have to be at least price competitive with petroleum-based fuels and conventional vehicles. For compressed natural gas and hydrogen, the alternative fuels would have to be made available with complementary vehicle and refueling infrastructure. To undertake the large investments necessary for the development and widespread availability of any alternative fuels, the fuel producers and distributors will have to be convinced that there eventually will be a profitable market for those fuels, including assurance that they will not be undercut by low-cost petroleum. The price of petroleum-based fuels would have to be relatively high and stable for investors to be confident in the profitability of alternatives. One policy that has promise for creating price stability in the oil market is a tax on petroleum that moves inversely with petroleum price and is levied only when petroleum prices fall below a target level, as discussed in Chapter 6. This tax approach ensures the price stability necessary to provide better signals to investors to invest more in efficiency or in alternative energy sources.

FINDING. Taxes on petroleum-based fuels can create a price signal against petroleum demand, offset the "rebound effect" induced by increasingly efficient vehicles, and help assure innovators, producers, and distributors that there is a profitable market for improved efficiency in energy use and for alternative fuels. The range of possible tax policies includes a fixed tax rate per barrel on petroleum that is a surtax on current taxes, or a tax that moves inversely with the oil price when the price falls below a target level, thereby stabilizing prices so that they are at or above the target. Fuel subsidies or quantity mandates are more difficult than taxes to use effectively. Subsidies require government funding, and sometimes complex decisions about who is eligible for the subsidies. Until alternative fuels become cost competitive with petroleum-based fuels, quantity mandates for alternative fuels would require fuel producers to cross-subsidize their money-losing alternative fuels from their profitable petroleum-based fuels. Creating and then maintaining the conditions necessary for successful cross-subsidization would be difficult, politically and otherwise, for the government. Yet without adopting one or more of these policy approaches, the lure of eventual profitability necessary to induce investment is absent, and so the investment is unlikely to occur.

POLICY OPTION. High and stable oil prices would be helpful in transitioning away from oil use in LDVs and meeting the 80 percent reduction goal by 2050. If fluctuations in oil prices and often low oil prices persist, it may be necessary to impose a tax on domestic use of petroleum-based fuels or set a price floor target for petroleum-based fuels.

Taxing petroleum or implementing a price floor to prevent the decline of petroleum price beyond a certain level would discourage its use and contribute to reducing VMT and increasing the use of fuel-efficient internal combustion engine vehicles (ICEVs) if petroleum-based fuel remained the dominant fuel. (See Box 7.1 and see Section 6.3.5, "A Price Floor Target for Motor Fuels," in Chapter 6). A reduction in petroleum use also would reduce the social cost of oil consumption. (See Box 5.5, "Social Costs of Oil Dependence," in Chapter 5.)

FINDING. The Renewable Fuel Standard contributed to reducing petroleum use by LDVs. As a result of the failure of cellulosic biofuels to achieve commercial viability and the ability of the EPA to waive the requirement, the volume of cellulosic biofuels mandated by the RFS has repeatedly been reduced. The RFS could become more effective if the EPA's authority to reduce the mandated requirement either is eliminated so as to maintain a guaranteed market for any cellulosic biofuels produced or linked to a requirement to fund RD&D for progress toward the improved viability of cellulosic biofuels.

POLICY OPTION. The committee supports continuation of the Renewable Fuel Standard because it has been modestly effective in displacing petroleum. The committee suggests periodic review of the RFS by Congress to assess whether the mandated volumes should be increased and whether other alternative fuels should be included in the mandate to encourage the use of alternative fuels and reduce the share of petroleum-based fuels in use for LDVs. The committee also supports further research and analysis for refinement of the means of assessing how fuels qualify as renewable.

7.3 POLICIES TO REDUCE GHG EMISSIONS ASSOCIATED WITH LDV FUELS

Policies that reduce the overall energy demand of LDVs through improving vehicle efficiency and lowering travel demand contribute to a reduction in GHG emissions. In addition, reducing GHG emissions requires policies that limit the net GHG emissions associated with the fuels used by LDVs. In considering fuel-related policies, it is crucial to distinguish between the fuels themselves—that is, the end-use energy carriers used directly by vehicles—and the primary energy resources (such as fossil fuels) and associated energy sector systems that supply end-use fuels. GHG emissions from fuel use can be limited through three basic approaches:

BOX 7.1
The Case for Fuel Pricing

The case for fuel pricing policies is based on economic theory as well as experience: for most goods, raising the price reduces the quantity demanded. One way to reduce petroleum use or greenhouse gas (GHG) emissions is to tax them. GHG emissions are environmental externalities, and their full societal costs are not reflected in market prices. As discussed in Box 5.5 in Chapter 5, a range of estimates exist for the damage that may be caused by GHG emissions. The committee chose a value at the high end of the range, $136.20 per metric ton of CO_2, because that is most consistent with the 80 percent GHG mitigation goal. There are excess social costs of oil dependence, as well, caused by the use of market power by oil producers, as well as increased public expenditures on defense (Greene and Leiby, 1993). As discussed in Box 5.6 in Chapter 5, a tax on the order of $10.50 to $38 per barrel with a midpoint of $24 in 2009 dollars would be needed to reflect the full social costs of oil dependence.

Fuel prices affect producer and consumer behavior with respect to the three parameters that affect petroleum use and GHG emissions: fuels, vehicles, and vehicle miles traveled. Experience both here and abroad indicates that producers and consumers indeed respond to fuel prices (Sterner, 2007; Dahl, 2012) but that fuel demand is relatively inelastic. For example, estimates of the elasticity of demand for gasoline range from only 0.1 over short periods when it is difficult to modify use, to about 0.3 to 0.5 over longer periods when there are more opportunities to change behavior. One study finds that a tax on gasoline that increases to about $2.00 a gallon by 2030 results in decreased gasoline use of about 25 percent over that same period (Krupnick et al., 2010). There is little experience with GHG pricing of transportation fuels and their supply chains, and so the overall GHG emissions response to including such pricing could be greater than the demand response alone.

There are also reasons why a fuel or GHG tax may need to be combined with other policies. Pricing gasoline to reflect its full costs will still not induce consumers to make optimal choices about fuel-efficient vehicles if they undervalue fuel economy (Greene, 2010). This point is discussed more fully in Chapter 5, but to the extent it is true, then a combination of pricing and vehicle standards will be important. The committee's scenario analyses suggest that significant ongoing fuel economy improvement is likely to play a very large role in meeting both the petroleum reduction and GHG emissions reduction goals (Greene, 2011; Allcott et al., 2012). That is why one of the committee's high-priority suggestions is to continue to strengthen vehicle standards for fuel economy and GHG emissions.

There are other reasons why pricing energy will be helpful in conjunction with such vehicle standards:

- Reducing VMT, including countering the rebound effect. Because fuel economy standards reduce the variable cost of driving, they encourage more driving, partially offsetting the fuel-use-reducing benefits of the standards. This phenomenon is called the rebound effect. Raising fuel prices counters the rebound effect and reduces the demand for fuel-consuming travel generally.
- Increasing demand for fuel-efficient vehicles. Higher fuel prices increase consumers' demand for fuel-efficient vehicles, thereby aligning the requirements faced by automakers under vehicle standards with the demands of consumers.

Any of these behavioral rationales for higher fuel taxation would represent a significant departure in U.S. fiscal policy. Traditionally, federal, state, and local fuel taxes have been justified only as a way to raise revenue for transportation infrastructure and maintenance. Federal U.S. gasoline taxes have not increased in nominal terms in almost 20 years; in real terms, they have declined dramatically, leading to crumbling roads, bridges, and tunnels. Other studies have documented a justification for higher fuel taxes in order to make up for this substantial shortfall in transportation funding (National Surface Transportation Policy and Revenue Study Commission, 2007). Thus, taxing fuels to reduce oil use and GHG emissions could have the important co-benefit of raising needed revenue for our transportation system. Although this behavioral rationale for fuel pricing is not traditional in U.S. policy, it has been used in Western Europe and other countries and is one reason for the higher levels of vehicle fuel economy and lower levels of per capita demand for automobile travel observed in those countries relative to the United States.

- By counterbalancing the end-use (vehicular) CO_2 emissions from carbon-based fuels with sufficient net CO_2 uptake elsewhere. Because this CO_2 uptake and the emissions associated with feedstock growth and processing (e.g., for biofuels) occur outside the transportation sector, the optimal policies are not those directed at the transportation sector per se, but rather measures to address net GHG emissions in fossil fuel extraction and refining, biorefining, agriculture, forestry, and related land-use management sectors involved in supplying carbon-based fuels.

(See also Chapter 6.) In the future, counterbalancing also might occur through geologic storage or biological sequestration techniques.
- By using physically carbon-free fuels such as electricity or hydrogen, which avoid release of CO_2 from vehicles themselves. These energy carriers must then be supplied from low-GHG emitting-production sectors. Therefore, optimal policies are not those directed at the transportation sector per se, but rather measures addressing electric power generation and other industrial sectors that produce carbon-free fuel.

- By capturing and preventing the release of the CO_2 produced during combustion or other utilization of carbon-based fuels directly on vehicles, or by avoiding the production of CO_2 during on-board energy utilization. Because no practical means of on-board CO_2 capture or avoidance are currently known, this third approach is not considered in this report.

This list demonstrates that it is impossible to have a complete policy for controlling auto-sector GHG emissions in isolation from policy to control emissions in other sectors, namely, those that supply energy and feedstock for fuel production. This principle is true whether the fuel is carbon-based or carbon-free. The extent to which policies are also needed to affect the choice of vehicular fuel depends on whether a change of end-use energy carrier is required. That question cannot be resolved on the basis of present scientific knowledge. As the committee's scenario analyses demonstrate, some technological approaches for meeting the task statement goals entail entirely new fuels and fuel distribution systems, but others (namely, the use of drop-in biofuels in high-efficiency vehicles) do not. In each scenario evaluated where the goals are achieved, a major change is required in the energy sectors that supply automotive fuel.

The committee recognizes that GHG emissions that occur in the non-transportation sectors involved in supplying energy and feedstock for fuel production need to be addressed to reduce net GHG emissions effects of the LDV sector. However, a thorough treatment of policies for addressing GHG emissions that occur in the non-transportation sectors is beyond the scope of this study. (See Appendix A for the statement of task.) Either an economy-wide GHG policy or a coordinated multisector GHG policy is likely to offer the most economically efficient and equitable way to achieve deep GHG emissions reductions across multiple sectors. Broadly speaking, the options for multisector GHG policy include direct regulation of GHG emissions under the Clean Air Act (CAA), carbon taxation, or a cap-and-trade system that blends elements of regulatory and fiscal policies by placing an economy-wide limit on GHG emissions and propagating a price signal to motivate emissions reductions across multiple sectors.

The EPA is beginning to pursue CAA regulation of GHG emissions; however, without new congressional authorization, the agency might not pursue targets that are stringent enough to support GHG emissions reduction of 80 percent by 2050. Carbon taxation is another way to motivate reductions. If the policy is of stringency comparable to that of setting a cap on energy supply sector GHG emissions at about 20 percent of the 2005 level by 2050, it would encourage GHG emissions reduction from other sectors (e.g., electricity and agriculture) that would contribute to reducing GHG emissions from the LDV sector. However, determining the tax level needed will be difficult. Given the large revenues that

would result (which could be helpful for federal finances), pursuing a carbon tax would entail engaging in a major fiscal policy discussion that affects many other aspects of national policy.

Although the near-term political prospects of cap-and-trade are poor, it may ultimately be favored over other options. It was the leading national GHG policy option in prior Congresses. Cap-and-trade once had some bipartisan support even though it fell short of sufficient majority support. California is implementing an economy-wide GHG cap-and-trade through its AB 32 program. The northeast Regional Greenhouse Gas Initiative is implementing a GHG cap-and-trade program for the power sector.

FINDING. Meeting the GHG emissions reduction target of this study requires addressing the upstream emissions that occur in the non-transportation sectors involved in supplying energy and feedstock for fuel production. Substituting hydrogen, biofuels, or electricity for petroleum-based gasoline in vehicles will result in net GHG emissions reductions only if these alternative fuels are produced using technologies and processes that emit few GHGs. Carbon capture and storage (CCS) is likely a critical technology for producing low-GHG hydrogen and electricity, but other options that directly produce electricity and can indirectly produce hydrogen through electrolysis exist (e.g., nuclear and renewable power).

POLICY OPTION. A policy that addresses GHG emissions from the energy sources and sectors that supply fuels used in LDVs is needed if GHG emissions from the LDV sector, including upstream emissions, are to be reduced enough to meet the 2050 goals. That policy can take the form of a set of measures that are specific to each sector that affects fuel production and distribution, or it can embody a comprehensive approach to addressing GHG emissions (e.g., a carbon tax or a carbon cap-and-trade policy).

7.4 POLICIES TO REDUCE THE RATE OF GROWTH OF VMT

As shown in the previous chapter, increases in vehicle miles traveled by LDVs have offset much of the potential reduction in petroleum use and in GHG emissions caused by improved fuel economy over the last several decades. If VMT increases at the rates projected in the "business as usual" scenario described in Chapter 5, the same is likely to be true in the decades ahead.[2]

---

[2]The "business as usual" and "reference" cases assume a slowdown in the rate of growth of VMT in the future. Nevertheless, in these cases, as well as in the committee's simulations, VMT continues to grow. This growth in VMT will offset some of the reductions in petroleum use and GHG emissions that otherwise would occur.

A range of policy options exists that have the potential to reduce VMT growth, but they differ widely in their likely impact. For example, policies to increase residential density are likely to produce limited results on a national scale. As discussed in Chapter 6, a previous National Research Council (NRC) report has found that a doubling in density of 75 percent of the new development by 2050, something that the report characterizes as "require[ing] such a significant departure from current housing trends, land use policies of jurisdictions on the urban fringe, and public preferences that they would be unrealistic absent a strong state or regional role in growth management," would reduce VMT by only 8 to 11 percent below what it otherwise would be in 2050 (NRC, 2009). And even this extremely optimistic degree of doubling of the density of new residential development would have to be accompanied by large increases in the amount of mixed-use development and in the quality and accessibility of transit. A major study of the potential impact of other much-discussed factors, such as pedestrian and bicycle strategies, has shown them to have only a small impact on national VMT (NRC, 2009).

Indeed, the policies found to have the most significant impact on VMT are those that raise the marginal cost of driving—for example, increasing fuel taxes. Other possible policies would be "pay at the pump" insurance, a means by which vehicle owners can pay for their car insurance through charges added to the price of gasoline, or a road-user charge. A road-user charge of $0.12 per mile would have an effect on the variable cost of driving roughly comparable in magnitude to the effect of current West European motor fuel taxes. The report *Moving Cooler. An Analysis of Transportation Strategies for Reducing Greenhouse Gas Emissions* estimated that a charge of this level would reduce 2050 VMT by 5 percent, and that just the VMT impact portion of a carbon tax levied at similar levels would reduce 2050 VMT by almost 8 percent (Collaborative Strategies Group, 2009).

FINDING. The policies that have the most significant impact on reducing the rate of growth of VMT are those that raise the marginal cost of driving. Policies other than those that raise the marginal cost of driving could result in significant reductions in the rate of VMT growth or even reductions in total VMT in certain individual urban areas, but they are not likely to result in significant reductions in GHG emissions or petroleum use at the national level by 2050.

POLICY OPTION. If reducing VMT growth is to be pursued to meet the study goals of reducing petroleum use and GHG emissions, policies that increase the marginal cost of driving should be considered.

7.5 POLICIES TO ENCOURAGE RESEARCH AND DEVELOPMENT, DEMONSTRATION, AND DEPLOYMENT

As discussed in Chapter 6, the federal government has implemented a range of policies intended to encourage the development and use of fuel-efficient LDVs and the alternative fuels to power them, with mixed success. Stages of advancement for new technologies are separated into research and development (R&D) (which involves basic and applied research on improvements to or evolution of the technology, including prototypes), demonstrations (which test the feasibility of developed technology, including significant impediments to commercial success), and deployment of the technology into the market at large scale.

The government's role in facilitating each of these stages varies with the type of technology, how far along in the advancement process the technology for either the vehicle or the fuel has progressed, and what policies are already in place. For example, new technologies for advanced ICEVs and hybrid vehicles powered by gasoline are continually developed, and regulatory policies such as CAFE and pricing policies such as feebates encourage the market adoption of fuel-efficient technologies and vehicle designs. Other powertrains and fuels, such as FCEVs, BEVs, hydrogen fuels produced with low net GHG emissions, and biofuels are at early stages of commercialization. BEVs have been introduced commercially, although sales are still low. Several companies have demonstrated FCEVs at small scale and expect to start introducing them commercially by 2015. However, significant technology and production progress is needed for cost reduction before these vehicles will be competitive at scale with existing ICEVs. Some alternatives to petroleum are at early stages of development, and demonstrations may be important in addition to R&D.

7.5.1 Research and Development

There is a strong case for R&D, whether public or private, to advance the intellectual infrastructure of the country for meeting technical challenges, as discussed in Chapter 6.

FINDING. Fuel cells, batteries, biofuels, low-GHG production of hydrogen, carbon capture and storage, and vehicle efficiency should all be part of the current R&D strategy. It is unclear which options may emerge as the more promising and cost-effective. At the present time, foreclosing any of the options the committee has analyzed would decrease the chances of achieving the 2050 goals. The committee believes that hydrogen fuel cell vehicles are at least as promising as battery electric vehicles in the long term and should be funded accordingly. Both pathways show promise and should continue to receive federal R&D support.

POLICY OPTION. The committee supports consistent R&D to advance technology development and to reduce the costs of alternative fuels and vehicles. The best approach is to promote a portfolio of vehicle and fuel R&D, supported by both government and industry, designed to solve the critical technical challenges in each major candidate pathway. Such primary research efforts need continuing evaluation of progress against performance goals to determine which technologies, fuels, designs, and production methods are emerging as the most promising and cost-effective.

FINDING. Current methods for the accounting of net GHG emissions associated with the production and use of transportation fuels involve numerous uncertainties. Reducing the uncertainties and developing robust accounting approaches are important for defining R&D strategies, guiding private sector investments, and developing effective public policies for reducing the net GHG emissions associated with fuels used by light duty vehicles.

POLICY OPTION. Because of the uncertainties associated with existing methods of accounting for the net GHG emissions impacts of the production and use of transportation fuels, especially for electricity, biofuels, and hydrogen, the committee suggests further efforts to develop accounting methods to account for GHG emissions that are applicable to the design of public policies for addressing these impacts.

7.5.2 Demonstration

The alternative vehicles discussed in Chapter 2 have demonstrated their performance readiness. Remaining challenges are cost reduction and further advancement through continued R&D, and potentially, successful deployment. Private industry may choose to demonstrate new technologies or new vehicle models or prototypes, but the need for further government involvement appears to be limited to areas of special government interest, such as validating the safety or performance of alternative vehicles.

For fuels, vehicles, and GHG management technologies that show promise of commercial readiness, appropriately scaled demonstration projects that are supported by both industry and government are likely to be important for validating feasibility, proving physical and environmental safety, and establishing cost-effectiveness. The results of such demonstrations could provide essential information for identification of which alternative fuel and GHG management technologies have long-term potential to both compete with gasoline in the marketplace and achieve GHG emissions reduction goals, and to establish readiness for deployment. Another appropriate role for the government is the coordina-

tion of integrated demonstrations of promising vehicles and fuel systems or stations.

FINDING. Demonstrations are needed for technologies to reduce GHG emissions at appropriate scale (e.g., hydrogen produced with low net GHG emissions and CCS) to validate performance, readiness, and safety.

FINDING. Integrated demonstrations of vehicles and fueling infrastructure are necessary to promote understanding of performance, safety, consumer use, and other important characteristics under real-world driving conditions.

POLICY OPTION. The committee supports the government's involvement in limited demonstration projects at appropriate scale to promote understanding of the performance and safety of alternative vehicles and fueling systems. For such projects, substantial private sector investment should complement the government investment, and the government should ensure that the demonstration incorporates well-designed data collection and learning to inform future policy making and investment. The information collected with government funds should be made available to the public consistent with applicable rules that protect confidential data.

7.5.3 Deployment

Many of the findings and policy options mentioned earlier in this chapter will encourage deployment of highly efficient or alternative vehicles and alternative fuels, and policy will be a critical driver of deployment. Policy options include CAFE and feebate policies for vehicles, performance standards, consumption mandates or pricing policies for fuels, and carbon control policies. Modeling results described in Chapter 5 show that such policies will greatly increase the shares of highly efficient and alternative vehicles over time. However, Chapter 5 also found that additional deployment policies will likely be needed for some alternative-vehicle fuel systems if they are to be part of the strategy to attain the significant reductions in petroleum use and GHG emissions discussed in this report. Additional policies such as subsidies or mandates for vehicles or fuel infrastructure investment will depend on the path of future technology, market conditions, and the urgency of the energy security and climate-change issues that these fuels are needed to address. The timing of additional deployment policies is critical and will depend on how close any one technology or combinations of technologies are to market readiness. At present, it is unclear which vehicle and fuel technology or technologies will have consumer acceptance and the best potential for lowest costs at scale to achieve the goals addressed in this study. Data on the costs of particular technologies will accumulate over time and will inform future policy decisions.

In addition, for alternative-vehicle fuel systems, the government, in partnership with industry, will likely have a role in coordinating the commercial deployment of alternative vehicles with the fueling infrastructure for those vehicles. Coordination of vehicle sales and provision of refueling infrastructure are more challenging for hydrogen than for electricity or natural gas because hydrogen requires a completely new, large-scale fuel production and delivery system. In contrast, natural gas and electricity already have a large, robust, and ubiquitous distribution system, and the additional deployment needed is an accessible dispensing infrastructure.

Assessments of the readiness of affected technologies and continuous assessments of the effectiveness of deployment policies are important. Such assessments would require metrics to be established to determine when to initiate a deployment effort, to assess progress during initial deployment, to guide adjustments based on the achieved results, and to determine when to terminate deployment efforts that are ineffective or have been overcome by events. Starting deployment prematurely will increase the chance of failure and costs, extend the time for support, and undermine public confidence. Yet prolonged delay in deployment risks failure to meet the GHG emissions reduction and fuel saving goals. Determining technical and market readiness is challenging and should involve an unbiased expert review of available data, and consideration of the viewpoints of applicable stakeholders. In particular, the analysis in Chapter 5 indicates that subsidies of particular vehicles and fuels as a deployment strategy may be important, but careful and periodic evaluations are needed to ensure their effectiveness.

FINDING. The commercialization of fuel and vehicle technologies is best left to the private sector in response to performance-based policies, or policies that target reductions in GHG emissions or petroleum use rather than specifc technologies. Performance-based policies for deployment (e.g., CAFE standards) or technology mandates (e.g., RFS) do not require direct government expenditure for particular vehicle or fuel technologies. Additional deployment policies such as vehicle or fuel subsidies, or quantity mandates directed at specific technologies are risky but may be necessary to attain large reductions in petroleum use and GHG emissions.

POLICY OPTION. The committee suggests that an expert review process independent of the agencies implementing the deployment policies and also independent of any political or economic interest groups advocating for the technologies being evaluated be used to assess available data, and predictions of costs and performance. Such assessments could determine the readiness of technologies to benefit from policy support to help bring them into the market at a volume sufficient to promote economies of scale. If such policies are implemented, they should have specific goals and time horizons for deployment. The review process should include assessments of net reductions in petroleum use and GHG emissions, vehicle and fuel costs, potential penetration rates, and consumer responses.

FINDING. For alternative-vehicle fuel systems, government involvement with industry may be needed to help coordinate commercial deployment of alternative vehicles with the fueling infrastructure for those vehicles.

The committee's analysis found that the timing and the scope of policy-related actions have a major influence on the successful transition to new vehicle and fuel technologies. If the policies are insufficient, ill-targeted, or improperly timed to overcome the cost barriers to making the transition, then the transition will not occur and the costs of the policy-related actions can be wasted.

7.6 THE NEED FOR AN ADAPTIVE POLICY FRAMEWORK

FINDING. Many uncertainties surround not only advanced vehicle, fuel, and energy supply technologies but also the response of the many LDV market actors to policies implemented for meeting goals such as those described in this committee's task statement. Therefore, policy makers will be well served to establish an adaptive framework that enables the set of measures enacted to be systematically adjusted as the world changes and as new information becomes available while staying on track to meet the long-term policy goals.

As found in Chapter 6, such a framework should not only anticipate the range of conditions that lie ahead but also be designed to be robust in the face of unanticipated developments. Aspects of such policy design include provisions for integrated and forward-looking analysis, policy development deliberations involving multiple key stakeholders, and performance metrics that are monitored to trigger automatic adjustments in parameters of the policy. To be effective, such a framework requires the establishment of clear, measurable, and durable goals. Because of the uncertainty about which technologies would emerge as most effective and cost-effective, and about how consumers will respond to those technologies and fuel delivery systems, new evidence and information will be key to developing the best policies. Chapter 5 (see Section 5.7, "Simulating Uncertainty About the Market's Response") illustrates the dilemma in setting policy in the absence of good information about key aspects of consumer preferences on the demand side, and learning and scale economies on the supply side of the market. This and other information would have to be provided by various sources, and its assessment will inform effective policy decisions.

FINDING. The policies and measures needed to achieve the petroleum and GHG emissions reduction goals stated in the committee's statement of task will be implemented by more than one federal agency, as well as coordinated with state and local jurisdictions. Moreover, as experience is gained and new information becomes available, adjustments will be needed and will be coordinated across the implementing agencies.

POLICY OPTION. To meet the petroleum-use and GHG reduction goals stated in the statement of task, the committee considers it desirable to define a federal light-duty vehicle petroleum and GHG emissions reduction policy with the following elements:

- **Establish overall goals (e.g., via congressional action).**
- **Assign relevant federal agencies having jurisdiction over LDV energy use and GHG emissions, in collaboration with the other relevant federal, state, and local agencies, to carry out periodic assessments of progress against the goals and to report the results. The assessments would include:**
  - **Quantifying progress to date and assessing the efficacy of the programs and policies in use for reducing petroleum use and GHG emissions;**
  - **Identifying the causes of emerging shortfalls in meeting the goals, and the steps being taken and planned to remedy those shortfalls, consistent with the authority of the implementing agencies; and**
  - **Identifying changes in implementing authority needed to remedy shortfalls and recommending those changes to Congress.**

If national policies are established to address these issues more broadly across the economy, then this LDV sector adaptive policy should be coordinated with, and appropriately incorporated within, the overall national energy and climate policy framework.

7.7 THE NEED FOR PUBLIC INFORMATION AND EDUCATION

FINDING. The committee considers that a vigorous program of public information and education is essential to the success of the other recommended policies and thus to achievement of the twin goals of reduced GHG emissions and reduced use of petroleum-based fuels. Increased research regarding public understanding and attitudes associated with these issues would inform the design of improved public information and education programs. Because the payoff of public education and information programs is long term and is typically measured in public benefit rather than direct financial return, it is critical that government be involved in developing and fostering such programs, because they tend to be underprovided by the private sector.

POLICY OPTION. If the United States is to achieve the goals of reduced petroleum use and reduced GHG emissions from the LDV fleet, then U.S. policy makers could develop public programs aimed at informing consumers of the goals to be achieved, the reasons such achievement is necessary, and the nature of the costs and benefits— individual and societal—to be derived from the policies being implemented.

As noted elsewhere in this report, the committee has differing views regarding the value of public promotion of specific alternative vehicle and fuel technologies, a difference of view that carries over into public information policy. Where there is agreement is in the value of informing consumers about the broad importance of the national goals, the connection with fuel economy and perhaps other objective vehicle environmental performance metrics to these goals, and the value of choosing highly fuel-efficient vehicles accordingly.

7.8 REFERENCES

Allcott, H., S. Mullainathan, and D. Taubinsky. 2012. *Externalities, Internalities and the Targeting of Energy Policy.* Cambridge, Mass.: National Bureau of Economic Research.

Collaborative Strategies Group, LLC. 2009. *Moving Cooler: An Analysis of Transportation Strategies for Reducing Greenhouse Gas Emissions.* Washington, D.C.: Collaborative Strategies Group, LLC.

Dahl, C.A. 2012. Measuring global gasoline and diesel price and income elasticities. *Energy Policy* 41:2-13.

Greene, D.L. 2010. *How Consumers Value Fuel Economy: A Literature Review.* Washington, D.C.: U.S. Environmental Protection Agency.

———. 2011. Uncertainty, loss aversion, and markets for energy efficiency. *Energy Economics* 33(4):608-616.

Greene, D.L., and P.N. Leiby. 1993. *The Social Costs to the U.S. of Monopolization of the World Oil Market, 1972-1991.* Oak Ridge, Tenn.: Oak Ridge National Laboratory.

Krupnick, A., I. Perry, M. Walls, T. Knowles, and K. Hayes. 2010. *Toward a New National Energy Policy: Assessing the Options.* Washington, D.C.: Resources for the Future.

National Surface Transportation Policy and Revenue Study Commission. 2007. *Transportation for Tomorrow.* Washington, D.C.: National Surface Transportation Policy and Revenue Study Commission.

NRC (National Research Council). 2009. *Driving and the Built Environment: The Effects of Compact Development on Motorized Travel, Energy Use, and CO₂ Emissions.* Washington, D.C.: The National Academies Press.

Sterner, T. 2007. Fuel taxes: An important instrument for climate policy. *Energy Policy* 35(6):3194-3202.

Appendixes

A

Statement of Task

The NRC will appoint an ad hoc study committee to conduct a comprehensive analysis of energy use within the light-duty vehicle transportation sector, and use the analyses to conduct an integrated study of the technology and fuel options (including electricity) that could reduce petroleum consumption and greenhouse gas emissions. As was accomplished with the NRC *Transitions to Alternative Transportation Technologies: A Focus on Hydrogen* study, the study will address the following issues over the time frame out to 2050:

- Assess the current status of light-duty vehicle technologies and their potential for future improvements in terms of fuel economy and costs including:
  —Advanced conventional ICE and hybrid-electric vehicles, including improved combustion and rolling resistance, and weight reduction (safety implications of lighter weight vehicles will be considered);
  —All-electric and plug-in hybrid electric vehicles;
  —Hydrogen fueled ICE and fuel cell vehicles;
  —Biofueled vehicles; and
  —Natural gas vehicles.
- Assess the status and prospects for current and future fuels and electric power that would be needed to power the vehicles. A variety of alternative fuels will be considered such as hydrogen, fuels derived from fossil feedstocks, and different biofuels derived from biomass feedstocks.
- Develop scenarios or estimates of the rate at which each of the vehicle technologies considered might be able to penetrate the market and what would be the associated costs, greenhouse gas emissions and petroleum consumption impacts out to 2050. This would also include the infrastructure needs either for

production of the vehicles or supplying the energy requirements for the vehicles. Costs would be put on a consistent basis to serve as a better index of comparing options. Scenarios will consider technology as well as policy options and consider the likelihood of achieving 50 percent reduction in petroleum consumption by 2030 as well as 80 percent reduction in petroleum consumption and greenhouse gas emissions by 2050. In addition to technology, potential reduction in vehicle miles traveled (VMT) will also be considered.
- Identify the barriers that might exist in transitioning to these vehicle and fuel technologies.
- Consider and compare, as appropriate, the results to those obtained in recent National Academies studies as well as in other outside analyses and make comparisons based on similar assumptions and cost and benefit calculations.
- Recommend improvements in, and priorities for, the federal R&D program activities to accelerate the development of the most promising technologies.
- Suggest policies and strategies for achieving up to 80 percent reduction in petroleum consumption and carbon dioxide emissions by 2050 through commercial deployment of the light-duty vehicle technologies analyzed in the study.
- Write a report documenting the analyses, conclusions, and recommendations.

To the extent possible the committee will consider issues relating to vehicle duty cycles, regional distinctions, and technology development timelines and will build on the recent work of the National Academies reports as well as other recent studies that have been conducted.

B

Committee Biographies

DOUGLAS M. CHAPIN (NAE), *Chair*, is a principal of MPR Associates, Inc., Alexandria, Virginia. He has extensive experience in electrical, chemical, and nuclear engineering, with particular application to nuclear and conventional power plant problems and functions, including numerous aspects of power plant systems and associated components. He has worked in such areas as instrumentation and control systems, nuclear fuels, fluid mechanics, heat transfer, pumps, advanced analysis methods, test facility design, and electrical systems and components. Dr. Chapin has worked on a number of efforts including the Japan/Germany/United States research program on loss of coolant accidents, served as project leader for the design, construction, and testing of the loss of fluid test facility, was a member of the Electric Power Research Institute's (EPRI's) Utility Review Committee on Advanced Reactor Designs, and worked with the Utility/ EPRI Advanced Light Water Reactor Program that defines utility requirements for future nuclear power plants. He was chair of the NRC's Committee on Application of Digital Instrumentation and Control Technology to Nuclear Power Plant Operations and Safety. He has served on a number of NRC committees, including the Committee on America's Energy Future, the Committee on Review of Department of Energy's (DOE's) Nuclear Energy R&D Program, and Board on Energy and Environmental Systems (chair). Dr. Chapin is a member of the National Academy of Engineering (NAE). He served as a member of the NAE's Electric Power/Energy Systems Engineering Peer Committee and as a member of the NAE's Committee on Membership. He is also a fellow of the American Nuclear Society. He has a B.S. degree in electrical engineering, Duke University, an M.S. degree in applied science, George Washington University, and a Ph.D. degree, nuclear studies in chemical engineering, Princeton University.

RALPH J. BRODD is president of Broddarp of Nevada, Inc., a consulting firm specializing in technology assessment, strategic planning and battery technology, production, and marketing. Dr. Brodd began his career at the National Bureau of Standards studying electrode reactions and phenomena that occur in battery operation. In 1961, Dr. Brodd joined the L.T.V. research Center of Ling Temco Vought, Inc., where he established a group in fuel cells and batteries. In 1963, he moved to the Battery Products Technology Center of Union Carbide Corporation, with technical management responsibilities for nickel-cadmium and lead acid rechargeable batteries, alkaline and carbon-zinc product lines, and exploratory R&D. He joined ESB (INCO Electroenergy, Inc.) in 1978 as director of technology. In 1982, Dr. Brodd established Broddarp, Inc., a consulting firm specializing in battery technology, strategic planning, and technology planning. He moved to Amoco Research Center as project manager of a rechargeable lithium sulfur dioxide battery project. He subsequently moved to Gould, Inc., to establish their Lithium Powerdex Battery Venture and then to Valence Technology, a venture group developing a solid polymer electrolyte battery system for rechargeable batteries for portable consumer devices as vice president, marketing. Dr. Brodd was elected president of the Electrochemical Society in 1981 and honorary member in 1987. He was elected national secretary of the International Society of Electrochemistry (1977-1982) and vice president (1981-1983). He is past chairman of the Board of Directors of the International Battery Materials Association (IBA). Dr. Brodd has more than 100 publications and patents. He received a B.A. degree in chemistry from Augustana College and M.A. and Ph.D. degrees in physical chemistry from the University of Texas at Austin.

GARY L. COWGER (NAE) is currently chairman and CEO of GLC Ventures, LLC—a management consultancy. He retired from General Motors Corporation as Group Vice President—Global Manufacturing, Labor Relations and Manufacturing Engineering. In this position he was responsible for all of GM's Global Manufacturing Operations. He held a variety of other senior positions at GM, including President of GM North America; Chairman—Adam Opel,

AG; Vice President for Operations, GM Europe; and President and Managing Director of GM de Mexico. Mr. Cowger has extensive experience in business, technology, engineering and manufacturing operations. He was responsible for the development and implementation of the GM global manufacturing system. He has also had extensive experience in benchmarking, target-setting, and the creation and application of organizational and production-based performance measures. Mr. Cowger is the past Chairman of the Board for Kettering University and holds other Board positions in private and public organizations. Mr. Cowger holds an M.S. degree in management from the Massachusetts Institute of Technology and a B.S. degree in industrial engineering from Kettering University (formally General Motors Institute).

JOHN M. DeCICCO is a professor of practice at the School of Natural Resources and Environment and research professor at the University of Michigan Energy Institute. Previous positions include senior fellow, automotive strategies, Environmental Defense Fund; transportation program director, American Council for an Energy-Efficient Economy; and staff scientist, National Audubon Society. His teaching and advising interests address energy use and greenhouse gas (GHG) emissions from transportation as well as broader aspects of sustainable mobility and energy use. His research seeks to further public understanding of transportation systems and GHGs, including the interlinked decision-making structures (both private market and public process) that underpin energy demand and emissions in the sector. He has published widely on analysis of the cost and improvements in emissions and fuel economy of advanced automotive technologies and in recent years has focused increasingly on the challenges of transportation fuels and GHG emissions. He has a Ph.D. in mechanical engineering from Princeton University, an M.S.M.E. from North Carolina State University, and a B.A. in mathematics from Catholic University of America.

GEORGE C. EADS retired from Charles River Associates in 2008 after serving 12 years as a vice president. He remains a senior consultant with the company. Prior to joining CRA, Dr Eads held several positions at the General Motors Corporation, including vice president and chief economist; vice president, Worldwide Economic and Market Analysis Staff; and vice president, Product Planning and Economics Staff. Before joining GM, Dr. Eads was dean of the School of Public Affairs at the University of Maryland, College Park, where he also was a professor. Before that, he served as a member of President Carter's Council of Economic Advisors, was a program manager at the RAND Corporation, served as executive director of the National Commission on Supplies and Shortages, as Assistant Director of President Ford's Council on Wage and Price Stability, and taught at Harvard University, Princeton University, and the George Washington University. He has been involved in numerous projects

concerning transport and energy. In 1994 and 1995, he was a member of President Clinton's policy dialogue on reducing greenhouse gas emissions from personal motor vehicles. He co-authored the World Energy Council's 1998 report *Global Transport and Energy Development—The Scope for Change*. He was Lead Consultant to the World Business Council for Sustainable Development's Sustainable Mobility Project, a project funded and carried out by 12 leading international automotive and energy companies. Dr. Eads is a member of the Presidents' Circle of the National Academies. He is an at-large director of the National Bureau of Economic Research. He received a Ph.D. degree in economics from Yale University. He has been on several National Academies committees, including the TRB study on Potential Energy Savings and Greenhouse Gas Reductions from Transportation, the TRB study on Climate Change and U.S. Transportation, and the America's Climate Choices study.

RAMON L. ESPINO is currently a research professor at the University of Virginia, where he has been on the faculty since 1999. Prior to joining the Department of Chemical Engineering, he was with ExxonMobil for 26 years. He held a number of research management positions in petroleum exploration and production, petroleum process and products, alternative fuels and petrochemicals. He has published about 20 technical articles and holds 9 patents. Dr. Espino's research interests focus on fuel cell technology, specifically in the development of processors that convert clean fuels into hydrogen and of fuel cell anodes that are resistant to carbon monoxide poisoning. Another area of interest is the conversion of methane to clean liquid fuels and specifically the development of catalysts for the selective partial oxidation of methane to synthesis gas. He has served on NRC committees dealing with R&D in DOE's fossil fuels programs, mitigation of greenhouse gases and other topics related to energy efficiency. He received a B.S. degree in chemical engineering from Louisiana State University and an M.S. and a doctor of science in chemical engineering from MIT.

JOHN GERMAN is a senior fellow for the International Council for Clean Transportation, with primarily responsibility for technology innovation and U.S. policy development. He has been involved with advanced technology and efficiency since joining Chrysler in 1976, where he spent eight years in Powertrain Engineering working on fuel economy issues. He then spent 13 years doing research and writing regulations for EPA's Office of Mobile Sources' laboratory in Ann Arbor, Michigan. Prior to joining ICCT four years ago, he spent 11 years as Manager of Environmental and Energy Analyses for American Honda Motor Company, with an emphasis on being a liaison between Honda's R&D people in Japan and regulatory affairs. Mr. German is the author of a book on hybrid gasoline-electric vehicles published by SAE and a variety of technical papers, including the future of hybrid vehicles, technology costs and benefits, consumer valuation of fuel savings, feebates, and light truck trends. He was the first recipient of the Barry D. McNutt award, presented annually by SAE for Excellence in Automotive Policy Analysis. He has a bachelor's degree in physics from the University of Michigan and partial credit toward an MBA.

DAVID L. GREENE is a corporate fellow of Oak Ridge National Laboratory, where he has researched transportation energy policy issues for the U.S. government for 35 years, a Senior Fellow of the Howard H. Baker, Jr. Center for Public Policy and a Research Professor of Economics at the University of Tennessee. Greene is an author of more than 250 publications on transportation, energy and related issues. He is an emeritus member of both the Energy and Alternative Fuels Committees of the Transportation Research Board and a lifetime National Associate of the National Academies. He is a recipient of the TRB's 2012 Roy W. Crum Award for distinguished achievement in transportation research, the TRB's Pyke Johnson Award, the Society of Automotive Engineers' 2004 Barry D. McNutt Award for Excellence in Automotive Policy Analysis, the Department of Energy's 2007 Hydrogen R&D Award and 2011 Vehicle Technologies R&D Award, the International Association for Energy Economics' Award for Outstanding Paper of 1999 for his research on the rebound effect, the Association of American Geographers' 2011 Edward L. Ullman Award, and was recognized by the Intergovernmental Panel on Climate Change for contributions to the IPCC's receipt of the 2007 Nobel Peace Prize. He holds a B.A. from Columbia University, an M.A. from the University of Oregon, and a Ph.D. in geography and environmental engineering from the Johns Hopkins University.

JUDI GREENWALD is the vice president of technology and innovation at the Center for Climate and Energy Solutions. She oversees the analysis and promotion of innovation in the major sectors that contribute to climate change, including transportation, electric power, and buildings. Ms. Greenwald focuses on technology, business, state, regional, and federal innovation. She served on the Resource Panel for the northeast Greenhouse Gas Initiative and the California Market Advisory Committee, and as a policy advisor to the Western Climate Initiative and the Midwest Greenhouse Gas Accord Advisory Group. She previously served as the vice president for innovative solutions at the Pew Center on Global Climate Change, C2ES's predecessor organization. Ms. Greenwald has nearly 30 years of experience working on energy and environmental policy. Prior to coming to the Pew Center, she worked as a consultant, focusing on innovative approaches to solving environmental problems, including climate change. She also served as a senior advisor on the White House Climate Change Task Force. As a member of the professional staff of the Energy and Commerce Committee of the U.S. House of Representatives, she worked on the 1990 Clean Air Act Amendments, the 1992 Energy Policy Act, and a number of other energy and environmental statutes. She was also a congressional fellow with then-Senate Majority Leader Robert C. Byrd, an environmental scientist with the U.S. Nuclear Regulatory Commission, and an environmental engineer and policy analyst at the EPA. Ms. Greenwald has a B.S. in engineering, cum laude, from Princeton University and an M.A. in science, technology and public policy from George Washington University.

L. LOUIS HEGEDUS (NAE) is the retired senior vice president, R&D, of Arkema Inc., and a visiting distinguished fellow at RTI International. Research programs at Arkema supported market applications in the automotive, petroleum, energy conversion and storage, electronics, and construction industries. Dr. Hegedus was previously vice president, Corporate Technical Group, at W.R. Grace. Research programs included catalysts for petroleum refining, chemicals, emission control, and fuel cells; technical and electronic ceramics; electrochemical products including polymeric membranes for electric storage batteries of various types; and construction materials and products. Prior to joining W.R. Grace, Dr. Hegedus was affiliated with the General Motors Research Laboratories where he managed research on the development of the catalytic converter for automobile emission control. Before his graduate studies, he was an engineer with Daimler-Benz in Germany. He is a member of NAE, and he is a recipient of the R.H. Wilhelm, Professional Progress, Catalysis and Reaction Engineering Practice, and the Management Division awards of the American Institute of Chemical Engineers (AIChE) and the Leo Friend Award of the American Chemical Society (ACS)-Chemtech. At the occasion of their 100th anniversary, AIChE named Dr. Hegedus as one of "Hundred Chemical Engineers of the Modern Era." He was a founding member of AIChE's Commission on Energy Challenges and has served on several panels of the NRC's Board on Chemical Sciences and Technology, including one on critical chemical technologies, one on the future of catalysis, and one charged with the international benchmarking of the U.S. chemical engineering competencies. Most recently, Dr. Hegedus served on panels of the National Science Foundation dealing with the manufacture of nanomaterials and with the development of rechargeable lithium battery technology. At RTI International, he co-edited and co-authored the book *Viewing America's Energy Future in Three Dimensions—Technology, Economics, Society*. Dr. Hegedus obtained his Ph.D. in chemical engineering from the University of California, Berkeley, and his M.S. in chemical engineering from the Technical University of Budapest, from which he also received an honorary doctorate.

JOHN B. HEYWOOD (NAE) has been a faculty member at MIT since 1968, where he has been the Sun Jae Professor of Mechanical Engineering and director of the Sloan Automotive Laboratory. His interests are focused on internal combustion engines, their fuels, and broader studies of future transportation technology, fuel supply options, and air pollutant and GHG emissions. He has published more than 200 papers in the technical literature and is the author of five books, including a major text and professional reference, *Internal Combustion Engine Fundamentals*. He is a fellow of the Society of Automotive Engineers. He has received many awards for his work, including the 1996 U.S. Department of Transportation Award for the Advancement of Motor Vehicle Research and Development and the Society of Automotive Engineers 2008 Award for his contributions to Automotive Policy. He is a member of the NAE and a fellow of the American Academy of Arts and Sciences. He has a Ph.D. from MIT, a D.Sc. from Cambridge University, and honorary degrees from Chalmers University of Technology, Sweden, and City University, London.

VIRGINIA McCONNELL is senior fellow in the Quality of the Environment Division of Resources for the Future (RFF), Inc. She is also a professor of economics at the University of Maryland, Baltimore County. Her recent work has centered on the evaluation of policies to reduce motor vehicle pollution, particularly on the role of pricing and other incentive-based policies. She recently completed a study on hybrid vehicles and the effectiveness of policies designed to increase the share of hybrids and electric vehicles in the U.S. fleet, part of a larger effort at RFF to assess a range of transportation and other policies to reduce oil use and GHG emissions in the United States by 2030. She was co-editor of the 2007 book *Controlling Vehicle Pollution* and has published on a range of transportation policy issues. In addition, she has served on a number of EPA and state advisory committees related to transportation and air quality. She is currently serving on a public policy panel to look at the prospects for Transport Electrification. She has been a member of several NRC panels in recent years, including the Committee on Vehicle Emission Inspection and Maintenance Program, the Committee on State Practices in Setting Mobile Source Emissions Standards, and the Committee for a Study of Potential Energy Savings and Greenhouse Gas Reductions from Transportation. Dr. McConnell received a B.S. degree in economics from Smith College and a Ph.D. degree in economics from the University of Maryland.

STEPHEN J. McGOVERN has more than 35 years of experience in the refining and petrochemical industries. Dr. McGovern has been a principal of PetroTech Consultants since 2000, providing consulting services on various refining technologies, including clean fuels projects and refining economics. He has assisted numerous refiners in the evaluation of gasoline and diesel desulfurization technologies, Catalytic Cracking and environmental issues. Dr. McGovern has provided technical advice to DARPA and commercial enterprises for the production of biofuels. Previously, he was with Mobil Technology Company, where he led various efforts in process development and refinery technical sup-

port. He has 17 patents and more than 20 technical publications and was a member of the NRC Committee on Economic and Environmental Impacts of Increasing Biofuels Production. He has lectured, published and consulted on refining technology, environmental and alternate fuels issues. Dr. McGovern is a licensed professional engineer in New Jersey and a past director of the Fuels and Petrochemicals Division of AIChE. He earned a B.S. degree (magna cum laude) and M.S. degree in chemical engineering from Drexel University and M.A. and Ph.D. degrees in chemical engineering from Princeton University.

GENE NEMANICH is a consultant specializing in chemical processes. Previously, he was director of hydrogen systems for ChevronTexaco Technology Ventures where he was responsible for hydrogen supply and developing and commercializing new hydrogen storage technologies. He has 31 years of experience with integrated oil companies, including Exxon, Cities Service, Texaco, and ChevronTexaco. He has also worked in the areas of refining, clean coal technology, oil supply and trading, and hydrogen systems. He represented Texaco in the California Fuel Cell Partnership in 2000-2001 and is a director of Texaco Ovonic Hydrogen Systems, LLC, a joint venture with Energy Conversion Devices to commercialize metal hydride hydrogen storage systems. He was one of seven industry leaders that helped prepare the DOE-sponsored Hydrogen Roadmap, and he has served as chairman of the National Hydrogen Association. He has a B.S. in chemical engineering from University of Illinois and an MBA from University of Houston.

JOHN O'DELL is senior editor with the Edmunds.com editorial team, where he originated online coverage of the environmental or "green" automotive segment, producing articles dealing with advanced and alternative vehicle policies, financing, technology, politics, alternative fuels, and related issues. Mr. O'Dell is regularly quoted by major newspapers, periodicals, wire services, and broadcast media as an expert on the growing green car and alternative fuels markets. Prior to joining Edmunds, Mr. O'Dell was a staff writer and editor at the *Los Angeles Times* from 1980-2007. He co-founded the consumer automotive section of the *L.A. Times*, Highway 1, in 1998, and was the paper's automotive industry reporter from 1998-2007. He also served variously as city beat reporter, county government writer, business reporter, and assistant business editor at the *Times' Orange County Edition* and was variously a city beat reporter, investigative reporter, political writer, and assistant city editor at the *Orange County Register*. Mr. O'Dell holds a B.A. in communications from California State College at Fullerton and completed the coursework there toward a graduate degree in communications with an emphasis in consumer economics. His career as a journalist has been marked by numerous awards for professional excellence in writing, research, and project development. He was part of the reporting teams that won Pulitzer prizes for the *Los Angeles Times* in 1992 for coverage of the Los Angeles Riots and in 1994 for coverage of the Northridge Earthquake.

ROBERT F. SAWYER (NAE) is the Class of 1935 Professor of Energy emeritus in the Department of Mechanical Engineering at the University of California, Berkeley. His research interests are in combustion, pollutant formation and control, regulatory policy, rocket propulsion, and fire safety. He served as chairman of the California Air Resources Board, chairman of the energy and resources group of the University of California at Berkeley, chief of the liquid systems analysis section at the U.S. Air Force Rocket Propulsion Laboratory, and president of the Combustion Institute. Dr. Sawyer has served on numerous National Research Council committees and was a member of the NRC's Board on Environmental Studies and Toxicology. He holds a B.S. and M.S. in mechanical engineering from Stanford University and a M.A. in aeronautical engineering and a Ph.D. in aerospace science from Princeton University.

CHRISTINE S. SLOANE retired from General Motors Corporation as the head of the global team for hydrogen and fuel cell vehicle codes and standards development. She coordinated development of GM policy and technical strategy across safety, engineering, and public policy requirements to ensure global consistency in GM interaction with government and professional industry organizations. She previously directed the GM interaction with the U.S. FreedomCAR program, which included R&D to advance fuel cell power systems, and earlier served as chief technologist for the development and demonstration team for Precept, GM's 80 mile-per-gallon five-passenger HEV concept vehicle. She has also been responsible for global climate issues and for mobile emission issues involving advanced technology vehicles. Her early research interests included air quality, and manufacturing and vehicle emissions. Dr. Sloane has authored more than 80 technical papers and co-edited one book. She has served on several boards of professional organizations and numerous National Academy of Sciences

panels and study groups. Dr. Sloane received her Ph.D. from MIT in chemical physics.

WILLIAM H. WALSH, JR., is an automobile safety consultant. He consults on vehicle safety activities with several technology companies to speed the introduction of advanced life-saving technology into the automobile fleet as well as substantive involvement in corporate average fuel economy (CAFE) rulemakings. He held several positions at the U.S. National Highway Traffic Safety Administration (NHTSA), including senior associate administrator for policy and operations; associate administrator for plans and policy; director, National Center for Statistics and Analysis; director, Office of Budget, Planning and Policy; and science advisor to the administrator of NHTSA. He also held the position of supervisory general engineer at the DOE's Appliance Efficiency Program. His expertise covers all aspects of vehicle safety performance, cost/benefit analyses, strategic planning, statistics analyses and modeling, and policy formulation. He serves on the Transportation Research Board's Occupant Protection Committee. He has a B.S. in aerospace engineering, University of Notre Dame, and an M.S. in system engineering, George Washington University.

MICHAEL EVAN WEBBER is the Josey Centennial Fellow in Energy Resources, associate professor of mechanical engineering, associate director for the Center for International Energy and Environmental Policy, and co-director of the Clean Energy Incubator, all at the University of Texas at Austin. Previously he was an associate engineer at RAND Corporation and senior scientist at Pranalytica, Inc. He holds four patents involving instrumentation. He serves on the board of advisers of *Scientific American* and is on the editorial board of several other journals. Dr. Webber is also a member of the Electric Utility Commission of the City of Austin and is active in a variety of other public and civic organizations. He has an M.S. and Ph.D. in mechanical engineering (minor, electrical engineering) from Stanford University and B.S./B.A. degrees with high honors from the University of Texas at Austin.

C

Meetings and Presentations

FIRST COMMITTEE MEETING
OCTOBER 21-22, 2010, WASHINGTON, D.C.

Overview of DOE's Vehicle Technologies Program: Potential for Light Duty Vehicle Technologies NAS Study
Patrick Davis, U.S. Department of Energy

Vehicle Technologies Program (VTP): Analysis Briefing for NAS
Phillip Patterson and Jacob Ward, U. S. Department of Energy

FY2011 VTP Energy Storage R&D
David Howell, U.S. Department of Energy

Analysis Methods from Recent Studies
Robert Fri, U.S. Department of Energy

Transportation Energy Futures
Austin Brown, National Renewable Energy Laboratory

SECOND COMMITTEE MEETING
DECEMBER 14-15, 2010, WASHINGTON, D.C.

Liquid Transportation Fuels from Coal and Biomass: Technological Status, Costs, and Environmental Impacts
Mike Ramage, Consultant

Alternative Transportation Technologies: Hydrogen, Biofuels, Advanced ICEs, HEVs and PHEVs
Mike Ramage, Consultant

Perspectives on Energy Security and Transportation: The Intersection of National Security and Economic Challenges
Robbie Diamond, Electrification Coalition

Biofuels: Technology Status and Challenges
Andy Aden, National Renewable Energy Laboratory

EPA's Light-Duty Vehicle GHG Technical Activities
Bill Charmley, U.S. Environmental Protection Agency

THIRD COMMITTEE MEETING
FEBRUARY 1-2, 2011, WASHINGTON D.C.

BP Energy Outlook 2030
Mark Finley, BP

Reducing Greenhouse Gas Emissions from U.S. Transportation
David Greene, Howard H. Baker Center for Public Policy and Steve Plotkin, Argonne National Laboratory

Critical Materials Strategy
Diana Bauer, United States Department of Energy

Toward a New National Energy Policy: Assessing the Options
Alan Krupnick and Virginia McConnell, Resources for the Future

FOURTH COMMITTEE MEETING
MARCH 21-22, 2011, WASHINGTON, D.C.

ARPA-E's BEEST Program: Ultra-High Energy, Low Cost Energy Storage for Ubiquitous Electric Vehicles
David Danielson, ARPA-E

Carbon Capture and Storage RD&D
Jay Braitsch, U.S. Department of Energy

Future Transportation Fuels Study, National Petroleum Council
Linda Capuano, Marathon Oil Company

Overview of Hydrogen and Fuel Cells
Sunita Satyapal and Fred Joseck, U.S. Department of Energy

The Mercedes-Benz Hydrogen Roadmap
  Sascha Simon, Mercedes Benz

FIFTH COMMITTEE MEETING
MAY 12-13, 2011, DETROIT, MICHIGAN

No open sessions were held during this meeting.

SIXTH COMMITTEE MEETING
JUNE 27-29, 2011, IRVINE, CALIFORNIA

Potential for Light-Duty Fuel Cell EVs, 2010-2050
  Ben Knight, Honda

Alternative Fuel Strategy . . . As Seen by a Policy Wonk, Regulator, and Academic
  Dan Sperling, University of California, Davis

Plug-in Electric Vehicles and their Impact to the Grid
  Reiko Takemasa, Pacific Gas and Electric Company

ADDITIONAL COMMITTEE MEETINGS

The committee met in closed session for deliberations and report writing and review on the following dates: August 10-11, 2011; September 12, 2011; October 5-7, 2011; December 14-15, 2011; January 25-26, 2012; March 29-30, 2012; and May 15-16, 2012.